THE MARLIN'S FIERY EYE

THE MARLIN'S FIERY EYE

AND OTHER TALES FROM THE EXTRAORDINARY WORLD OF MARINE FISHES

Joe E. Meisel

Foreword by Rod Fujita

Comstock Publishing Associates
an imprint of
Cornell University Press
Ithaca and London

First published 2025 by Cornell University Press

Printed in the United States of America

Library of Congress Cataloging-in-Publication Data

Names: Meisel, Joe E., author.
Title: The marlin's fiery eye and other tales from the extraordinary world of marine fishes / Joe E. Meisel.
Description: Ithaca : Comstock Publishing Associates, an imprint of Cornell University Press, 2025. | Includes bibliographical references and index.
Identifiers: LCCN 2024020248 (print) | LCCN 2024020249 (ebook) | ISBN 9781501779442 (paperback) | ISBN 9781501779466 (epub) | ISBN 9781501779459 (pdf)
Subjects: LCSH: Marine fishes—Behavior. | Marine fishes—Conservation.
Classification: LCC QL620 .M45 2025 (print) | LCC QL620 (ebook) | DDC 597—dc23/eng/20240809
LC record available at https://lccn.loc.gov/2024020248
LC ebook record available at https://lccn.loc.gov/2024020249

To Catherine

The sea is the vast reservoir of nature.
—Jules Verne

In all things of nature there is something of the marvelous.
—Aristotle

Curiosity is, in great and generous minds, the first passion
and the last.
—Samuel Johnson

CONTENTS

FOREWORD

YES, FISH ARE FOOD. But they are so much more.

In *The Marlin's Fiery Eye and Other Tales from the Extraordinary World of Marine Fishes*, Joe E. Meisel introduces us to fish as wild beings, living out their own lives. Like us, they seek shelter, look for food, and reproduce. But they do these things in wildly different ways, shaped by hundreds of millions of years of evolution in an environment that is radically different from our own. To paraphrase Charles Darwin: there seems to be no limit to the beauty and infinite complexity of the adaptations of living beings to their environments.

I had my own awakening to the lives of fishes and other ocean wildlife. I grew up near the sea, and my relationship with it as a boy was defined by what I could see from the sandy beaches. The ocean would reveal rare glimpses of its life, in the form of a dolphin undulating through the waves or grunion writhing on the beach during spawning season. But for the most part, the murky waters off my hometown obscured the life in the water. Then I discovered tidepools.

Tidepools lifted the ocean's veil a little more. Strange and beautiful life forms like anemones and hermit crabs were easy and fun to watch in the pools during low tides. As I gained more understanding of the ocean in my academic studies, the relationship changed. The ocean became the object of my fascination as I sought to learn everything I could about it from books and my teachers. Decades of experiencing the ocean in various ways, from meditating on the shore to scuba diving, deepened the relationship. For a time, I lived on a platform about 50 feet above an isolated coral reef. From there I could watch the movements of animals from dawn to dusk. I did several dives a day to check transects and run experiments. After a while, I felt my relationship with the reef shift again. I started to recognize individual barracuda and their hunting grounds. The same group of reef squid would come over to check my work periodically. I started to feel a sense of community as the life and the rhythms of the reef became more familiar.

One of the lessons I learned, and one that's reflected in *The Marlin's Fiery Eye and Other Tales from the Extraordinary World of Marine Fishes*, is that with familiarity and belongingness comes love and compassion. And with love and compassion comes protectiveness and reciprocity—a new kind of relationship with the ocean and its wildlife.

We humans are story tellers. The stories we tell ourselves about who we are give rise to our values, beliefs, and actions. Some of the stories we tell about what corporations or nations are for give rise to amazing feats of cooperation, while others result in exploitation, strife, and war. One story that some of us used to tell about nature is that nature's bounty is inexhaustible. Once that was proved to be false, some of us told another story about how nature exists to serve us, but that we needed to exploit natural resources sustainably. It now seems clear that the sustainability story is not going to have an entirely happy ending. The destruction of life-sustaining habitats and emissions of pollutants that are altering the life support systems of the entire planet have continued, with catastrophic effects.

The Marlin's Fiery Eye and Other Tales from the Extraordinary World of Marine Fishes includes many fascinating stories about how people have discovered, used, and come to understand fish over the centuries. But it also contains stories about people caring for the ocean and its life, as well as the amazing stories about the fish themselves. Readers get an extended look at a world teeming with fish leading remarkable lives that is mostly invisible to us, most of the time. These are the plot elements of a new story about a reciprocal relationship with nature, one in which we are responsible for repairing the damage that we have caused. A story in which fish are food but also our partners on this planet, inspiring respect, awe, and a sense that we part of something larger than ourselves.

Rod Fujita
April 24, 2023

ACKNOWLEDGMENTS

I OWE A DEBT TO ALL THE EXPERTS I interviewed, who gave selflessly of their time and knowledge, and whose contributions have, I hope, greatly improved the accuracy, completeness, and tone of this book. Thank you to Bjarte Bogstad, Megan Corazza, Fabio Di Dario, Becca Franks, Rod Fujita, Lexa Grutter, Jennifer Jacquet, Chris Lowe, Karen Maruska (who sadly passed away in 2023), Brian Perkins, Monty Priede, Jorge Ramirez, Jennifer Schmidt, and Robert Steneck.

Imagination only gets one so far, but artwork brings creatures to life. Mariah Jane Robinson created all of the drawings in this book. I am deeply grateful to her for the superb illustrations that not only accurately depict biological details but also imbue fishes with an artistic beauty they richly deserve.

Thanks to Leta Landucci, for her early research forays that uncovered many of the stories featured in this book.

I single out Kitty Liu at Cornell University Press, in gratitude for her unflagging support and belief in this book, despite setbacks, delays, and floods in the kitchen. I'm grateful as well to India Miraglia, Mary Kate Murphy, and Lucy Treadwell for their attention to detail during preparation of the manuscript.

To my close friends, thank you for setting unparalleled examples of lives lived well: Zeke and Michelle for courage and selflessness, Ritt and Victoria for unflagging hard work and love of music, Andy and Jenny for entrepreneurial spirit and goofiness, Eric and Nina for steadfast allegiance to art, Brendan and Mira for internationalism, Didi for devotion to great writing, and John for a love of literature and all things weird.

I am grateful to the many Madisonians who prove that socializing is one of life's great pleasures and invaluable balms: Julie, Paolo, Mary, Gaby and Brian, Keith, Jennifer and Mark, P. J., Danielle and Rich (taken from us far too soon), Elizabeth and Dan, Peter, Carol, Cheryl, Mike, Dennis and Margaret, Heather and Drew, Dino and Christine, and James and Shamane. An extra special thanks to Ian, for miles of cycling, pints of ale, and hours of conversation; his hometown of Grimsby inspired the opening of the chapter on cod.

To the staff of the Ceiba Foundation, Haley, Yasi, María José, Doménica, Marcos, Joffre, Jessy, Mariela, Steve, and Susan, thank you for putting up with the sporadic abdications of duty that writing this book unfortunately demanded.

I am immeasurably grateful to my family; qualities in my character are wholly due to their contributions, and shortcomings are my fault alone. To my mother, who told her sons they could achieve anything, and to my father who proved it by his own example. To my brother Chris, a superb teacher who sets a commendable example of balancing work and pleasure. To Colleen, for her independence of thought and Dr. Doolittle-esque love of animals. To my nephews Finn and Beckett, whom I have watched grow into thoughtful, kind, and brilliant young men. To my aunt Teena, for proof that dedicating oneself to art will always provide rewards. To my St. Louis family, Pat, Molly, Jim and Kim, Susan and Ed, and the laughter that underpins every conversation. To my western family, Brad and Kim, Mary and Chuck, for setting examples of how you can always do it your own way. And to my German family: Michaela for her unequivocally unique style, Willi for his delight in nature, and Ursula for her wide-ranging knowledge and for a trip to Norway where this manuscript first set sail.

Finally, I thank Catherine, the most accomplished, dedicated, and indefatigable person I know. With her I share a love of biology, conservation, travel, teaching, curiosity, and an undying enthusiasm for the natural world both above and below the waves. She is my partner, my love, and my inspiration.

THE
MARLIN'S
FIERY EYE

INTRODUCTION

YOU GLIDE EFFORTLESSLY THROUGH the balmy and crystalline waters of a tropical coral reef, propelled by coordinated flaps of your wing-like fins. Peering overhead, you catch your reflection on the silvery surface: a female blue-chin parrotfish looks back at you. Large in size and sturdy of build, you are merrily painted with alternating bars of yellow and blue. You wield a powerful set of beak-like front teeth, to expertly scrape algae from rocks. Occasionally, you take a vigorous bite of coral, and extract the soft residents by crushing their hard exterior with a constellation of flattop teeth that stud your muscular throat. At night you spin a cocoon of mucus, a sleeping bag of sorts, that protects your scaled surface from nautical parasites and prevents nocturnal predators from picking up your scent as you doze. By day you cruise the reef, your bright colors on full display, illuminated by the ballet of sunshine refracted through the waves. Now and then you take a brief break at a cleaning station, where attentive wrasses and perhaps an industrious shrimp provide the full salon grooming treatment, plucking tiny, bloodsucking parasites and encrusting algae from your scales. Emerging fresh and clean, you detect a scent in the water, a faint but unmistakable odor that signals courtship is underway.

Excitedly, you scurry to a protected area amid boulders and corals and join a handful of other females, each competing for the attention of one dominant male parrotfish. You females far outnumber the solitary male, a large fellow with a pronounced hump on his forehead, who holds court over a sort of harem; this is polygyny at its most pronounced. Diluting his command only slightly are a few inferior males, smaller and less adept, who hang around the margins of the dance floor, hoping to steal a kiss, as it were. Suddenly, in a demonstrative rush, two females turn their chins toward the sky and rocket upward to the surface. Not wanting to miss your chance, you lash your tail and chase exuberantly after them. A few meters beneath the waves, in spectacularly coordinated maneuvers, you and the surging females jettison eggs into the water at the precise moment when the top male knives through the crowd, releasing a white cloud of sperm. With luck, a few dozen eggs are successfully fertilized, carrying into future generations the genes of several females and the single dominant male. Occasionally, one of the satellite males will streak through the crowd and surreptitiously release his own gametes; now sperm competition in open water determines whose genes fertilize whose eggs.

But your eggs are shed, your die is cast, and the fate of your offspring is now entirely out of your hands, or fins. Exhausted by your efforts, you retire to a still and shady location under a slab of encrusted rock. The coffee break may have saved your life. Because above you, just as you disappear from view, a silent killer slashes into the crowd of spawning fish. An oceanic whitetip shark, speeding onto the reef from dark blue waters nearby, seizes the dominant male with saw-like teeth and drags him into the depths to his death. This brutal abduction from the seraglio leaves the harem without a master, the crowd of stunned females without a mating partner. A void in the hierarchy has been created. Then, something truly amazing happens.

You sense an advantage to be had. If you were to take the place of the dominant male, every offspring of every fertilized egg of every female would carry your genes. No sooner does this revelation dawn than your body begins a remarkable transformation. The color of your skin changes, and you begin to display new hues and strikingly different patterns. Your yellow and turquoise stripes fade, supplanted by brilliant bluish-green on the flanks, lavender markings on your fins, and a spiderweb of bright blue lines around your mouth and eyes. The graceful curve of your forehead begins to swell into a comical bump shape, as though you had been clobbered by a cartoon club. Externally, you are swiftly becoming quite distinct from the other females; soon, you will not recognize your reflection. But internally, an even more shocking change is taking place. Something miraculous is about to happen, something landlubbers can scarcely imagine, yet it is just one of a thousand wondrous mysteries that play out beneath the waves every single day.

You, our protagonist parrotfish, were born a hermaphrodite, possessing the genes and reproductive capabilities of both genders. You began life as one but are about to transform into another. As your forehead swells and the colors of skin and scale transform, your estradiol levels begin to decline, and your testosterone levels climb. As this modified hormone balance courses through the circulatory system, your body begins to respond. Your ovaries shrink and your testes enlarge. Shortly, you lose the capacity to produce eggs, but this loss is offset by a newfound ability to release countless sperm: you have become a male. In light of the gender-swapping of many fish, the terms "female" and "male" now seem ill-fitting. Instead, you have transitioned from an "initial" phase to a "terminal" phase adult. The conversion complete, you are ready to seize control of the harem of other females—initials—as the dominant breeder. Now, when spawning occurs, all the eggs will be fertilized by you, thanks to the hormone-induced activation of your latent sperm-producing powers. All the offspring will carry your genes, as the little larval fishes are swept through the ocean by currents. Today you are a terminal-phase, harem-holding blue-chin parrotfish, and the father of millions.

MYSTERIOUS AND MIRRORED WATERS

> All that we do is touched with ocean, yet we remain on the shore of what we know.
> —Richard Purdy Wilbur, *For Dudley*

The ocean is more mirror than lens. It reflects back to us our aspirations and dreams, our struggles, our failures and victories. It reveals souls to poets, portrays sunsets for painters, and echoes the stars floating overhead. Yet scarcely do we see what lies beneath the surface. One can gaze upon the sea for hours, marvel at glimmering lights dancing on the ripples, at the moon suspended amid flecked foam. We can observe patiently, penetratingly, but never catch so much as a glimpse of the life below the waves. We barely know the fish that see those same waves, those same stars, that same moonrise, but all from below.

Thousands upon thousands of species call the sea home. Indeed, a single drop of seawater can contain more than a million organisms. Most are tiny algae and other invertebrates, but every drop can also transport hundreds of microscopic larvae that will grow into fishes, crabs, squids, and more. Fishes are astoundingly diverse, with 33,700 described species comprising half of all vertebrates on planet Earth.[1] They live in the sunlit shallows and in the black depths. They hunt in open waters and over the continental shelf. They hug the bottom or fly above the surface, squeeze into crevices, bury in sand, or flash through the great three-dimensional watery space that covers most of our planet. There are as many surprising strategies for capturing prey as there are for eluding predators. There are as many extraordinary exceptions as there are rules. And there are as many entertaining tales as there are fish in the sea.

Still, as land-dwelling humans, we are largely blind to the anthology of stories whose main characters are marine fishes. For us, fishes are restricted to tanks in an aquarium, or digital representations in popular films, or breaded fillets. If you asked your sister or brother to draw a pollock, or a haddock, could they? Given that most people have probably eaten pollock, and two-thirds of the planet is aquatic, it is shocking how little we know about fish.

A minority of us, albeit a sizable one, enjoy sport fishing in lakes, streams, and the sea. Anglers know quite a bit more about the lives of the fish they seek with hook, line, and lure. Ask anyone who fly fishes, they command a detailed knowledge of the insects that are hatching, the fish that feed on them, and how to trick them into taking a manufactured fly instead of the genuine article. But the wet feet and waders and bass boats only get anglers slightly closer to fish; we remain forever above the surface. We only infrequently spy our quarry before they are on the line. Those fleeting peeks through murky waters or tumbling rivulets afford limited glimpses into the lives of fish. We know not how they court, where they shelter at night, why they choose one pool over another, nor how they perceive their world. While there is much we do understand, there is far more that eludes our ken.

Even fewer of us are fortunate to enter the sea with a snorkel or scuba tank, where we can briefly observe fishes in their natural environment. Our gear is like a temporary backstage pass to the world behind the curtain of the waves. Divers slip beneath the surface, where they can watch fish with the same intensity that bird-watchers apply to their feathered friends. Spellbinding behaviors are revealed: a moray lurching from its narrow cave to nip at prey; eagle rays swatting the sandy bottom as though unfurling picnic blankets, flushing sediment to reveal hidden crabs; a richly colored damselfish tenaciously guarding a patch of eggs glued to a small rock face.

Amid all the diversity, the colors, the movement and sound, however, it can be easy to forget just how limited are our underwater experiences. People rarely spend more than an hour beneath the waves before returning to the surface. Almost all diving is done by day; even on a rare night dive, your perspicacity is limited by the strength of your flashlight. Most scuba and snorkel trips are near the shores of continents or islands, so we rarely venture into the great open waters. We also explore only the shallowest fraction of the ocean: depth limits for recreational diving are around 150 feet, while the average depth of the ocean is over 12,000 feet. Consider, then, the enormity of what we do not see, compared with the infinitesimal sliver of marine life into which we peer. The dark depths, the open ocean, the seamounts and undersea ridges, the abyssal trenches, all are hidden from us, as are the creatures who therein reside.

WAVES OF DIVERSITY

Fishes were among the first vertebrates—animals with backbones—to appear on Earth, half a billion years ago. They survived each of the planet's five cataclysmic extinction events and multiplied in diversity during the rich years that followed. Many of the fishes we know today descend from recognizable ancestors in the Devonian, a geologic period that began over 400 million years ago and was characterized by an equable climate, both terrestrial and marine. In the sandy and silty bottoms of warm tropical seas, the so-called Age of Fish had begun.

From primitive scaled creatures with stubby, fleshy appendages arose a lineage of fish that possessed true fins, stiffened with bony spines. Three pairs articulated to the body: pectoral (the parrotfish's wings) from the flanks; pelvic and anal fins emerged aft, from the belly. Another fin, the dorsal, topped the back like a sail; a caudal fin tipped the muscular tail to provide thrust. Fish bearing such fins were magnificently successful at exploiting complex undersea environments. Across the breadth and depth of the oceans new species bloomed at an astonishing rate, giving birth to many of the fish families swimming the seas today. Hot spots of diversity erupted in coral reefs and rocky submarine peaks, those places with complex topography that nurture rich communities of algae and inverte-brates. The explosion of fishes found its epicenter in the island-speckled western Pacific, where the ocean's great age, its stability, and a galaxy of shallow environments offer unrivaled conditions for biodiversity.

Fishes live together, not alone, and are engaged in a constant evolu-tionary ballet with one another. Schools of sardine and herring developed intricate choreography to feed, and to avoid being eaten. Meanwhile, mackerel and tuna and other speedy predators adopted creative responses to hack this group defense and snatch a mouthful. Fishes help maintain their homes, like clownfish who fertilize an anemone abode, or humbugs who fan oxygen over the coral that shelters them. Colleagues as dissimilar as a grouper and an eel work together closely, flushing prey from narrow tunnels and engulfing them in a huge mouth. Little fishes enthusiastically groom larger fish; huge fishes tow petite companions across oceans. One fish even strikes up a touching relationship with a blind shrimp, who lacks sight but boasts a weapon powerful enough to defend them both. Such relationships, each a dance of coevolution, built diversity upon diversity as watery eons rolled through our planet's seas.

While the continents were undergoing geological upheaval and cli-matic turmoil, the oceans remained comparatively stable. Water heats and cools more slowly than air and is buffered—though not completely—from drastic fluctuations in temperature, acidity, and more. Rising and falling sea levels alternately drowned and desiccated coastal regions, rains drenched interior forests or left them as deserts, glaciers advanced like bulldozers, but oceans were left largely unaffected. During millennia of stability, fishes were free to adapt ever more exquisitely to their envi-ronments. Still, global extinction spasm sporadically rocked the oceans, sometimes wiping out as much as 90 percent of all species. And yet, the marine realm retained enough organisms in the wake of those decimating events to rebuild its diversity swiftly, often in just a few million years.

Our seas are exceptionally resilient, quick to recover diversity, eco-logical function, abundance, and productivity, if given the chance. This resiliency is at the heart of modern ocean conservation strategies. On islands where excessive fishing has devastated reefs and fish populations, the creation of protected "no-take" zones has proved immensely success-ful. With heartwarming speed, fish can recover to previous levels: total abundance surges, average size increases, and reproductive potential swells exponentially. These protected areas more than pay for themselves, through the spillover of juvenile and adult fish who boost productivity in neighboring marine sites. Fishers who initially opposed marine protected areas swiftly become converts, their staunchest supporters and most strident defenders. The recuperative power of the seas, their ability to heal themselves if insults like plastics and pollution are reduced, if excessive fishing is scaled back, is a reason for profound optimism. For every challenge that confronts oceans, proven solutions exist, and evidence abounds that the marine world, a vital contributor to life on this largely blue planet, can be kept healthy and productive.

The seas cover about two-thirds of the surface of the Earth, but because they are three-dimensional and deep, they represent 99 percent of all habitat on the planet. Imagine rolling together all the rainforests, marshes, deserts, pine woods, grasslands, deciduous forests, and every other terrestrial ecosystem into a ball; that sphere packed with plants and animals would be dwarfed by the colossus of water that is the marine realm. Immense storehouses of abundance and extravagant productivity, the oceans harbor some 2 billion tons of fishes, six times the weight of all the people on Earth. But fish abundance is not evenly distributed. The relatively unproductive centers of oceans are sparsely populated, while marine abundance flourishes in waters that are cold and rich in nutrients: over continental shelves where rivers discharge, in the oxygen-rich polar seas, and at the delivery ends of deepwater conveyer belts known as upwellings. These places are home to colossal fish schools that sustain the planet's biggest fisheries, whose nets and lines procure 100 million tons of seafood every year.

FEEDING A PLANET

> Perhaps I should not have been a fisherman, he thought.
> But that was the thing that I was born for.
> —Ernest Hemingway, *The Old Man and the Sea*

Since the first human strode across a dune to view the great sea, men and women have hunted fish. Our ancestors used lines, poles, hooks, spears, weirs, and nets to capture the silent animals that slipped stealthily through the waters. To this day, fish provide some 3 billion people on Earth with a substantial portion of their daily protein.[2] The bounty of the oceans granted meat that could be eaten directly, or smoked, salted, or dried for consumption during long, cold winter months. Advanced civilizations, such as Indigenous Americans of the Pacific Northwest, developed rich cultures bursting with art, storytelling, weavings, and dances all made possible by ready access to protein: breathtaking runs of millions of salmon. Coastal communities around the globe have subsisted on ample harvest of shellfish, crabs, barnacles, and other intertidal animals that could be plucked without resorting to a boat. Cultures on South America's coast invented an advanced monetary system using coins carved from the wine-colored shells of *Spondylus*, a local mollusk. In the Far East, algae have been gathered or farmed for millennia, providing valuable calories and essential minerals and vitamins.

Formerly, the prodigiously productive oceans met our needs with barely a shrug. Fish and crab and clam populations rebounded as fast as we could harvest them. Our population and our demand for seafood have grown, however, and fisheries have intensified: larger ships, longer voyages, bigger nets, stronger lines, an industrialized approach to a noble art. There is room still for the sport fisher, the hobbyist, and the coastal communities that ply their trade in boats of wood and fiberglass, risking their lives on the pummeling waves to bring a meal to families and neighbors. But massive and heavily equipped industrial operations are fishing the seas so thoroughly that hook-and-line small timers are being driven out, heading for home with boats nearly empty.

In response, the world has taken a dramatic step forward, rearing immense harvests of farmed fish in aquaculture facilities based on land and in the ocean. Tremendous financial and scientific investment has been directed to this effort. The yield from all aquaculture operations, both freshwater and marine species, now equals the total production of global fisheries. While capture of fish upon the ocean waves is sure to continue, there is a pleasing echo of history in humankind's evolution away from hunting wild game and toward the stewardship of animals reared in farms. If managed well, aquaculture can alleviate pressure on wild fish populations while satisfying global demand for seafood. We can quite literally have our fishes and eat them too.

SETTING OUR COURSE

> The sea, once it casts its spell, holds one in its net of
> wonder forever.
> —Jacques Cousteau, *Life and Death in a Coral Sea*

This book is an exploration of the sublime spectacle of exotic behaviors and curious adaptations exhibited by saltwater fishes. Thanks to nature programs and magazines, zoos and ecotourism, we know plenty about elephants, giraffes, chimpanzees, toucans, and tortoises. We have learned some of their fascinating life history stories: box tortoises pull into their shells for defense, elephants march for miles to communal graveyards, chimpanzees can use tools and learn sign language. These animals have become familiar to us. But that familiarity does not extend to the denizens of the oceans. We can hardly recognize a frogfish, describe the development of a flounder, or comprehend the size of a full-grown manta ray. Fishes are a mystery as deep as the sea. Did you know that marlins have an exotic heating organ to warm their eyes, by as much as 25 degrees, so they can see better while hunting in cold waters? Or that eels can use a compass, but only on a one-way journey? In the chapters ahead we will embark on an underwater safari, to discover some of these spellbinding and awe-inspiring creatures of the saltwater world.

The diversity of marine fish species is bewildering, and we cannot attempt to cover them all. Instead, this book divides the ocean into broad zones: the open ocean, the seafloor, and the shore. In each, chapters portray groups of fishes that behave in similar ways, such as open water predators like tuna and swordfish, or that live in particular environments like coral reefs, seamounts, or the deep abyss. Each chapter focuses on what makes those fishes unique, on their body shape and arrangement of fins, and how those forms represent finely tuned adaptations to their lifestyles and habitat. For example, tuna are streamlined, torpedo-shaped, and their fins fold into narrow slots so perfectly they lie absolutely flush with the body surface. These are adaptions to a life of pursuit hunting: efficiently cruising hundreds of miles of water in search of prey, then rocketing into schools of smaller fish to feed. In contrast, fishes that live on coral reefs typically are flattened from side to side, like a dinner plate stood on end, to permit slipping into narrow crevices for safety. Their fins may be long, or broad, or tipped with defensive spines, since they do not rely on them to swim great distances; their mouths commonly are projected forward on a long snout, the better to pluck tiny invertebrates from rock faces. Marine fishes present an enchanting array of forms, and each provides clues to the fish's life story, environment, and evolutionary history.

Form and function are inextricably linked, in fish as in many things. We will explore how fish behaviors are facilitated by body shape, fin placement, skin coloration, and internal morphology. Parrotfish—your sexually flexible alter ego of a few pages ago—largely subsist on a diet of stony corals and hard, calcium-rich algae. To harvest these encrusting food sources, their front teeth are sturdy, broad, and protrude almost like a beaver's. With them a parrotfish can take a bite of coral, itself a colony of soft polyps surrounded by a protective cement rampart. But like a baseball player with a mouth full of sunflower seeds, they must crush the cement to release the polyps before swallowing them. The interior of a parrotfish's mouth is studded with bizarre, molar-like teeth, which even extend down into a muscular throat. Those mouth molars and muscular pharynx pulverize the cement skeleton, releasing the soft, nutritious polyps for digestion. Rich in alabaster-colored calcium carbonate, the cement dust is then excreted onto the reef: parrotfish excrement is a principal component of the gorgeous snow-white sand of tropical beaches.

Because the oceans are such an alien environment to us, the adaptations and behaviors found there may be astounding to land dwellers. Fish are free from the size limits imposed on terrestrial animals by gravity— little wonder that the largest animals on Earth, great whales and giant squid, are exclusively marine. The oceans are deeper than Mount Everest is tall, and curious fish forms have evolved in the abyssal depths where they must adapt to the crushing pressure, extreme cold, and acute scarcity of food. Water holds far less oxygen than air, but sound travels faster and farther: damselfish and toadfish communicate habitually with whistles and chirps. Light penetrates only the first few hundred feet of seawater, and certain colors like red drop out almost immediately; however, chemical signals disperse superbly in water, so many deep-sea creatures are blind to imagery but communicate by pheromones. Each of these adaptations opens a window into the environment in which the fishes live and evolved.

Most fishes exhibit life history characteristics that place special constraints on the fisheries that pursue them, and the diners who eat them. While certain species are fast growing, like anchovies, and can be caught in great numbers, they undergo mysterious population fluctuations that befuddle agencies tasked with setting catch targets. Others, like orange roughy, are exceptionally slow growing and late to reach breeding age.[3] While early catches were bountiful, now restraint is required to ensure young fish can safely grow and reproduce. Large predators like swordfish and tuna have a disagreeable tendency to accumulate mercury in their muscles.[4] Coldwater fishes like cod and pollock can be spectacularly abundant, but they are sensitive to minute fluctuations of temperature and food availability. If overfished too dramatically, as unfolded in the northwest Atlantic, their populations can cross an invisible tipping point, never to return. Without knowing the life history of each fish, fishers cannot successfully catch them, but neither can the public contribute intelligently to their lasting stewardship.

The final chapter will dive into the challenges facing global fisheries. We will investigate the role of aquaculture in meeting demand for seafood, survey the current state of sustainable fisheries, and explore the swelling tide of ocean conservation. There are a great many successes to relate, in both marine conservation and in fisheries management. We should be buoyed by the breadth of proven solutions, and the abundance of data demonstrating their effectiveness. The island nation of Palau, in just one example, calculated the value of a live shark to ecotourism at $2 million, dwarfing the $100 a dead shark might yield the fishing industry.[5] Palau elected to ban shark fishing within its expansive national waters, benefiting not only the marine realm but also diverse local businesses and the underwater tourists who sustain them.

The unseen world of marine fishes may be unfamiliar to us, but their lives are extraordinary, surprising, even arresting. The oceans are full of diverse stories and spectacles, and you are about to enter the theater where they unfold. To truly understand the fishes of the sea, to learn about their forms and fins, their changing colors, the exhausting migrations they undertake, and the ways they greet and eat one another, is to enter a watery and wondrous world that is as spellbinding as a Shakespearean drama. Take a seat; the curtain, or at least the tide, is about to rise.

Part I
BIG BLUE

IN THE MIDDLE OF THE PACIFIC OCEAN, late in the year 1820, a colossal sperm whale rammed the *Essex*, a 240-ton whaling ship from Nantucket, crushing the bow and sending the wooden vessel gurgling to a watery grave. The high seas battle was dramatized by Herman Melville in his classic novel *Moby Dick*. After the whale struck, Captain Owen Pollard, first mate Owen Chase, and the ill-fated crew took refuge in three shallow-draft dories. These twenty men had the misfortune of being marooned in one of the most remote and lifeless places on the planet: the open ocean of the southern Pacific. Its seemingly hospitable, equatorial waters are so devoid of fishes and marine life that the castaways very nearly starved to death. In the weeks that followed, the men were unable to catch even a scrap of food to sustain themselves. More than half perished. In a gruesome plot twist, they even resorted to eating the deceased, in a wretched attempt to stave off starvation. Earth's marine world, in places and at times so bountiful that schools of fish can be measured in miles and by tons, here was so barren as to sentence Pollard and Chase to death by famine. How, they must have thought, can such warm tropical waters, so transparent and welcoming, afford us so little to eat?

Life in the oceans, as on land, starts with sunlight. The sun's rays filter into just a thin skin of the sea, down to only a few hundred feet. Living in those illuminated surface waters are tiny algae which, like the lowly grasses of the Serengeti, sustain a multitude of animals and the ecosystem that knits them together. Sunshine powers photosynthesis, the biochemical magic trick that turns carbon dioxide into sugar and starch (and oxygen). Photosynthesis fuels algae that sprout into a fast growing, and nutritious, floating lawn on which all other marine life graze. Such plant-like algae are dubbed phytòplankton, and their photosynthetic output is so immense that they manufacture nearly three-quarters of the oxygen in Earth's atmosphere.[1] Miniscule animals called zooplankton voraciously consume phytoplankton, creatures like copepods (tiny shrimp relatives, occasionally sold as "sea monkeys"), and both mini-animals and pseudo-plants are gulped and strained by a ravenous bevy of fishes.

Virtually the entire web of marine life rests on this proverbial lawn of plankton. Where algae struggle to thrive, the ocean is clear and fishing is meager; but where seawater is a thick, luxuriant broth, fishes abound. Schools of anchovy, sardine, and herring achieve spectacular abundance by feeding on zooplankton. Powerful tuna, marlins, and sharks hunt those schools. Gentle behemoths like manta rays and whale sharks are content to feed directly on the soup of plankton, sieving them from the water just as the giant whales do.

As the crew of the *Essex* discovered, the distribution of ocean productivity is agonizingly uneven. Much of the sea is barren, and marine life booms in only a select few regions, and on intermittent occasions. Essentially, if plankton can thrive, then marine life flourishes. Just like land plants, phytoplankton rely on sunlight and carbon dioxide for photosynthesis. They require nutrients to flourish, including nitrogen, phosphorus, and iron. And they need oxygen to power their own metabolism, growth, and reproduction. Where, or when, any of these requirements are lacking, the sea is a desert; where they are abundant, it is a banquet.

Sunlight only reaches the upper lamina of the ocean, about 5 percent of all the seawater in the world, leaving the vast depths cloaked in darkness and robbed of productivity. Carbon dioxide is widely available, and increasingly so in recent decades, but oxygen is most abundant only in cold seas. This happens because warm water accelerates the atomic dance of dissolved oxygen molecules, which more frequently bump one another from the aquatic dance floor and into the atmosphere. Most of the great fisheries of the world—sardines, halibut, herring, cod, and more—are found in cold, plankton-rich waters far from the equator.

Nutrients are plentiful near coastlines, where they are transported by rivers like the Mississippi, a watery goliath that injects around 2 million tons of nitrogen annually into the Gulf of Mexico.[2] Nitrogen and phosphorus fertilize enormous phytoplankton productivity that sustains immense fisheries, but the flow of nutrients soon sinks below the sunlit zone just a few dozen miles offshore. Not just river nutrients, but also the inert bodies of deceased plankton, fish, and all manner of marine life fall to the depths, succumbing to the relentless pull of gravity. Plankton counter this force by packing themselves with buoyant oils that float them near the sunlit surface; breakdown of those oils liberates a unique odor (dimethyl-sulfide) widely recognized as the characteristic smell of the sea.[3] Once it sinks beneath the surface, all that nourishment slowly settles to the ocean bottom, in whose inky depths it is greedily awaited by abyssal fishes.

There is only one force on earth that can lift marine nutrition off the seafloor, and that is an upwelling, a vertical current that carries nutrients to the surface like an escalator. Upwellings are driven mostly by global wind patterns, with the trade winds wafting steadily from the east throughout the tropics and the westerlies gusting in the opposite direction in temperate latitudes, both blowing surface water away from the margins of continents. Deep water is drawn upward to fill the void, and a rich brew of fallen plankton, river sediment, phosphorus, fish feces, and other nutrients rises to provoke booms of plankton and explosions of fish populations. Along with the polar seas, the upwelling zones off the west coasts of North and South America, and Africa, are the richest and most productive on the planet.

Unfortunately for Captain Pollard and his crew, their ship had sunk about as far from marine productivity hot spots as one can get. The middle of the south Pacific Ocean boasts sunny skies and warm breezes, but low oxygen levels, no nutrient inputs, and no upwellings. They were stranded in the marine equivalent of the Sahara Desert. The sperm whales hunted by the *Essex* had gathered in these waters not to feed, but to give birth: the equatorial latitudes offer a balmy and tranquil nursery for newborn calves who would not survive infancy in icy polar waters. Only once they are generously fattened by their mothers' rich milk can the young leave the tropics and swim poleward, to feed in the fruitful but frigid seas found there.

Other than polar and coastal waters, much of the world's ocean realm is cursed with low productivity. Upwellings, for example, where nearly half the fishes on Earth are caught by hooks and nets, account for only one percent of all ocean waters.[4] The remainder of the seas present slim pickings. In the barren zones, predators must be exceptionally mobile, cruising vast empty spaces in search of food that is few and far between.

To meet this challenge, tuna and marlins and other top carnivores have evolved into pinnacles of hydrodynamic efficiency that boat designers and Olympic swim teams can only dream of approaching. Powerful and graceful, these finned predators literally crisscross the globe to find outbreaks of phytoplankton and gorge themselves on the bountiful schools of fish that attend them.

One of the most spectacularly fertile places in all the world, and a dining destination for millions of predatory fishes, is the massive upwelling system off the coast of Chile and Peru. There, a combination of polar and deep-ocean waters stimulates a level of productivity that beggars the imagination. Stretching over 2,500 miles, the Peruvian upwelling sustains the world's greatest population of anchovies and supports the largest fishery on the planet.[5] We may regard these small, oily fishes merely as a salty pizza topping, but in the ocean they are highly evolved machines for transmuting plankton into muscle. If Captain Pollard's ship had tussled with the sperm whale near this upwelling, he and his men would have encountered schools of anchovies so dense they could have practically walked to shore on the backs of more than a trillion fish. So copious is their abundance that annual harvests by the Peruvian fishing fleet comprise a fifth of all fish landings on the planet.[6]

For schooling anchovies, the danger does not relent beyond the edge of a knotted net. All manner of predators devour these hand-sized delicacies, plump with oils from feasting on phytoplankton. Bluefin tuna, striped marlins, mackerels, swordfish, and even dolphins commute from far-flung corners of the sea to fatten themselves. But menace can also arrive in the form of a furred sea lion striking from below, or a feathered seabird plunging from above. The immense schools of anchovies sustain sprawling colonies of seabirds, from penguins and boobies to gannets and pelicans. Collectively, those fowl have been dining on anchovies for millennia, and carrying their catches back to hungry beaks in millions of nests that liberally speckle the coastline and nearby islands. Powered by bountiful anchovy feasts, those humble seabirds were not just rearing chicks, but amassing an invaluable resource that would solve one of the greatest problems in human history.

1 ALL TOGETHER NOW

Anchovy, Sardine, and Herring

ON JULY 23, 1839, the aptly named ship *Heroine* glided into the busy harbor of Liverpool, England, carrying a seemingly innocuous cargo that would change the course of history. Thirty dusty sacks were aboard, collected from the coastline of Peru and ferried heroically across the high seas (along with bundles of cinchona bark, a malaria cure). Those sacks carried the concentrated bounty of the world's most productive ocean upwelling, and they were destined to nourish an upheaval of their own, an agrarian revolution that would feed millions of people. Within five years, a booming trade would mushroom from this modest sample, with thousands of pounds, then hundreds of thousands of tons passing through the very same port.[1] The sacks were packed with brown gold: dried bird guano from South America, the richest fertilizer the planet had ever seen.

Famed explorer and naturalist Alexander von Humboldt noted in 1802 that Peruvian fields were fertilized with seabird guano, and three years later he brought a few samples home to Europe.[2] Some of the packets reached the brilliant Cornish chemist Humphrey Davy, who found the guano exceptionally rich in phosphorus and nitrogen, both elements critical to plant growth (Davy also coined the term "laughing gas" after noting that experiments with nitrous oxide left him giggling). The discovery arrived at a critical time in the history of European agriculture. Yields were declining across the continent just as populations soared. Farmers had long relied on spreading animal manure onto overworked fields, but famine everywhere hovered menacingly and something new was desperately needed.

That something new was imported fertilizer. While horses and hogs and cows provided abundant manure, its nitrogen and phosphorus came from the same farms, often the very same fields, where they were applied as fertilizer. Guano would allow European farms to spread on their croplands the end product of millions of seabirds—gannets, cormorants, boobies, pelicans, and more—eating anchovies for thousands of years, half a world away. The very term "guano" was derived from the ancient word *huano,* meaning dung or droppings in Quechua, the language of Indigenous South Americans with a long tradition of fertilizing their farms with animal manure. A Peruvian proverb heaps praise on dung, saying, "Guano, though no saint, works many miracles."[3]

The productivity of Peruvian schools of anchovy, known locally as anchoveta, was so vast that seabird guano deposits had built up to depths of 100 feet or more.[4] Europeans were soon slicing, shoveling, and blasting their way into these deposits, shipping tons to Europe to be processed into commercial fertilizer. Crop yields boomed in response, and the population of the continent surged, now that starvation was at bay. Fortunes were made, so vast that a few years later the Guano Wars between Spain and its former colonies erupted over who would control the precious South American trade. The ensuing search for unclaimed deposits prompted the US Congress to pass in 1856 the Guano Islands Act, which effectively authorized Americans to seize unoccupied islands anywhere in the world: "When any citizen of the United States may have discovered . . . a deposit of guano on any island, rock, or key . . . said island, rock, or key may, at the discretion of the President of the United States, be considered as appertaining to the United States."[5]

Seabirds prefer to nest where no predators appertain to attack defenseless nestlings, and where food is plentiful. These birds could scarcely find a better place for their hungry chicks than next door to an oceanic upwelling. Phosphorus and nitrogen are stirred to the surface by immense rising currents, provoking a boom of plankton. Phytoplankton capture sunlight, and zooplankton like copepods happily dine on the tiny algae, setting the table for an anchovy banquet. One female anchovy can release more than 10,000 eggs in a season, and mass mating events are timed so the eggs hatch in synchrony with plankton explosions.[6] Tiny larvae emerge, at first bearing a yolk sac, a tidy little bag lunch that sustains them for the first few days. But the yolk is swiftly expended, and the proto-anchovies must begin feeding by their fourth day out of the egg; if they cannot find food, the larvae will starve in as little as two or three days.[7] But if the anchovies can tuck into a copepod buffet they will grow quickly, graduating to juvenile-hood in a little over a month.

Gawky youngsters soon form colossal schools, their hyperactive metabolisms transmuting plankton into pounds of fish. In a few months a swirling bounty swims just beneath the surface, within reach of the seabirds who plunge into the glinting waters to gather food for themselves and their nestlings. During one bumper year in the 1950s, Peruvian seabirds caught some 2 million metric tons of these silvery little fish, equivalent to the weight of six Empire State Buildings.[8] About one-tenth of that gulped fish was excreted, by the adults and their flightless nestlings, onto growing mounds of nitrogen-rich guano waste. Century after century, phytoplankton were transformed into manure, and the piles rose ever higher. By the time these breeding grounds were targeted by European and American exporters, the guano had grown into veritable monuments to fertility.

Guano's arrival prompted discoveries in Europe beyond the direct benefits to crop yields. In 1844, German chemist Julius Unger isolated a chemical compound from seabird dung that would prove to be a fundamental building block of life; he dubbed it guanine in recognition of its source. Years earlier, crystals rich in guanine had been extracted from shimmery fish scales by a Frenchman who labeled the resulting substance *Essence d'Orient* because it could be turned into artificial pearls (formerly acquired by trade with Asia). Even today the lustrous, pearly shine of shampoo and nail polish is often contrived by these glossy crystals. Some silvery spiders even divert nitrogen to produce guanine crystals that give their bodies a ghostly, reflective shine.[9] The reflectivity of guanine is also vital to the vision of deep-sea fishes who rely on a mirror-like layer behind their retina to see in near-total darkness.

But the most important discovery of all came a century after Peruvian guano first reached Europe. Guanine was pinpointed as one of four nucleotide bases, the building blocks of DNA, whose secrets were unraveled by Francis Crick and James Watson.[10] Along with the other bases—adenine, thymine, and cytosine—guanine forms the backbone of DNA's twin strands, and staccato sequences of these four bases code for all the proteins in every plant and animal on Earth. This humble molecule, named for fertilizer squirted from anchovy-eating seabirds, comprises one-quarter of the alphabet in which the script of all living things is written.

Anchoveta (*Engraulis ringens*)

THE SCHOOL CAFETERIA

Anchovies, a family of nearly 160 species, are distinctive from their small schooling brethren: sardines, herring, and menhaden. Reaching just over a foot in length (though many species are shorter), they resemble the others in their silvery scales, slender frame with dorsal fin set about halfway down the back, and acutely notched tail. But they are readily distinguished by their face: an underslung jaw, downward-pointing mouth, and pronounced snout have earned them the uncomplimentary epithet of "pig-like." That snout contains the poorly understood rostral organ, a dense agglomeration of sensory antennae that puzzle scientists to this day. The perplexing organ is packed with neuromasts, sensory cells that may give anchovies a refined ability to sense minute water movements ahead of them; in ancient fishes like coelacanth, however, the rostral organ confers the ability to sense electrical fields, like an on-board voltage detector.[11] This sensory apparatus likely helps anchovies detect the jittery swimming of their prey, skittish zooplankton like copepods, giving the fish an extra ability to anticipate their quarry's frantic escape moves. Anchovies, however, are not celebrated for being predators. They are famous as prey, abundant and delectable. For a tiny fish, the attack of a predator must be terrifying. Five hundred pounds of fins and teeth hurtle toward you at blinding speed, and in the open water there is nowhere to hide.

Anchovies play an invaluable role in marine food webs, whether they like it or not. They are the linchpin that connects the luxurious productivity of ocean plankton to the magnificent diversity of predatory fishes. On land there is nothing comparable: a small animal that grazes on tiny particles and is a mainstay food source for hundreds of larger species. It would be as if jaguars, lions, tigers, and eagles all subsisted solely on a diet of mice, and if those tiny rodents were so numerous as to carpet the plains with murine abundance (incidentally, a cluster of mice equivalent to a school of fish is charmingly called a "mischief of mice"). Unsurprisingly, anchovies are not particularly fond of their role as everyone's favorite menu item in the ocean's cafeteria. On their own, they can neither hide from an inrushing tuna or swordfish, nor can they outrun these muscular marauders. They can attempt the C-start maneuver, violently folding their body in half when a predator attacks, then rocketing into a 90 degree turn with an explosive contraction of the flank muscles. This escape reflex is aided by oversized neurons in the hindbrain known as Mauthner cells that can transmit impulses in as little as 10 milliseconds, ten times faster than the proverbial blink of an eye.[12] But a small fish is soon outpaced, so they have instead embraced a collaborative tactic to better their odds: swimming in mesmerizing, swirling schools that bob and weave around the thrusts and jabs of hungry predators.

Schools come in three flavors. Loose aggregations of fish that swim together, but without highly coordinated movements, are known as shoals. More tightly packed groups oriented all in the same direction are called polarized schools: changes in pace or course are mimicked instantly by their neighbors. Every turn, surge, dive, and about-face seems to be made by the entire school simultaneously, like a well-rehearsed ballet company. Fish may transition smoothly between these two flavors of schooling, as they continuously weigh the anti-predator benefits against the cost of feeding near schoolmates who would happily steal your lunch. But when the threat of attackers is imminent, a third form emerges. The panicking school retreats into itself, spaces between the fish shrink, and the densely swirling mass coalesces into a spinning sphere called a bait ball. The name was applied by fishermen who dip nets into such balls to catch bait for their hooks, an example of how a defensive strategy against one threat, predatory fishes, can leave anchovies at risk from another, fishing boats.

A catalog of benefits is delivered by schools to their enrollees, both offensive and defensive. If a predator strikes, each school member can just cross their fins and trust in the dilution effect: attackers are more likely to chomp any of a thousand schoolmates than you, the ultimate example of safety in numbers. But predator detection is a key advantage of schools: a few thousand eyes are more likely to see an incoming marauder. Characteristic movements by the sentinel reverberate swiftly through a polarized school, an alarm that travels up to seven times faster than the predator itself.[13] This communication advantage is known as the Trafalgar effect, after the battle of the same name during which Admiral Nelson used signal flags to inform his fleet of enemy tactics.[14] Schools and bait balls can also confuse their attackers, whether they are nearby or far off. From a distance, tightly packed clouds of anchovies may resemble much larger organisms, perhaps a shimmery whale, something a mackerel or tuna might not dare attack. If those predators do approach more closely, a different confusion effect sets in, the dizzying challenge of focusing on a single individual. When there are thousands of anchovies eddying around you, it is nearly impossible to lock on to just one to pursue and strike.

It is easy to see why small, defenseless fish so readily form schools and swim in such tight companies. But unlike a ballet, a polarized school has no choreography. Rather, each fish detects the movement patterns of his or her nearest neighbors and reacts accordingly. The school itself is referred to as an emergent phenomenon, one that arises from individual fish obeying a surprisingly simple set of dance rules. Schooling fish rely on their hyper-responsive vision to trigger those rules, but another sensory system is responding, one that is peculiar to fish and works only in a watery world.

Seawater is about 800 times denser than air, and it transmits waves of pressure far more effectively. Marine researchers working above the Arctic Circle in the 1960s could detect small underwater explosions at a distance of more than 700 miles, and there is evidence that a whale singing in Newfoundland can be heard by another in the Bahamas, over 2000 miles away.[15] Fish can hear quite well, but they also can detect vibrations in water that are below the threshold of sound. A unique network of pressure sensors, whose origin dates back more than 400 million years, allows fish to perceive and respond to minute movements of water produced by neighbors, prey, or attackers. The network is known as the lateral line system, because it manifests in most fish as a fine stripe down each flank. It is an unfortunately pedestrian name, akin to calling our own sense of ear-mediated hearing the pink flap system.

The streak of the lateral line is on closer inspection composed of a meandering row of fine dots. Each is an opening to an internal water-filled canal studded with tiny, finger-like projections called neuromasts. Vibrations in water outside the fish reverberate into the canals and tickle a dunce cap of fine hairs perched atop each neuromast. The slightest deflection of these hairs triggers a nerve impulse that registers the velocity, and direction, of the vibration. Fish neuromasts can detect the faintest of water pressure waves, and they are as much as twenty-two times more sensitive than our own fingertips.[16] Their resolution is also astonishing: blind cave fishes, who rely wholly on lateral line signals, can distinguish objects just four one-hundredths of an inch apart.[17] So acute is this system that a fish can feel the movement of its neighbors and follow the rules that convene a school, even if it is fitted experimentally with blinkers covering its eyes.[18] Fish can school using only vision, if their lateral lines are severed, but their ability to synchronize the speed and acceleration of their neighbors suffers.

Many of these experiments were performed on fishes that shoal and school in small numbers, a handful to a hundred. But anchovies and their cousins, in whose massive schools millions of swirling and turning fish are enrolled, require even more acute detection of changes in their neighbors' speed and direction. In these small schoolers, the lateral line systems have been enhanced in myriad ways to achieve extraordinary sensitivity, rivaling even Spiderman's.

First, the sensory canals that normally stretch along the flanks are concentrated around the head, where they are better able to read the movements of schoolmates up ahead, and where they can transmit signals that much more quickly to the nearby brain.[19] The canals also branch and braid around the eyes and are linked to the inner ear, where they may feed signals into the auditory system. Some species evolved connections to the swim bladder, a gas-filled sac inflated for buoyancy control, but which can amplify the vibrations resonating from outside, like a taut drum.[20] Anchovy's distant relatives, the longfin herrings, even have modified ribs, with small cavities inside them; the chambers' small size may extend the range of frequencies that can be detected.[21] Many species also possess mysterious inflated canals inside the skull behind the eyes, called bullae, that also connect to the swim bladder; it is not known whether the bullae send signals to the ear, or to the sensory canals, or both.[22] It is clear, however, that the linked network of canals and hollows helps transmit and magnify vibrations, like the sound of a saxophonist echoing through the tunnels of a subway platform. Small schoolers have packed a lot of technology into their tiny heads, which operate as some of the keenest motion detectors on Earth. And yet, their predators have a few tricks up their proverbial sleeve as well.

CRASHING THE SCHOOL DANCE

I can see no limit to the amount of change, to the beauty
and infinite complexity of the co-adaptations between all
organic beings, one with another.
—**Charles Darwin**, *Origin of Species*

When one fish evolves a behavior, like schooling, that protects them against such marauding predators as swordfish, the hunters do not just fold up their tents and go home. Instead, they counter with novel hunting strategies of their own to disrupt the defense. The schoolers then respond in kind, and the cycle of coevolution between assailant and quarry takes another turn. Schools today are the refined result of millennia of adaptation and counter-adaptation, but their defense can be breached by even one new innovation. And predators are innovating all the time.

Many predators hunt in packs, the better to disrupt schools of prey. Spotted weakfish (*Cynoscion nebulosus*) prey on menhaden, small schoolers in the herring family, and have adopted a coordinated attack strategy. Menhaden meet the thrust of a single weakfish with a school response known as the fountain maneuver, in which the ball of menhaden flees the weakfish, then splits and turns to swim past it, then turns again as a reassembled school behind the bewildered attacker.[23] But spotted weakfish have been observed gathering in small phalanxes and charging as a compact line. The school, when faced with multiple predators, becomes fragmented and discombobulated and can no longer communicate and respond smoothly (the Trafalgar Effect is negated).[24] Indo-Pacific sailfish (*Istiophorus platypterus*) also hunt in groups, with one powerful attacker after another hurtling through schools of sardines, like a relay race. Though not all charges yield a mouthful of prey, nearly every sailfish manages to injure a few sardines as it plows through the school slashing its sword-like bill. The accumulating injuries wear down the school's defenses, as wounded fish tire or fail to swim in synchrony; once this happens, the sailfish are increasingly able to pick off sardines from the battered school.[25] Other fishes, like tuna or silky sharks (*Carcharhinus falciformis*) will gather in massive hunting parties of a hundred individuals or more and literally herd and harass their prey into tight bait balls. Once the prey, quite often anchovies, are swirling in a compact sphere, the hunters will slam through the middle of the school, again and again. While you might think the center to be the safest, predators have responded to this strategy, and several species specialize in aiming like an arrow at the heart of the target. Their repeated body blows result in a splintered school, and some well-fed tuna.

Even without a squad to assist the hunt, open ocean predators have developed a toolshed of techniques to overcome the defenses of small schoolers. Sailfish, when not hunting in a group, rely on a creative method that takes full advantage of their long, distinctive bill. A sailfish creeps slowly to the wall of a bait ball, then delicately inserts its bill into the churning fish, so gently that they swim along unperturbed. The sailfish then selects a likely target and taps it lightly.[26] This must come as something of a surprise to the sardine. One moment you are swimming happily among thousands of mates, the next a 1000-pound predator with a 3-foot spear is poking you in the flank. If the disoriented sardine loses concentration and disengages ever so slightly from the school, all advantage is lost, and the sailfish takes it in a single gulp.

Another solitary hunter is the thresher shark, a distinctive species whose tail sports an impressive, scythe-like upper lobe nearly as long as the entire rest of the body. Threshers scour deep blue seas for schools of sardines, but if they attack with a direct charge, the school pulls the fountain maneuver and the shark comes up empty. Instead, a thresher will swim alongside the school, fold its body almost in half, and let fly with a powerful slap from the tail. So abrupt and powerful is the tail's whipping motion that the tip reaches speeds of nearly 50 miles per hour.[27] Sardines clobbered by this rubbery lash are stunned or killed outright; the thresher wheels quickly once the slap has done its work, to scoop up its hapless victims.

The evolutionary struggle to defeat schools may have even played a role in the coloration of ocean predators. Researchers from the University of Cape Town noticed a similarity between school-hunting animals as unrelated as striped dolphins and African penguins.[28] Both move quickly through the water, both attack schooling fishes, and both are marked by distinctive black-and-white stripes running from stem to stern between black backs and white bellies. The conspicuously streaked penguins had been observed swimming rapid circles around small schools which then mysteriously disbanded: the school depolarized, leaving the schoolmates swimming higgledy-piggledy, unprotected, and ripe for the taking. In an ingenious laboratory experiment, the researchers menaced schools of anchovies with various penguin-like facsimiles. The anchovies, swimming in a close-knit school, reacted not at all to a translucent plastic model. When rushed by two-toned, black-and-white version, the school split in the fountain effect but largely continued their coordinated swimming. But, when a realistically striped replica was sent into battle, the school depolarized in two of every three trials and could not continue swimming in unison. The researchers theorized that the visual stimulation from the fast-moving stripes may overload the anchovies' nervous system with information: perceiving the stripes took up so much sensory bandwidth that the little fish were unable to adequately process their neighbors' positions, and the school collapsed. In other words, the penguins were hacking the anchovy security network.

Just as the strategies of predators continue to evolve, so too have the tactics of the small prey who are themselves hunted by schooling fishes. Copepods, amusing little crustaceans that row through the water with two tiny oar-like arms, are one of the favored targets of small schoolers. Sardines, anchovies, herring, and more dine voraciously on them. The response by copepods has been to develop a lurching swimming style: they jump sideways in the water, by slashing just one of their oars, to escape an onrushing herring's maw. In response, juvenile herring coordinate the distance between each fish so the hunting school forms a grid matching the distance a copepod can jump.[29] A little crustacean sidesteps one oncoming herring only to land precisely in the path of the next, literally rowing from the frying pan into the fire.

Another, positively brilliant defensive innovation evolved in plankton as far back as 400 million years.[30] When set upon by a school of hungry fish, some plankton glow with a blue-green light. This bioluminescence, powered by the interaction of the devilishly named compounds luciferin and luciferase, is nearly the same as the light of fireflies (beetles, in truth). When activated by the roiling waters of an attacking school, the plankton's glow illuminates marauding fish like security lights around a warehouse. Now the tide of battle has turned: the erstwhile predators become prey themselves, as the light summons even larger fishes to feast on them. Mackerel, tuna, sharks, and more home in on the glowing waters where they are sure to find anchovies, herring, or sardines. Plankton plying this so-called burglar alarm strategy take advantage of the adage "the enemy of my enemy is my friend" to gain some relief from the torment of their schooling hunters.

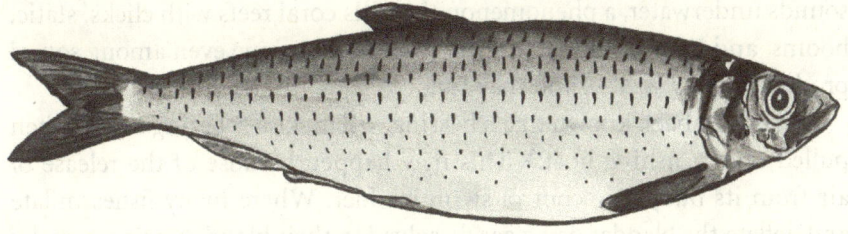

Atlantic herring (*Clupea harengus*)

FOLLOW THE RED HERRING

> There is no family of fishes and no group of aquatic
> animals that contributes so largely to the support of the
> human race as the herrings.
> —Hugh M. Smith, *King Herring*

Although schooling behavior evolved in anchovies, herring, and their relatives primarily to limit their own losses to predation, there is one predator on Earth that has turned the stratagem against them. For several thousand years, fishing boats have sought out schools of fish, encircled them with ever-larger nets, and hauled them out of the water to feed people and industry. Enormous catches are converted to fertilizer, fish oil, and protein for a hungry planet. And just as vast populations of anchovy made Peru and Chile into fishing powerhouses, the vast schools of herring that ply the northern seas built mighty empires in Europe and led others to their downfall.

Herring, despite all the trouble they have provoked, are handsome and unassuming little fish on first glance. They lack the pig-nosed snout and underslung chin of their anchovy cousins, sporting instead a slightly protruding lower jaw that hinges in front of, rather than behind, their eye. Herring resemble other small schoolers in their silvery color, forked tail, and elongated body form. They reproduce quickly, and in large numbers, and usually swim in schools. They favor cooler and more temperate waters like the North Atlantic, a distribution that has for centuries made them a target of European fisheries. They do exhibit a few odd characteristics that distinguish them, including the production of sound. Many fishes make sounds underwater, a phenomenon that fills coral reefs with clicks, static, booms, and honks. But herring are somewhat unique even among sound producers: they sneeze, and they fart.

It has long been known that a herring will make a sneezing sound when pulled onto a fishing boat.[31] This may happen because of the release of air from its buoyancy-control swim bladder. Where many fishes inflate and deflate the bladder using gas dissolved in their blood, herring actually gulp air from the surface to puff up the swim bladder. Several fishes who favor nearshore waters do this as well. To prevent air escaping the bladder by seeping across its walls, the lining is impregnated with none other than guanine, the compound that made small schoolers famous, and which has the added value of slowing gas diffusion. When a herring is caught and hauled rapidly to the surface, the tightly sealed swim bladder can expand rapidly due to the reduced water pressure, akin to taking a party balloon on an airplane. The only route to relieving this pressure is by expelling air through their mouth and nostrils, which may explain why some herring give a wet, explosive blast not unlike a sneeze.

There is another route for releasing pent-up air from the swim bladder, and it may explain the unique and slightly comical explosive blasts herring emit underwater. For years researchers detected repetitive ticking sounds made by herring. High resolution cameras revealed that some species expel a bubble from the anus that coincides with the ticking.[32] To make matters even more amusing, the acronym for the fast repetitive ticking, or FRT, sounds an awful lot like "fart." Whoever said scientists were humorless had not spent time with FRT-specialists. Those sounds, however, almost certainly have a use beyond entertaining researchers and fishing boat crews. Herring most likely are communicating with their schoolmates, producing what ornithologists refer to as a contact note, something that says, in effect, "I'm over here." In the dark, Pacific herring (*Clupea pallasii*) emit more FRTs than in the light, suggesting the ticking sounds play a role in maintaining group cohesion in the absence of vision.

Despite, or perhaps because of, these odd and amusing characteristics, herring have been much beloved across Europe, provided an invaluable source of food for centuries, and served as a wellspring from which numerous colorful expressions have been drawn. When eaten fresh, they have a strong flavor due to their high oil content. But in older times, pungent fish were the food of the poor, and a plateful would send rich lords and ladies into paroxysms of nose-clutching. The expression "dead as a herring" referenced the indubitably deceased nature of a stinking fish. That powerful smell was occasionally used to distract hunting dogs, putting them off the scent of the fox or quail, a practice that gave rise to the term "red herring," meaning something that draws attention away from the matter at hand. Sherlock Holmes mysteries, for example, relentlessly introduce seemingly panicky or guilt-stricken characters who serve as mere distractions while the true killer lurks in plain sight.

A herring became "red" when it was preserved in salty brine, a common preservative that allowed the highly seasonal catch to be shipped across Europe or stockpiled for leaner months. When butterflied and brined whole, a herring becomes a kipper, that famous fishy addition to English breakfast or high tea. Sometimes, kippers were made by simply gutting a herring and hanging it on a line to dry, a method much used when salt became scarce. Someone in England who had been framed might say they were "done up like a kipper," where the American equivalent would be "hung out to dry." And in London, when Parliament banned salmon fishing in the River Thames for several months each year, locals relied heavily on dried herring: the expression "kipper season" came to mean any period when work is scarce and belts must be tightened.

Herring have even influenced European geography, with numerous cities having been sited just to take advantage of plentiful spawning and feeding grounds, counting Copenhagen and Amsterdam among them. Other regions rose to prominence thanks to salt mines that supplied brine to preserve the catch. Thus the map of Europe to this day is imprinted with the atlas of the herring industry. So sought after was the wealth tendered by herring fisheries that conflicts inevitably broke out, between boats and towns and even nations. The 1429 Battle of Herring pitted French forces against a British caravan bringing supplies to English soldiers, then laying siege to the city of Orléans.[33] The caravan was loaded with herring and outwitted the attackers by literally circling the wagons before springing a counter-offensive. Those fish would go on to fuel a victory that marked the turning of the tide against the French in the Hundred Years' War, and incidentally Joan of Arc's first defeat in battle.

Hundreds of years later, changing sea temperatures and currents have conspired to shift the spawning and feeding grounds of many commercial species of fish and have provoked renewed hostility between fishing nations. In 2013 the Faroe Islands announced its intent to triple their allotted herring quota because the fish were becoming more abundant and staying longer in their national waters. The United Kingdom and the European Fisheries Council responded angrily to this unilateral corner- ing of a shared fishery and imposed import bans that crippled the tiny nation's industry. Negotiations eventually resolved the issue, but Senior Fisheries Biologist for the Faroe Islands Kjartan Hoydal noted wryly, "cli- mate change will break up all agreements—there will be chaos."[34]

The value of herring was not always measured just by the pounds of stored food it provided, or by its contribution to national treasuries. Her- ring for many years constituted an important source of oil, fertilizer, and even explosives. Before the advent of widespread electrical grids, Norway powered street lights with herring oil pressed in large factories that dated back to the 1800s. This habit, so valuable in a northern country with long dark winters, persisted in many isolated towns until the 1950s. After the oil was squeezed out, what remained of the fish was minced into fertilizer for farming, an echo of the Peruvian guano industry that long before had buoyed the fertility of overworked fields. By the mid-1900s fish rendering was so lucrative in Norway that it consumed 95 percent of the national catch.[35] Industry factories often tortured nearby residents with the over- powering stench of volatile compounds like trimethylamine, otherwise known as essence of rotting fish. But it was a different kind of volatility that made fish oil famous.

In 1847, Italian chemist Ascanio Sobrero discovered nitroglycerine, a compound that could be synthesized from nitrogen-rich fish oils. One of his fellow students was Alfred Nobel, who later built an immense for- tune on the production of dynamite and used that wealth to bankroll the world-famous prize awarded annually in dozens of disciplines. Back in Norway, fish oil pressed in the Lofoten Islands was converted to nitro- glycerine explosives during World War II, but the facilities were seized by the Nazis after they invaded the defenseless nation. One of the first war- time acts undertaken by the British armed forces after the heroic rescue at Dunkirk was bombing the Lofoten factories to prevent the Germans from controlling nitroglycerine and using it against the Allies.

As a general rule, only small and plentiful fish are destined for oil and fish meal rendering. Large species like tuna and salmon fetch better prices as fillets, steaks, and whole fish. But the expensive and tedious labor of gutting, deboning, and filleting pint-sized herring, sardines, or anchovies makes it much cheaper to process them by the ton. Thus fish meal and oil-rendering industries focus on small species that reproduce copiously and aggregate in easily captured schools. Those fishes produce huge numbers of eggs, and if conditions are favorable the offspring grow quickly to reproductive maturity. A single female herring can produce more than 30,000 eggs per year;[36] a sardine yields over 100,000 and may spawn every year for a decade or more.[37] The trick, however, is whether there will be enough food for all those fishy offspring once that multitude of eggs hatch. If phytoplankton production is high and there is abundant zooplankton to eat, the juveniles will survive their post-hatching hunger, and the population will boom in the following year. But, if conditions are unfavorable and the sea's buffet line is empty when the larvae emerge, they are doomed to starvation and next year's population will plummet.

Boom and bust cycles of abundance are typical of small schooling fishes. One year the ocean boils with huge schools, the next it is quiet and still. Fisheries have had to adapt, sometimes painfully, to the cyclical nature of the seas. One of the strongest drivers of ocean productivity is seawater temperature, and even modest fluctuations can cause populations to plummet, or shift, as seen in the case of the Faroe Islands. In Peru, fisheries biologists spotted an intriguing trend in the catch data. Poor harvest years were typically sandwiched between better years, and the collapses seemed to occur at regular intervals. It was the first clue that planet Earth was experiencing intermittent vacillations in ocean temperature, which we now know as the El Niño phenomenon.

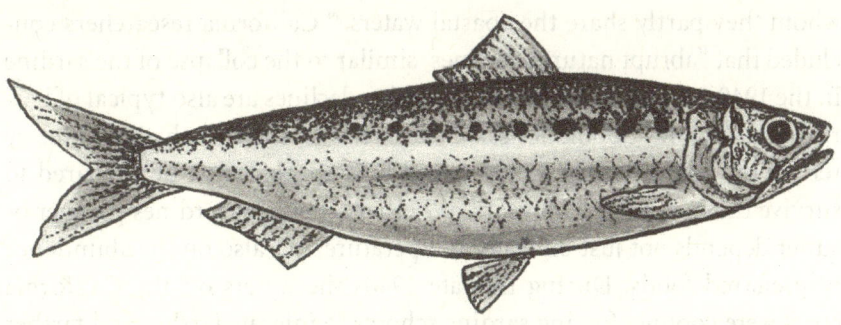

Pacific sardine (*Sardinops sagax*)

A TIN CAN PACKED WITH GOLD

> In the morning when the sardine fleet has made a catch,
> the purse-seiners waddle heavily into the bay blowing
> their whistles. The deep-laden boats pull in against the
> coast where the canneries dip their tails into the bay.
> —John Steinbeck, *Cannery Row*

Little sardines were once big business in northern California, making Monterey famous as the state looked to match the success of profitable canning operations in Europe. After an encircling net called a lampara boosted catch rates of these small schoolers, the state's fishing boats landed more than 27,000 tons in 1917 and reached a peak of 790,000 tons by 1937.[38] But by 1952 the sardine population had plummeted, and the fishery utterly collapsed. The industry's autopsy produced more vitriol than verifiable fact, with one side placing the blame squarely on unregulated overfishing and the other pointing a finger at unpredictable environmental shifts. The reality likely lies somewhere in the middle. California fishers had vehemently resisted all efforts to restrain their industry, despite warnings by the Fish and Game Division that quotas must be established to protect the catch. The fishery had boomed during a period in human history when it was inconceivable in some quarters that a natural resource could be exhausted, despite considerable evidence to the contrary (just ask the great whales).

Warm waters entice sardines northward, but when the seas chill they drift south and become outnumbered by cold-tolerant anchovies with whom they partly share the coastal waters.[39] California researchers concluded that "abrupt natural declines, similar to the collapse of the sardine in the 1940s, are not uncommon." Sudden declines are also typical of herring and anchovy populations, so it comes as no surprise that an industry reliant on catches of these boom and bust species must be prepared to survive the inevitable lean years. Whether schools of sardines prosper or suffer depends not just on water temperature, but also on the abundance of preferred foods. During the late 1940s the waters off the California coast were cooling, forcing sardine schools to migrate farther and farther south, until they were out of the reach of cannery boats. Counts of sardine scales in seafloor sediments show a roughly sixty-year cycle between abundance and scarcity, one that has characterized the population for some 2000 years.[40] The availability of delectable zooplankton, including tiny shrimps and nimble copepods, also teeters between bountiful and scarce as water temperatures swing from warm to cold. When the seas are productive, sardines can be found in dense schools during the day, feeding in the upper water column. Those schools typically disperse at night, however, when each sardine charts its own course through the dark waters; at dawn, the school reassembles.[41]

Visual hunters like tuna, mackerel, seabirds, sea lions, and the like are the primary predators on sardines and their diminutive ilk. The collective defense offered by a school is highly effective during the day, but once those predators punch out for the evening, there is less need for the protection granted by enrollment.[42] Zooplankton are also hunted by sight, with their predators—the anchovies and herring and sardines—eyeing these little motes as they bob beneath the sunlit waves and plucking them like cherries from a bowl. Consequently, zooplankton have evolved a feeding pattern of their own, in which they sequester in the dark depths by day, sometimes beyond 300 feet down, then ascend to the surface under cover of night to browse on a salad of phytoplankton. Photosynthetic algae, of course, must remain near the surface where life-giving sunlight penetrates. Thanks to the migration, hungry sardines can tuck into a more bountiful zooplankton meal near the surface at night than by day, often by several orders of magnitude.[43] Zooplankton's daily pattern of vertical migration was first detected by Navy sonar experiments that found a dense layer lurking in the deep. Fears that the blob on their screens were enemy submarines prowling off the coast were dismissed when scientists discovered this so-called deep scattering layer was an innocuous carpet of copepods and other organisms that rose by night to feed and dove at dawn to safety. One can imagine the sonarman's relief upon learning that the menacing subs packed with torpedoes were in fact innocuous crustaceans no larger than a flea.

In the early 1900s, when the waters off northern California were rich in both plankton and sardines, the cannery industry surged. Thousands of people were employed to crew fishing vessels, unload the catch, and process tons upon tons of sardines before they went into the tin. Early on, a small side business emerged to handle the discarded heads and innards, which were boiled down (rendered or reduced, in industry parlance) into fish oil and fish meal.[44] Rendered oil was used for soap, linoleum, and other products, while the meal was destined for chicken feed. A cottage industry even arose to turn sardine scales into pearl essence, the shimmery liquid that heralded the discovery of guanine.[45] But, as the labor cost of hand-prepping sardines for canning soared, the fishery shifted increasingly to rendering until it effectively swallowed the entire catch.

Sardines, like anchovies and other small schoolers, are rich in oil. It is part of what gives them their strong fishy taste, but it also dispenses a shoal of health benefits. The oils are a direct reflection of their diet. Zooplankton eat phytoplankton that often rely on buoyant oils to prevent them from sinking. The zooplankton accumulate oil with every bite, and when gulped themselves they pass those oils on to a sardine. One small serving of sardines provides loads of vitamin B_{12}, which supports a healthy nervous system. These little oily fish are rich in niacin (B_3) and B_2, along with many essential minerals. And they are packed with a now-famous group of oils known as omega-3 fatty acids. A health craze focused on omega-3 oils has swept the globe, making vast fortunes for suppliers, and spurring formerly quiet fisheries to pull an ever-increasing, and potentially unsustainable, number of small schoolers from the sea. Indeed, the very evolution of humans, of our brain capacity and even our speech, may be the result of an ancient history of eating seafood. And the poster child for the omega-3 craze, once fantastically abundant off the shores of North America, is the most important fish that nobody has ever heard of, the menhaden.

A FISH FOR EVERY STALK

> An acre thus dressed will produce and yeald so much
> corne as three acres without fish.
> —Thomas Morton, *The New English Canaan*

Five or six million years ago, in a marsh in eastern Africa or perhaps on its southern coastline, apes began adding seafood to their diet.[46] Shellfish were easily collected, fish were abundant, and crustaceans could be plucked from the bottom. Wading into the water to reach this new and plentiful food allowed the apes to stand on two feet instead of four, buoyed as they were by the warm salt water. Life became a little easier, and more apes adopted the aquatic menu. Over time, their brains were nourished by the oils that little fish and mussels added to their diet. They may even have begun to dive, holding their breath to reach shellfish on the bottom, the newfound breath control encouraging speech to develop. Ample food supported larger gatherings, and stable societies began to emerge. Modern humans were born from apes because of seafood.

So goes the Aquatic Ape hypothesis, first presented to the world in the 1960s, and hotly debated ever since.[47] Proponents argue that the grassy savannas, often thought to be the cradle of civilization, lack sufficient omega-3 fats to unleash the rapid brain development that sparked the meteoric rise of modern humans. Opponents point out that much of the evidence is merely a coincidence of timing, leaving unresolved whether seafood made humans, or humans emerged and later chose a marine menu. Nonetheless, if one turns the clock of paleontology forward a few million years to the days of Neanderthals, even more evidence emerges. Paleolithic hominids living in northwest Spain discarded piles of shells, fish bones, and crab legs that would be the envy of any bib-wearer at a Louisiana seafood boil.[48] Grottos around Pinnacle Point in South Africa show stunningly advanced cave paintings from contemporary Neanderthals who also fed richly on abundant seafood. The reliability and abundance of those foods likely supported a more stable society, with sufficient free time to dedicate to art and other pursuits, a progression nicknamed the cognitive revolution.[49] Across the planet, the coincidence is striking. The earliest complex human societies on earth all emerged where fish and shellfish were plentiful: coastal China and New Guinea, Mesopotamia, coastal Peru and Mexico, and the upper Mississippi River valley.[50]

Today, the health benefits of omega-3 oils are trumpeted by medical experts and industry advertising, echoed in a tide of books and magazines, and breathlessly amplified by celebrities, chefs, and social media stars. Despite the faddish hype, regular doses of fish oil do confer a wide range of genuine benefits. Of some 7000 people studied in the Netherlands[51] and southern France,[52] those who ate fish at least once a week were significantly less likely to develop dementia. The Mediterranean diet, first detailed in 1975, curbs heart disease, cancer, and diabetes, extends average life span, and reduces healthcare costs. The benefits are linked to the diet's substitution of seafood for livestock, and replacement of butter by olive oil, which more closely resembles fish oil in its fundamental chemistry. Such correlations with marine-rich diets have since been refined, focusing on several omega-3 fats known as EPA and DHA (shorthand for their tongue-twistingly convoluted chemical appellations). These fats must be obtained from our diet, as the human body cannot synthesize them; they abound in the red muscle flesh of oily fishes like sardines and herring thanks to oil-rich plankton. EPA and DHA have been proven to promote fetal nerve cell development, tamp inflammatory reactions, improve cardiovascular function, and even slow the advance of Alzheimer's disease.[53] When grandmothers a century ago forced kids to endure a nightly spoonful of cod liver oil, they knew instinctively what they were doing.

Menhaden are attractive little fish, deeper-bodied than slender sardines, their silvery-purplish flanks overlain with a golden sheen and decorated with small black beauty marks. They were once so abundant in the Gulf of Mexico and North America's eastern seaboard that in 1871 Captain Nathanael Smith described seeing a school "two miles wide and forty miles long."[54] Adorably referred to as "bunkers," and the juveniles even more endearingly as "peanut bunkers," menhaden of all sizes are ravenously pursued by a medley of predatory fishes, from striped bass to sharks, porpoises to tuna. In a mad rush to escape becoming someone else's lunch, they would occasionally throw themselves from the sea in such numbers that naturalist Mark Catesby reported, in the mid 1700s, a beach "covered with them at a considerable Depth, and three miles in length along the Shore."[55] Menhaden's prodigious reproductive capacity, and their ability to sustain so many predators, stems from their reliance on an equally plentiful food source, phytoplankton. Their algae-rich diet makes them plump and oily, prompting Catesby to describe menhaden as "excellent Sweet Fish, and so excessive fat that Butter is never used in frying."[56] So abundant are, or were, menhaden that for many years their annual catch outweighed landings in America of all other fishes combined. Ponder that statistic for a moment. In 1955, for example, 1700 million tons of menhaden were netted, while the total catch of pollock, halibut, salmon, swordfish, tuna, flounder, perch, cod, snapper, striped bass, mackerel, and every other commercial fish taken in the United States added up to barely half that haul, at 975 million tons.[57]

The oily nature of menhaden and a preponderance of small bones limited their commercial value as a food fish, until their worth as a source of oil and fish meal was discovered. Boiling a few fish in small kettles gave way to processing plants as early as 1850. Production rose as the collapse of the whaling industry left nations scrambling for a new source of oil, and by 1874 fish oil manufacture had outpaced whale oil by at least 50 percent. As in Norway, the discovery of petroleum displaced the fish oil industry for a time, but surging interest in the health benefits of omega-3s spurred the rendering factories to renew production and reach all-time highs. Signs of overfishing emerged as early as 1871, when a report by the newly founded US Commission on Fish and Fisheries concluded that populations had dropped to one-tenth of their levels in 1801. While a menhaden reduction industry association confidently announced that "the plenitude or scarcity of sea fish is wholly independent of the operations of man," at least one fishing captain lamented, "for the last few years the entire coast line, from end to end, has been run over almost every day by the fishing vessels . . . have we not killed the goose that lays the golden egg?"[58]

The scarcity of menhaden reverberated through the predators who relied on them, attracting the attention and provoking the ire of the sports fishing community. When striped bass, weakfish, bluefish, and others began to disappear from coastal waters, a coalition emerged to fight the rendering industry. Lawsuits, legislation, countersuits, and protests stretched on for decades, and the fight continues to this day. Eventually, all states north of Virginia banned entry by the reduction fleet. Even within Virginian waters the average weight of captured menhaden fell to a mere 6 ounces: juveniles were all that remained. In 2005, the Atlantic States Marine Fisheries Commission placed a cap on rendering catches of menhaden; however, in a cruel twist revealing the cloakroom power of the industry, that cap was set equal to the average annual catch, and thus represented no reduction whatsoever. Finally, after widespread public outcry the same Commission cut harvests by 20 percent in 2010. By 2017 the menhaden, their predators, and the ecosystem they compose were showing distinct signs of recovery, evidence that with proper fisheries management the fish oil business, sports fishers, and the marine ecosystem can coexist.[59]

As far back as the early 1800s, the oil-rendering plants were faced with a problem: what to do with the flattened menhaden bodies after the oil had been pressed from them? The answer had been provided two centuries earlier, when the pilgrims who settled in New England were saved from starvation by a Patuxent man, and an unknown fish. Tisquantum, whose name was shortened and anglicized to simply "Squanto," may have been the last of the Patuxent. Its members, and nearly all of the broader Wampanoag tribal confederation to which they belonged, had been ravaged by diseases presumably brought by the colonizing English. The pilgrims were poorly equipped for their adventure, knowing almost nothing of farming, carrying few seeds, and virtually no tools. Tisquantum took mercy on them, a charitable display given the suffering his people had endured, and introduced the white settlers to a fish he called *munnawhatteaûg*, a name from the Algonquin language meaning "that which manures." The name, soon shortened to "menhaden," was apt. Tisquantum taught the ersatz farmers to plant corn seeds in a mound of earth, and to dress that mound with a few menhaden as fertilizer. Soon, corn crops flourished, and the colony was saved from starvation by a man whose own people had been decimated.

Meanwhile, out at sea, menhaden, sardines, herring, and anchovies continued to convert plankton into vast schools of small, oil-rich fish. Seabirds continued to feast on their abundance, as did sea lions, sharks, and swordfish. The erratic and capricious nature of ocean productivity, however, meant those seabirds who had built towering mountains of guano would suffer inevitable years of tragedy. The famine that threatened the ill-equipped pilgrims could also devastate animals during hard-luck seasons. Amid the 1982–1983 El Niño event, some 8 million boobies, pelicans, and cormorants in Peru, a staggering 85 percent of the population, died from starvation.[60] Under the waves, predatory fishes who rely on small schoolers are bound to come up empty when shifting conditions convert last year's bountiful buffet into this year's food desert. Over time, mighty hunters of the open ocean, like tuna and sailfish, evolved to roam vast waters tirelessly and efficiently, pursuing far-flung schools thanks to body plans, musculature, and sensory systems all finely tuned to scour the seas for an elusive feast.

2 HOT BLOODED

Tuna and the Open Ocean Predators

TAKE A TABLE at any sushi restaurant and you will be greeted by molded mounds of rice, elegantly arrayed on a slab of wood, with gleaming slices of raw fish perched atop like glittering jewels set in a crown of white gold. A culinary feast awaits, and the color, texture, and flavor of each morsel provide clues to the hidden world of these extraordinary animals. A slice of Atlantic bluefin tuna (*Thunnus thynnus*), the world's most prized sushi fish, smolders with a deep burgundy-red hue. Bluefin meat is suffused with myoglobin, a protein that stores oxygen for slow-twitch muscles essential to the long-distance, marathon-like swimming of these powerful predators. A pigment rich in iron like its cousin hemoglobin, myoglobin dyes tuna muscles a deep crimson in the same way that iron in paint colors barns their classic red.

Savoring your first bite, you note a buttery smoothness that hints at delicate layers of fat interwoven with muscle. Those rich sheets of fat are the gas tank of a fine automobile, storing energy to fuel long journeys between remote hunting grounds. The ruby-red meat is firm, reminiscent of exquisite rare beef. Fish, unlike land animals whose stringy musculature is sheathed by rubbery collagen, have muscles made of repeating stacks of short fibers separated by delicate connective tissue. These segmented muscles power the side-to-side contractions of swimming and yield flaky fillets. Roving predators like tuna add robustness to those layered muscles, building up a unique firmness through constant use in pursuit of prey. The combination of flavor, color, and texture makes bluefin the most expensive fish on earth: a single bluefin sold not long ago for 3 million dollars, an eye-watering price of more than 5000 dollars per pound.[1]

Second on the platter is a slice carved from the flanks of a bigeye tuna (*Thunnus obesus*), another strapping hunter who swims in cool ocean waters and routinely dives below 1000 feet to feed. The eponymously large eyes allow this tuna to spot prey in those inky waters, like an owl hunting mice by starlight. Deep waters are also frigid, and both bigeye and bluefin warm their muscles to attain peak performance, sometimes to as much as 40 degrees above the surrounding water.[2] Blood-rich musculature is the furnace generating this heat, furiously burning fat and myoglobin-stored oxygen. The resulting bigeye meat is deep red, and almost as luxuriously flavored as bluefin.

You proceed to a precisely carved rectangle of yellowfin tuna (*Thunnus albacares*), admire its hue of pale rose, and savor the delicate flavor and modest firmness. Yellowfin infrequently dive below their favored shallow and balmy seas, but still have meat tinged with myoglobin. They also undertake occasional cross-ocean migrations powered by fat stores and are hefty animals; both traits give their muscles the supple consistency of soft butter. Next your chopsticks pluck a morsel of skipjack tuna (*Katsuwonus pelamis*), a small-bodied cousin similarly at home in sun-warmed waters. Unlike yellowfin, however, skipjacks rarely dive to hunt the depths. They largely lack slow-twitch muscles, relying instead on fast-twitch fibers to power lightning lunges at prey. The predominance of sprinter's muscles yields sand-colored steaks uncommonly served as sushi but widely eaten as the main species in canned tuna.

Last in line is albacore tuna (*Thunnus alalunga*), another species more famously eaten from a can, the so-called "chicken of the sea." This genius marketing moniker, dreamed up by Frank Van Camp, convinced millions of households to eat a fish previously considered unpalatable by linking it to a popular white meat. Albacore are also smaller-bodied, shallow swimmers, with muscles dominated by fast-twitch fibers. They harbor even less myoglobin and their meat is more pale—chicken to skipjack's turkey. Both these smaller tuna store less fatty oils than their larger, colder water relatives, and as a result the flesh has a mild and agreeable flavor.

Exiting the beachside restaurant, you glance across the sand to the rippling, starry surface of the Pacific. You wonder with some guilt about the tuna that you sampled, residents of that same ocean, and ask how much do you know of their daily lives? Today you were the top predator, a

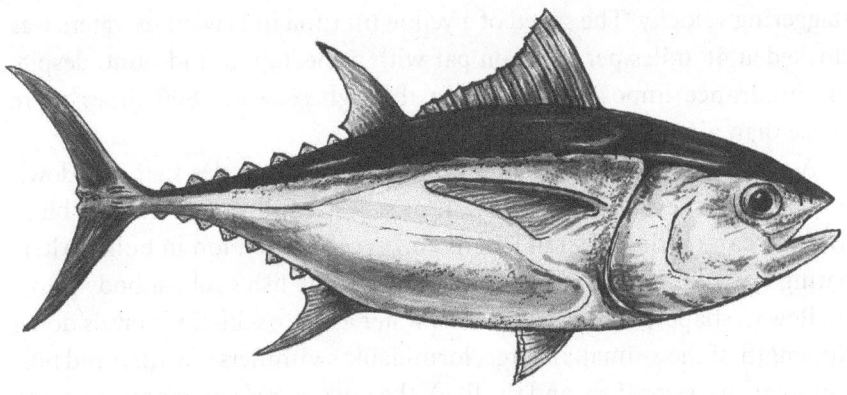

Atlantic bluefin tuna (*Thunnus thynnus*)

beneficiary of poles and hooks, lines and nets. But beneath those ceaseless waves and beyond the eyes of landlubbers, commanding and tireless tuna wield their own tools to pursue, snare, and devour prey for themselves.

PREDATORS OF THE HIGH SEAS

Tyger, tyger, burning bright, in the forests of the night.
—William Blake, *The Tyger*

When we think of wild animals stalking prey, closing with a burst of speed, jaws snapping to consummate the kill, our typical image is of a lion, a jaguar, or a tiger. But tuna, and other high seas species like speedy mackerel and bonito, colorful mahi-mahi, and titanic marlin and swordfish, are every bit as imperious as predators. They are brawny animals, their large muscles and long-distance journeys imposing substantial metabolic costs that must be met by regular, successful hunts. Prey must be found, not an easy task in the enormous, desert-like empty spaces of the sea. As far back as 60 million years ago, tuna and their ancestors slipped through the waters of the world's oceans in pursuit of a meal.[3] Over millennia they evolved into exceptional swimming machines, specialists in pelagic hunting far from land. Nearly every feature of these large-bodied predators, both external and internal, is tuned to achieve tireless endurance and staggering velocity. The speed of a yellowfin tuna in Hawaiian waters was clocked at 46 miles per hour, on par with a cheetah in mid-hunt, despite the hindrance imposed by pushing through seawater 800 times more dense than air.[4]

A tuna's primary engines are powerful flank muscles that run down both sides of these enormous aquatic missiles. A full-grown Atlantic blue-fin tuna can reach 15 feet in length and exceed half a ton in bulk.[5] Alter-nating waves of muscular contractions warp the fish's robust body into a shallow C-shape, pushing against the water as the oscillation travels down the length of the animal. In large, formidable swimmers like tuna and bill-fish (marlins, swordfish, and the like), the tail and rearward portion of the body provide the propulsion, while the forward half remains nearly stiff, stabbing through the water like a spear. They cover hundreds of miles in search of seas rich with prey, using an internal magnetic compass to help navigate the vast, empty expanses.[6] Tuna even accomplish celestial navi-gation thanks to a semi-transparent and light-sensitive "pineal window" atop their skull.[7] Such tigers of the sea push relentlessly through mile after mile of seawater, but conserve their energy for the final hunt. Blue mar-lins (*Makaira nigricans*), considered among the world's fastest fish, travel mostly at a leisurely pace of just 1 mile per hour for extended periods of time.[8] Once they accelerate to strike, however, these fearsome giants can exceed 40 mph.[9] Many open ocean predators literally travel around the world as they transition from juveniles feeding on local fish larvae to adults motivated by larger but more remote prey, and by the drive to reproduce. A black marlin (*Istiompax indica*) tagged and released near Baja California was later recaptured in New Zealand, a swim of more than 6000 miles.[10] By comparison, the English Channel, considered by many to be the acme of long-distance swimming, is only 21 miles across; a bather would have to recross the channel 284 times to equal the marlin's feat.

Most strong swimming fishes, particularly those that cover great dis-tances like billfish and tuna, are torpedo-shaped: their conical nose parts the water like a submarine, and their tails narrow severely. Water closes smoothly behind them, without making eddies that would cause drag like a race car trailing a parachute. To further cut down on drag, their scales may be small and smooth, as in tuna, or wholly absent, as in adult swordfish. Marlins have scales covered by a glossy layer of skin to further ease their passage through the water. The skin can change colors, to blend in with the background or to express emotion. Striped marlin (*Kajikia audax*) are known to flash silvery blue bars to the intensity of neon lights when they are excited. Many fishes possess multicolor vision, allowing them to perceive subtle communications sent by the color of their skin.[11] Mackerels, slender relatives of tuna, owe their pearlescent shine to miniscule but nearly iridescent scales loaded with guanine and sheathed by a thin layer of skin. The very word mackerel means "pimp" in medieval French, possibly because their shiny appearance lends them an aura of a slippery fellow sporting excessive jewelry. Even today, Tokyo slang slingers use the term "mackerel gals" to refer to women in glittery clothes.[12] Hiphop artists in the United States bent the term into their own lexicon, the hippest glossing themselves as "mack daddy."

Anything that juts out will brake an animal as it passes through the water. Ironically, in a drag race, drag is your worst enemy. Fins pose a conundrum. They are necessary for steerage, aiming the fish up or down, and serve as brakes when a hard turn is required. Pectoral fins, located foremost on the body, provide most of this control. But in tuna, their scimitar-shaped fins can nestle into shallow divots in each flank, so while the fish is cruising the pectorals are plastered to the body: stroke your hand over them and you will feel not the slightest bump or bulge. The dorsal fin is able to disappear completely, folding exquisitely into a deep slot like a Swiss Army knife. And in both tuna and mackerel, just forward of the forked tail, the body is decorated with two ranks of short serrations, delightfully called finlets. These finlets also reduce drag, in a way that today is being exploited by competitive swimmers. When water flows over a completely smooth surface, it slips imperfectly, tumbles, and tiny eddies develop; their backward motion sucks at the skin and slows progress. The finlets, however, trap a thin layer of water between them and the body. Ocean water now slides alongside that trapped layer, provoking less turbulence and reducing drag. Today, swimsuit manufacturers are making competition suits out of material with tiny teeth that mimic this effect in the never-ending quest for more speed in Olympic waters. But all these adaptations for high speed, in fish anyway, are for naught if those swimming predators cannot find prey to chase.

On the high seas, most of the life is found near the surface. Solar-powered phytoplankton are fed on by a diverse ecosystem of zooplankton: miniature crustaceans, diminutive jellyfish, precocious fish larvae, and more. The rich broth of the sunlit waters, a soup bubbling with life, is where small schoolers like anchovies and herring come to feed if they dare to run the risk. Meandering in well-illuminated waters, engrossed in pluck-ing tiny organisms for a morning meal, small fishes join a mortal game of roulette with their lives on the line. Predators like tuna and mackerel and marlins will find them, sooner or later, and launch violent attacks on the schooling prey. Most of these predators use a keen sense of smell to track groups of prey from a distance: yellowfin tuna can detect a single scent molecule in five trillion molecules of water.[13] Once the predators follow the odor to the quarry, they rely on sharp vision to make the final kill. In response, many smaller prey species pursue a simple strategy: feed by dark of night, hide by light of day. But where to hide? The big blue sea is featureless, lacking the cracks and crevices exploited for safety by their reef-dwelling cousins. The answer is to drop below the sunlit zone, descending as deep as 1000 feet, where they shroud themselves in the dimmest of waters, awaiting nightfall when they can rise again to feed in safety.

A wink is as good as a nod to a blind horse, but for a predatory fish in the dark, an empty mouthful of cold water is nowhere near as good as a meal. Evolution has provoked an arms race between how deep predators can hunt and how far down prey have to drop to find safety. First, like most animals that hunt in dim light—nocturnal mammals, frogs, owls—deepwater predators have enormous eyes. Their huge retinas have more power to detect scarce glimmers in coal-black waters, allowing them to spy the dim outlines of prey. In 2012 an eye the size of a softball washed up on the shores of Florida that once belonged to a swordfish (*Xiphias gladius*), a species that makes repeated deep dives to hunt lanternfish and other dark water species. Even more impressive is the 11-inch eye of a giant squid, who also stalks fish that inhabit the aquatic twilight, but is itself a favorite meal of sperm whales. Their colossal eye may serve as an early warning system, detecting incoming whales just beyond their sonar range of about 400 feet.[14]

Bigeye tuna, who dive deeper and more routinely than any other tuna species, have eyes half again larger than their closest relatives. Further-more, those oversized eyes have high levels of guanine in the tissue lying just behind the retina. Guanine acts like a mirror: photons that slip past the light-sensing retina without detection are bounced back for a second, reverse pass, effectively doubling the ability to see in dim light. This reflec-tive layer, dubbed a *tapetum lucidum*, is possessed by many nocturnal ani-mals and is responsible for eye-shine, the green or orange iridescence that bounces back to the viewer when a light or camera flash strikes the eyes. Sweet-natured house cats look satanic in snapshots because of the reflection from their night-adapted eyes. Bigeyes combine this doubly sensitive retina with oversized peepers, allowing them to stalk prey in deeper and dimmer waters than any other tuna. Managing the frigid cold in those inky depths, however, is another story.

PLUNGING INTO THE FRIGID DEEP

> They say the sea is cold, but the sea contains the hottest
> blood of all.
> —D. H. Lawrence, *Whales Weep Not!*

For centuries our understanding of tuna, marlins, swordfish, and other ocean giants was limited to the experiences of fishermen. Virile anglers like Zane Grey waxed poetic about the speed and power of these fishes and reported herculean battles on the high seas once the hook was set. But in the late twentieth century, a surprisingly different picture emerged. Stanford University marine biologist Barbara Block has spent years attaching telemetry tags to these giants, recording their depth, position, and the water temperature for weeks or months at a time. "We try to use the same chips that are in your computer, the same devices that allow you to talk to satellites and cell phones," Dr. Block explains, "and put them on big animals like white sharks and tunas, and we follow them across the globe. What we're trying to do is figure out how do big animals live in the ocean ecosystem."[15] The picture that emerged has shattered old stereotypes. Block's tags revealed that large ocean predators lead a fairly sedentary existence, swimming languidly near the surface, lolling and basking in the sunshine, rarely exceeding even a couple of miles per hour. "The fact that the cruising speeds of all these fish are exceedingly low suggests that slow swimming is a way to minimize the cost of locomotion over long distances."[16]

With improved battery life and dozens of collaborators, an ambitious program tracked these animals swimming across half the globe, and startling patterns emerged. "We got a glimpse for ten years of how the Pacific Ocean worked," Block marvels. "What we discovered was there was a pulsatile movement of the animals according to seasons. Animals you thought would wander everywhere were basically going away and coming home, going away and coming home . . . It was a finely-tuned periodicity much as you'd expect on the plains of Africa in which animals were going through large migrations on a seasonal scale."[17] Block's team also made a serendipitous discovery, that tuna undertake repeated staccato dives into deep waters, staying but briefly before returning to the shallows.[18] Why, they asked? The answer can be given in one drawn-out syllable: BRRR!

Three problems plague fishes descending into frigid waters to reach delicious and lethargic prey without chilling themselves to the bone in the process. Swimming muscles are debilitated, their contractions just half as quick with the 18 °F temperature drop common in dives to 3000 feet.[19] The chill robs swimming musculature of the burst energy needed to capture prey. Cold water also affects the efficiency of the most important muscle, the heart, making it pump less vigorously. Reduced blood flow slows oxygen supply to the body, further diminishing swimming vigor. Third, sensory systems respond less well in a cold environment, and in particular the retina of a fish's eye becomes less sensitive overall and less capable of detecting rapid movements like prey gliding across the field of vision. Deepwater hunters must adopt behaviors that limit their exposure to frigid water, and they have evolved key modifications to better tolerate repeated forays into the icy deep.

While fish generally are cold-blooded, large predators like swordfish, marlin, tuna, and mahi-mahi are homeotherms, capable of keeping themselves warm in all but the coldest waters. Alternating contractions of their flank and tail muscles burn fats and sugars in the bloodstream, a cellular blaze that warms the body and revs up swimming performance, giving them an advantage over the cold, sluggish fishes they are hunting. To conserve that valuable heat, their bodies have evolved over millions of years to slow its loss to the ocean waters. People in the cold feel it first on their extremities, with fingers, toes, and noses growing numb in an icy wind. In warm-blooded fishes, much of the heat generated by muscles is lost from the fins and tail, rather than toes. As we have seen, dorsal fins vanish into narrow slots, an anatomical disappearing act that also insulates those fins against the cold. To abate the loss of heat from their tails, tuna and their compatriot hunters rely on a plumbing trick called counter-current exchange. Warm blood traveling aft from the body flows in arteries lying directly alongside veins returning from the tail. The former warm the latter, so blood reaching the tail is pre-cooled and little overall heat is lost to the sea. This heat retention network can be dialed up or down by the fish. Wild swimming bigeye tuna diving into frigid waters use counter-current exchange to slow heat loss, but when ascending to the shallows they disengage the heat retention network, allowing balmy waters to warm the body more rapidly.[20]

Beyond physiological mechanisms, fishes can modify the amount of heat they generate through behavioral adaptations, in effect setting the furnace thermostat higher. When large ocean predators enter cold waters during a hunting dive, they shift from slow, wide swings of the tail to faster, tighter wiggles that produce more heat.[21] The more truncated stroke is akin to shivering, and it generates warmth at the expense of forward propulsion. That thermostat adjustment trick works only within a narrow range: albacore tuna can maintain their body temperature in waters above about 52 °F, but in colder seas, muscle heat alone cannot defend them, and their bodies began to chill. Once that happens, the only alternative is to return to shallower, sun-warmed waters, like Minnesota kids coming in from ice skating for a hot chocolate in the warming house. Tuna in frigid waters hunt by trampolining, explaining the bounce dives discovered by Barbara Block. They make short forays to depths where the temperature is too chilly for their internal furnace, staving off the cold as long as they are able with shiver-like swimming, and capturing as many slow-moving prey as they can manage; then they rise to the surface to warm up before another dive. Off the cost of Baja California, Kurt Schaefer and his team recorded yellowfin tuna diving repeatedly to depths below 800 feet, presumably hunting, and then returning quickly to the surface to warm up.[22] One yellowfin made as many as thirty dives in a day, plunging and rising about every twenty minutes. That's a lot of hot chocolate for one fish.

The bounce diving strategy only works when the proverbial warming house is open, that is, when the shallows are heated by the sun. Tuna make these repeated dives only during the day, but stick to the surface at night. Albacore tuna, who call both tropical and temperate waters home, bounce dive only in equatorial latitudes where the hot tropical sun provides welcome warmth after hunting at below 1000 feet.[23] In colder, temperate latitudes, albacore do not make regular dives, and as a consequence they subsist on a less diverse diet of mostly shallow-water prey. Bigeye tuna, who are larger bodied and more capable of thermoregulation, can routinely venture to depths beyond 1500 feet, but again only during daylight hours.[24] Blue marlin follow much the same pattern: in the Gulf of Mexico they repeatedly make daytime dives to just below the cooling zone found 200 feet down, then rise and loiter in balmy surface waters to recover.[25] This behavior, widespread among the high seas predators, confirms the enormous value of the prey schools lurking below the thermocline. Cold, however, is not the only challenge predators must surmount to hunt effectively in the deep; they must also contend with the risk of asphyxiation.

Muscles have simple requirements: like a campfire they need fuel and oxygen to perform. Fuel comes in the form of hunted prey, but without oxygen campfires do not burn and muscles cannot contract. Cramping sets in, fuel cannot be hunted, and death can follow. As in humans, oxygen is delivered to fish muscles by hemoglobin picked up by gills and shipped through the bloodstream. Tuna have some of the largest gill surface areas[26] and highest hemoglobin levels of any fish, both facilitating rapid and ample uptake of oxygen to power their carnivorous assaults.[27] They are challenged, however, because the ocean's oxygen minimum zone lies around 1000 feet, precisely where diving tuna go to hunt. As blood passes across a tuna's gills it cools rapidly, and that chilled blood goes straight to the heart. In land animals, hemoglobin's affinity for oxygen climbs as temperatures drop: cold blood is more "sticky" to oxygen. Normally beneficial—flapping wings and jumping legs are warm and need more oxygen—this relationship would starve a tuna's heart of the life-giving gas, as too much would remain lodged in the slushy bloodstream. Amazingly, bluefin and other deep divers show a reverse temperature dependence: their blood unloads oxygen more when it is chilled than when it is warmed.[28] Remarkable among animals, this mechanism ensures plenty of oxygen is supplied to the heart so that it can beat strongly despite the cold and pump oxygenated blood swiftly to the rest of the tuna's body. In the flanks where muscles are hard at work, carbon dioxide builds up as a by-product, much like a campfire belches smoke. That carbon dioxide causes blood to release oxygen, an acidity-mediated response named the Bohr effect a century ago. In tuna, the Bohr effect, rather than warmth, signals the hemoglobin to release oxygen to the swimming muscles. Diving tuna truly fit a poignant description penned by the University of Manchester's Holly Shiels: "a warm fish with a cold heart."[29]

The final key response by diving predators to the demands of hunting in cool waters is to protect their sensory systems. When icy water chills a fish venturing into the depths, it cools not only swimming muscles but also, critically, the brain and the eyes. A colder brain is less responsive and more slowly processes signals, like an older computer that sluggishly opens large, high-resolution photos. The retina of an eye at low temperature is less sensitive to light than in a warmer eye, and crucially it is less able to resolve fast-moving objects. Marlin and swordfish have excellent vision, and like tuna they frequently hunt in deep and dimly lit waters. They commonly stalk prey at depths beyond 2000 feet, but swordfish have been detected as far down as 9400 feet, where the waters are inky black and temperatures hover near freezing.[30] To counter the cold, these formidable hunters possess a unique heating organ that wraps their enormous eyes and rather small brains like an electric blanket.[31] Its brown-colored tissue is richly supplied with blood and packed with energy-producing cellular machinery that raises the eye temperature by 25 °F. As a result, the warmed retina enjoys a sevenfold improvement in what researchers call the flicker fusion rate, the speed at which the eye can take discrete pictures of the world whizzing past.[32] A delicious squid sprinting across the field of vision of a cold eye with a slow flicker rate would appear as an indistinct blur, but a warmed eye resolves the prey into crisp, stop-motion images. A fiery-eyed marlin can instantly measure the speed and direction of the harried squid and intercept its path for a welcome cold-water lunch.

Striped marlin (*Kajikia audax*)

SWORDFISH, MARLINS, AND SAILFISH

> A sword never kills anybody; it is a tool in the killer's hand.
>
> —Seneca, *Letters to Lucilis on Morals*

On a cool August morning in 1886, captain Franklin Langsford guided the fishing schooner *Venus* out beyond Halibut Point, a rocky headland some 30 miles northeast of Boston, on a quest for swordfish. It would be his last day at sea. Just before midday, the crew sighted a 300-pound swordfish swimming near the surface and hastily launched the ship's dory. With a powerful harpoon thrust, Langsford speared the fish, attaching a length of line and a buoy. The captain returned to the boat for lunch, leaving the swordfish to wear itself out. But the fish did not tire. Having dined, Langsford boarded the dory, grasped the buoy, and began to reel in the line. The swordfish pulled mightily, the taut line twanged, the captain hauled, poised with a lance to land a mortal blow. But in an instant that would change the course of Langsford's life, the swordfish turned and charged the boat. The line went slack, and the captain, leaning back in a small boat pitching on the waves, fell over backward. In that moment the swordfish, perhaps in a fit of vengeful rage, perhaps merely driven by blind instinct, hurled itself at the bottom of the dory and thrust its prodigious spear through the hull, piercing the wood by some two feet. Poseidon was not kind to Langsford that August day, because the captain had fallen in the dory on precisely the spot where the swordfish struck, and he was stabbed by the tip of the spear deep in his pelvis. "I think I am hurt, and quite badly!" he exclaimed, accurately diagnosing his predicament.[33] Although he survived the initial wound and made it back to land, Langsford perished three days later. The duel, fought on a watery field of honor, ended fatally for both combatants.

The sword of a swordfish, and of its close relatives the marlin and sailfish, is a bony extension of the skull. It is sheathed in skin and projects forward to constitute as much as one-third the total body length. This unique form evolved at least 100 million years ago and has persisted ever since. All three use their swords to slash through schools of prey, as we have seen, stunning or wounding a few unlucky individuals they seize and devour. The slash and dash hunting technique was first inferred from the contents of billfish stomachs: copious squid and fishes bearing distinctive gashes from having been literally put to the sword. Later, underwater cameras recorded the balletic but violent performances. The penchant by small fishes to shoal under floating objects, where they can be hunted by billfish, provided videographers an ideal site to film these behaviors. The aggregating popularity of such objects, for hunted and hunters, may explain bizarre stories of boats having been speared by marlins and swordfish who, unlike the victim of Frank Langford's harpoon, had no reason to attack a vessel. They may have simply been slashing and lunging their way through a thick, swirling school of fish that gathered beneath the boat and accidentally stabbed the hull. Unintentional or not, such impalements can be disastrous. In 2018 a Filipino fishing vessel was struck by a marlin and promptly sank, stranding five fishermen on a makeshift raft for days. The skipper Jimmy Batiller told rescuers, "It hit the bottom of our boat leaving two big holes. We suspect it was chasing a smaller fish. It swam around the sinking boat for a while, apparently disorientated."[34] Fortunately, men and fish survived their ordeal.

Billfish, like marlins, are nearly the acme of the food chain in the ocean, but not quite. They themselves are preyed upon by toothed whales and may use their bills not only for offense but also for defense against such assaults. This phenomenon was personally observed by one unfortunate diver, the oceanographic filmmaker Mark Ferrari. He was in the water when a wounded swordfish fleeing an attacking pod of false killer whales charged toward him. The injured goliath, maybe in a desperate attempt to escape, or possibly mistaking man for whale, sped at Mark and stabbed him powerfully in the chest.[35] Unlike Captain Lansford, Ferrari survived the terrifying impalement, despite losing a prodigious amount of blood. Billfish blades have been found buried in the blubber of captured whales around the world, albeit infrequently. Swords also have been recovered from even more unusual locations, from sea turtles to floating bales of rubber to undersea cables.[36] One swordfish observed below 2000 feet by the *Alvin* scientific submarine in the 1960s actually assailed the vessel, stabbing it with its bill; sadly, the sword became stuck, and the unfortunate swordfish perished as a result.[37] There are enough of these odd stories, tinged with fear, retaliation, or curiosity, to suggest we do not yet fully comprehend the ways that billfish wield their peculiar implements.

Between the three types of billfish the tools differ in form and size. A swordfish is appropriately named, as its weapon is a flattened blade; these are also the largest armaments among all billfish, reaching up to 5 feet in length. Marlin and sailfish brandish rapier-like implements, however, round in cross section. The former is named after a marlin-spike, the round and pointed metal tool used by sailors to splice ropes. Holes in boats, like the one skewered off Halibut Point, can sometimes be attributed to a specific animal, depending on the shape; the same is true for broken bills left behind in whales or sea turtles.

Swordfish and their relatives also differ in their flamboyant dorsal fins. Swordfish sport a relatively narrow fin that rises steeply from their back, from just behind the gill slit, and ends just as abruptly. In contrast, the aptly named sailfish boasts a fin stretching down the length of its back nearly to the tail, forming a long and broad sail when fully unfurled. Marlins are intermediate, with the topmost fin emerging fore of the gill opening and tapering swiftly to a shallow sail that stands only a few inches above the back. While the narrow fin of a swordfish is permanently upright, both marlins and sailfish can erect or furl their sails. When moving through the water, their sails are folded into narrow slits, like a tuna, to reduce drag and improve swimming efficiency. During hunting, there is evidence that an extended sail stabilizes the fish by steadying the side-to-side motion of its bill while swimming and permitting a stealthier approach to schooling prey.[38] That stabilizer may also give the sailfish a firm buttress to brace against when vigorously slashing its sword, generating bill accelerations that outstrip the escape speed of most prey. When not hunting, the sail's expansive surface area renders it an excellent solar collector. Billfish have been observed loitering at the surface, using the extended sail to warm themselves after a hunting foray into the frigid depths, capturing the sun's heat and funneling it to the stout body.

All billfish are enormous compared with most fishes in the sea, their bulk helping them stay warm during hunting dives. More cylindrical in shape than tuna, they similarly bear sickle-shaped tails. Females are considerably more hefty than males: full-grown adult swordfish can reach 15 feet in length and exceed 1400 pounds.[39] Blue marlins attain similar bulk but stretch to more than 16 feet.[40] Sailfish are more diminutive, reaching 8 or rarely 11 feet, but they still represent a large and fearsome hunter if you are a sprat on the run. All three are powered by immense muscles, making up the majority of their weight and requiring prodigious consumption of prey for fuel; small fishes and squid are favored as main courses. Unlike tuna, billfish tend to hunt singly or in small groups rather than in large schools, which dramatically affects the way they, in turn, are hunted by humans.

Fishing nets are designed to catch schooling fish, whether tiny anchovies or beefy and gregarious tuna, but they cannot reliably be employed for solitary targets like marlins and swordfish. Historically, these giants were caught on hook and line, in one-on-one battles with sports fishers. Commercial fisheries, in an effort to boost catch rates, have multiplied that approach, trailing staggeringly long lines behind fishing boats with thousands of baited hooks. One long line may stretch for 100 miles in the ocean, the distance from Philadelphia to New York City. They are dragged through productive waters for a few hours, then hauled in to harvest the fishes that took the bait. Predictably, this indiscriminate technique also ensnares a distressing amount of bycatch. Sharks, rays, sea turtles, and even seabirds like albatrosses are attracted to the bait, snagged by the hooks, and drowned. Deployment of tinsel-like streamers above the surface, and C-shaped hooks below, has eased the mortality of seabirds and marine animals, but the practice is inherently unselective and destructive. Whether line-caught or netted, billfish unfortunately share with tuna the twin traits of being economically valuable and slow to reproduce, a combination Bruce Collette at the National Marine Fisheries Service calls "double jeopardy."[41] Such species are pursued intensely by fisheries because the meat fetches a high price, but their late maturity means they are unable to replace themselves quickly enough to sustain overenthusiastic harvests. Consumer demand, unfortunately, has yet to submit to biological realities. And the hazards of swordfish and marlin harvesting affect not only the fishes themselves. Once those expensive steaks reach a diner's table, an unseen danger to the consumer lurks inside.

AT THE END OF THE CHAIN

> Even in the vast and mysterious reaches of the sea we are brought back to the fundamental truth that nothing lives to itself.
> —Rachel Carson, *Silent Spring*

In 2009, the actor Jeremy Piven took an unexpected leave of absence from his leading role in the hit Broadway show *Speed the Plow*, citing muscle fatigue and neurological dysfunction. The cause? His blood mercury level tested at more than six times the safe limit, likely caused by his obsessive consumption of sushi. Playwright David Mamet, who penned the script quipped, "So my understanding is that he is leaving show business to pursue a career as a thermometer."[42] But mercury poisoning is no laughing matter. In the 1860s when London scientists synthesized compounds of methyl mercury, the element's organic form, two lab technicians died from accidental poisoning. Their deaths must have been ghastly. Nearly a century later, the Chisso chemical company in a coastal Japanese town called Minamata discharged methyl mercury for decades into the waters of the bay.[43] There it was absorbed by fish and shellfish that were eaten by residents, particularly the families of fishermen. The consequences of this diet of pollution were disastrous. Adults were stricken by numbness, dizziness, and lack of coordination. Children suffered from severe physical deformities, acute neurological disability, and agonizing deaths. Over a thousand people succumbed, and many more were tormented for the rest of their lives by debilitating symptoms. The term "Minamata Disease" became synonymous with methyl mercury poisoning.

Epidemiologists were slow to diagnose this so-called disease, in part because they could not imagine how the miniscule concentrations of methyl mercury measured in the bay's waters could provoke such profound effects. After an extensive investigation they discovered that mercury concentrations were being magnified in the seafood eaten by local families. Trace but persistent levels in the water soared to around 20 parts per million (ppm) in the bay's mollusks and fish, then skyrocketed to 700 ppm in the hair of residents.[44] For comparison, the Environmental Protection Agency standard for "safe" methyl mercury levels in seafood is just three-tenths of a ppm; Minamata families were eating fish nearly a hundred times more poisonous.[45] Researchers eventually concluded that the marine food chain, a noble succession of organisms that progressively concentrate sunlight into dinner table fillets, had been corrupted as an amplifier of toxicity.

When a flying fish is swallowed by a mahi-mahi, the larger fish eats not just its victim, but also all the prey that flying fish has ever consumed, and the prey of that prey, and so on. Like nesting dolls, every mouthful contains a multitude. But in each step, only a small fraction of the dinner is permanently incorporated into the diner. The mahi-mahi's metabolism burns much of the flying fish calories, exhaling carbon dioxide across its gills. Mercury, however, cannot be used to power muscles, and thus is never exhaled. It also has a nasty habit of lodging in fatty tissues, rather than being purged as waste. As months stretch into years, each tiny morsel of mercury eaten by the mahi-mahi accumulates in its tissues. If that mahi-mahi is eaten by a larger predator, like a great white shark, its entire storehouse of mercury is passed on undiminished, and the indigestible toxins become biomagnified. These open ocean hunters, graceful and powerful predators like tuna and mahi-mahi and swordfish, can harbor a chemical curse in their streamlined bodies.

Older and larger fishes tend to have higher mercury levels, because time yields greater amplification, and big bodies are built from many meals. Anchovies, on which Peru built a seafood empire, manifested mercury levels of just two-tenths of a part per million in tests by the US Environmental Protection Agency, while big-bodied and long-lived marlins bore twice that amount.[46] Similarly, New Jersey researchers studying mercury in canned tuna found levels in "white" tuna were nearly four times higher than in cans of "light" tuna, because the white albacore tuna are larger, about twice the size of light skipjack tuna, and are harvested several years later in their lives.[47] Faster-growing animals like swordfish also accumulate toxins more swiftly, since their high metabolisms require more frequent meals. Toxicologists in Italy showed that mercury levels in Mediterranean swordfishes more than doubled with a twofold increase in fish length and urged buyers to stay away from the largest specimens.[48]

In the oceans, it is not only mercury that should cause us worry. Many pollutants that reach the sea will biomagnify in marine animals if they cannot be broken down. These include metals like cadmium and lead, complex industrial compounds like persistent organic pollutants (such as PCB and dioxin), and powerful insecticides like DDT. This last chemical (its full name would traverse two Scrabble boards: dichlorodiphenyltrichloroethane) was made infamous thanks to Rachel Carson's landmark book *Silent Spring*, her heartbreaking requiem for birds silenced by accumulation of the poison.[49] Fish, too, are debilitated by the toxins they ingest. Luís Vieira and his colleagues in Portugal have shown that common gobies (*Pomatoschistus microps*, a denizen of estuaries) are unable to swim as vigorously when exposed to mercury or excess copper in their water, leaving them more susceptible to hungry predators and more likely to pass their toxins.[50] Our pollution puts fishes, as well as our own bodies, at risk.

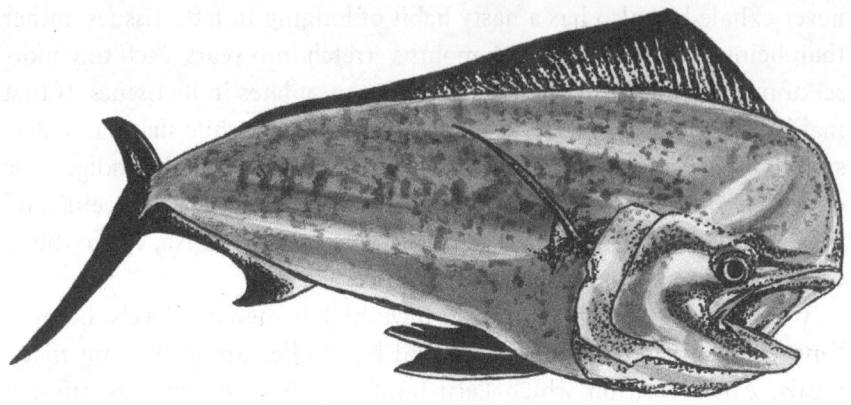

Mahi-mahi (*Coryphaena hippurus*)

MAHI-MAHI, JACKS, AND MACKEREL

Mahi-mahi (*Coryphaena hippurus*) are among the most beautiful, and most odd-looking, of the large predators who stalk the big blue ocean. Their common name derives from the doubled Hawaiian word for "strong," reflecting their robust build and vigorous swimming ability. A glittering gold coloration, however, is the origin of their Spanish name, *dorado*, or golden one. A deep but narrow body tapers gracefully to the scissor-shaped tail characteristic of speedy aquatic hunters; however, it is the astonishing head, with its high, bluntly domed forehead that stands out. Adult males develop this prominent bony crest as they age, giving them a unique, battering ram-like appearance. Parading a pronounced hump, while unappealing in the ogre of Notre Dame, is apparently quite alluring in large male fishes: salmon, parrotfishes, wrasses, and others develop conspicuous, bulbous head shapes as signals of their virility to would-be challengers contemplating a duel, and to females contemplating a family.

Females and youngsters lack this macho crest, having more normal, rounded heads, but all grown mahi-mahi are equally splashed with some of the most dazzling colors in the sea. Draping the body, atop a wash of aqua green, is a brilliant, lustrous golden color, an unmistakable combination, like a broad beam of sunlight drenching fresh grass. Glittering turquoise spots speckle the flanks, while long dorsal and anal fins of rich cobalt blue frame the sleek body, the former tracing almost the entire length of the back. The brilliant colors may provide a degree of camouflage, as these predators hunt primarily in the sunlit, turquoise waters of tropical and subtropical seas. Sadly, the gleaming colors fade rapidly after they are caught, but underwater they are a stunning sight to behold.

From Hawaii and myriad Pacific islands to tropical countries across the globe, mahi-mahi is widely sought as a food fish. The large bodies yield ample steaks of firm, pale flesh that is delicious: lighter in flavor than tuna or swordfish, but with more savor than bland whitefish. Add to this mild, sweet flavor the fact that mahi are excellent targets for fisheries, and it is easy to see why they are fast becoming a top seafood choice. They reach reproductive maturity in as little as four months and thus accumulate fewer toxins like mercury before capture. Mahi are well known to congregate under floating objects, like tuna, a trait exploited by fishers who dot the seas with all manner of clustered jetsam to draw them to their nets and hooks. They breed abundantly, in the wild and in captivity, rapidly replacing individuals reeled in and sold to restaurants. For these reasons mahi-mahi stocks are routinely ranked by seafood assessment organizations as safe and sustainable.

Despite these advantages, mahi-mahi was not an immediate hit in all places. Perhaps the wildly misleading common name "dolphinfish" was responsible; more likely, people were reticent to eat something unfamiliar. Fisherman Ed Ries relates a tale from the 1930s, when he had to resort to a bit of creative flim-flam to flog a dozen tasty mahi-mahi he had hooked. "We stowed the fish in the rumble seat of a Model A Ford," he begins, "and took off for L.A. to peddle our catch. To our dismay the small markets and restaurants on our route refused to buy. None had ever seen a dolphin or heard of the Hawai'ian taste treat, mahi-mahi." But necessity, as they say, is the mother of invention. Inspiration struck Ed when he realized the fish tails closely resembled those of more familiar, and delicious, yellowtail jacks; Ed, in turn, struck the mahi-mahi. "Chop, chop and back into the ice went the dolphins, minus their funny-looking heads. They were quickly disposed of at yellowtail prices."[51]

Yellowtail jacks (*Seriola lalandi*), the objects of Ed's deft impersonation-by-beheading, are members of a large group of fishes collectively called jacks that includes amberjack, trevally, pompano, and the confusingly named jack-mackerel, which is not at all a mackerel. All share a deeply forked tail, streamlined shape, and ravenous appetite for predation with their cousins, the tuna. Conversely, humans have a ravenous appetite for jacks: nearly all species are much sought-after for the table. They differ from tuna by being smaller, usually one to three feet in length, and more compressed: flattened from side to side, like a book standing on a shelf. Most are silvery, occasionally with bars or a yellowish wash; in a few species, like the giant trevally (*Caranx ignobilis*), males turn a deep brown, almost black, when they are of advanced age. African pompanos (*Alectis ciliaris*), on the other hand, are splendidly decorated as young-sters, with long, tinsel-like threads trailing from the tips of their dorsal and anal fins. The threads disappear by adulthood, suggesting they serve to make the juveniles less appetizing to predators with no wish to get floss stuck in their teeth.

Jacks are predatory and have developed some unique strategies to flush and capture their prey. Giant trevally, well-named for their hefty, almost brutish appearance, have been known to ambush seabirds loitering on the ocean's surface. Mesmerizing videos from the Seychelles, widely seen in David Attenborough–narrated documentaries, show the formidable fish lurking just below the waves, then bull-rushing an oblivious tern and seiz-ing it in powerful jaws. A few nimble birds take panicky flight at the last instant, spurring the trevally to burst upward into the air itself, leaping as much as a foot out of the water, salt spray glistening in the sun, to snatch a fowl on the wing and drag it down to a watery grave. But in the enormous, mostly empty expanse of the oceans, finding prey can require some cre-ative strategies, and some fish, it would seem, do know jack about hunting.

Frank Parrish, from the Pacific Islands Fisheries Science Center, made a surprising discovery about jacks when he fitted Hawaiian monk seals with underwater video cameras to observe the mammals while feeding. He recorded numerous instances of predatory fish accompanying the seals in what he termed "escort" behavior. The most frequent escorts were giant trevallies, presumably taking a break from plucking birds out of the sky, followed by several other species of jacks. "They were clearly attracted to the [seal's] intense bottom-searching activities," which flushed prey bur-ied in the sand or hidden inaccessibly among rocks.[52] "Jacks routinely positioned their mouths within inches of the seal's nose to maximize their chances of snatching prey items flushed by the bottom probing of the seal . . . [and] to capture prey before the seal could catch it."

While many predatory fishes are solitary hunters like swordfish and giant trevallies, some species hunt in packs, relying on their numbers. Complex routines are choreographed by mackerel as they gather in large schools with mouths open wide and pass through the ocean like a giant spider web, capturing copepod and other morsels as the ensemble glides forward. Mackerels resemble tuna, but are more slender, a javelin to the tuna's torpedo, and they shine a glittering silver. Many of the thirty or so species have dark patterns on their backs, ranging from loose speckling to wavy ripples to zebra-like barring. Both the shiny scales and the barcode-like markings are thought to help these animals traverse the upper ocean waters in tight schools without so much as a scrape or fender-bender. A fine example is the horse mackerel (*Trachurus trachurus*), who sports bright mirror-like patches on the tail and cheek, like reflectors on a boat trailer. The reflectivity of the patches shifts dramatically with the tiniest of movements of the head or tail, further signaling to other school members that this fish is adjusting course mid-gallop.[53] Taken together, the mix of visual signals from stripes and reflectors helps the school instantly detect, and match, changes in direction and speed, producing the coordinated movement we see as a polarized school.

Like tuna, mackerel are steady swimmers, relying on slow-twitch fibers to produce sustained travel. Their muscles are reddish and keep their bodies warmer than the sea, but they rely to a greater degree on fat as fuel. Mackerel undertake arduous migrations from the high seas to spawn in nearshore shallows, sometimes in staggering numbers. These fish are astonishingly fecund: a single female may release as many as 2 million eggs in a season.[54] During the pilgrimage and industrious spawning sessions, mackerel power their muscles with stored oils, rather than stopping to feed. To a diner, these fat stores are noticeable as fish oil, a slippery, energy-rich substance that make mackerel the quintessential "fishy" fish. They smell strongly, even unpleasantly to some, and the meat has a similarly potent flavor. Mackerel sequester oil in their muscles, and in cavities around the gut: a quarter or more of a fish's weight can be composed of oil.[55] The oils are beneficial to us, as they are offer important sources of vitamins A, B_{12}, and D, as well as the current poster child for dietary health, omega-3 fatty acids[56].

Mackerel, however, have been unappetizingly dubbed by some Japanese chefs "the fish that spoils while still alive."[57] Fishes high in oil tend to rot more rapidly because the fat-rich tissues break down quickly; in turn they yield by-products called lipid peroxides responsible for the distinctive rancid odor of rotten fish. The expression "stinks like a dead fish" should more technically be "stinks like a lipid peroxide," but it hardly rolls off the tongue as poetically. Better would be to reference the noxiously fetid names given to some of the foul by-products, like "putrescine" and the particularly stomach-churning "cadaverine." Decay of oils into putrid volatiles may be further accelerated by the abundance of myoglobin in these dark meat fishes, as the oxygen they supply supercharges the breakdown of fats and the production of malodorous stenches. Because of such rapid spoilage, mackerel have traditionally been salted, smoked, or dried as soon after capture as possible. But thanks to their abundance, particularly near coastlines where people with rods and reels and hungry bellies reside, they have always been a popular food fish.

Even an abundant fish can be mismanaged, however, and Atlantic mackerel (*Scomber scombrus*) provide a worrying illustration of the consequences of multiple countries managing a single, mobile stock. Once sustainably harvested in the North Sea, their schools have been shrinking dramatically, unable to withstand intense fishing pressure levied by overlapping fleets. Geoffroy Lean of the *Daily Telegraph* describes how climate change, migration, and optimistic fishing quotas all combined to drive northeastern stocks of Atlantic mackerel down to a fraction of their pre-industrial levels. "It began more than four years ago, when Iceland started increasing its landings of the fish, unilaterally upping its quota . . . The Faroe Islands followed suit, raising their own quota sixfold . . . Added to what other countries are legally catching, some 900,000 tonnes of mackerel are being landed a year from a fishery that can only stand to yield 500,000 tonnes. That cannot go on for long without disaster."[58]

There are reasons beyond sound resource management to be cautious about overeating mackerel. These fishes are common sources of scombroid toxicity, one of the most widespread types of seafood poisoning. The word scombroid refers to the taxonomic family of mackerel (and tuna), but the illness can be provoked by nearly any fish that is improperly chilled or preserved. It is instigated by a chain reaction in which bacteria on the fish's skin convert histidine—a harmless amino acid present in the flesh—into histamine, a toxin that triggers allergic reactions.[59] Symptoms are variable, with victims typically suffering moderate stomach upset, dizziness, or a bright red rash, usually appearing an hour after the meal, but disappearing within one day; in extreme cases it can cause anaphylactic shock, or even more rarely, death.[60] Fortunately, treatment with antihistamines is highly effective since scombroid toxicity is an allergic response, not a true poisoning.

Mackerel are known to harbor surprises even more off-putting than just allergens and pungent odors. Many is the time an angler has boated a mackerel and set about cleaning the fish, only to find the insides literally crawling with parasites. These range from tiny crustaceans to worms and flukes that do not kill the fish, using it instead as a stopover on a circuitous life cycle that normally includes several hosts. Anisakis, a roundworm that infests many fish species as well as marine mammals, can cause nausea or vomiting in people if the fish is eaten raw. Even more nauseating is the description by the Centers for Disease Control and Prevention of how unfortunate diners may learn they have just eaten the roundworm: "Some people experience a tingling sensation after or while eating raw or undercooked fish or squid. This is actually the worm moving in the mouth or throat."[61] Fortunately, freezing or cooking fish eliminates the risk of nearly all parasites; this is why most sushi, although eaten raw, is deep-frozen first.

Mackerel are so heavily parasitized that one study of nearly 2000 horse mackerels in the eastern Atlantic registered a total of forty-five different kinds of squirming colonizers.[62] But much can be learned from such diverse infestations. Parasites are often specific to restricted geographic regions, and the assemblage in a single fish can serve as a kind of bar code revealing the fish's past: its birthplace, the waters where it has swum, and the stock to which it belongs. By parsing the map of parasites in numerous fish, researchers discovered that Spanish mackerel (*Scomberomorus commerson*) in northern Australia was divided into five distinct populations, with about 5 percent of fish moving in their lifetime between the stocks.[63] So, parasites are not just disgusting and dangerous, they can be quite helpful . . . so long as you do not swallow them.

Some fish have the power to provoke revulsion even without the help of parasites. Escolar (*Lepidocybium flavobrunneum*), a deepwater cousin of mackerel, has gained fame as "the most dangerous fish in the world," not because it is a menace to bathers, but for the peril it poses to diners.[64] That peril has a name: keriorrhea, a sanitized clinical term for a revolting condition of uncontrollable, malodorous, waxy diarrhea. Escolar shares with mackerel the tendency to store energy in fats, but mostly in the form of indigestible wax-esters. Also found in carnauba wax—principal ingredient in shoe polish, car waxes, and furniture polish—wax-esters cannot be processed by our digestive systems. Food writer Harold McGee primly describes the greasy results: "The wax esters therefore pass intact, their lubricating properties undiminished, from the small intestine into the colon, where a sufficient quantity will defeat our normal control over the ultimate disposition of food residues."[65] While eating a few ounces may provoke no ill effects, larger portions may send the unfortunate diner rushing for the restroom. So foul are the results that escolar has been banned in Japan since 1977, where it is listed as toxic.[66] Elsewhere, the fish is sometimes distressingly marketed as "white tuna" by sushi restaurants, a deliberate mislabeling that can catch customers off guard, so ask before you eat.

THE TYGER'S FUTURE

> The solution often turns out to be more beautiful than the puzzle.
>
> —Richard Dawkins, *Unweaving the Rainbow*

In tuna, marlins, and the other tigers of the sea, every detail has been fine-tuned over millennia for a singular purpose: cruising the vast expanses of the great blue oceans in search of prey to hunt. The streamlined body shape, the acute senses, their tolerance for deepwater cold, and their unrivaled swimming endurance all contribute to a transfixing and majestic work of predatory art. People will fly across the world to see a cheetah hurtling over the Serengeti, or a pride of lions tear into an antelope, but they are barely aware of swordfish and tuna as equally extraordinary hunters. Granted, it is difficult to arrange an underwater safari to witness marlin and tuna in action. But like John Cage's symphony of pure silence, an unwatched tuna is nevertheless magnificent. These open ocean hunters deserve our appreciation, our awe, and our protection. What they habitually inspire, however, are our tableside appetites.

Barbara Block extends the comparison to terrestrial carnivores: "We wouldn't go into Africa and eat the lions, zebras and elephants, in most cases. We are basically doing that in the ocean. We are not looking at wildlife in the ocean as anything but food, and we could leave to our children an ocean without these animals."[67] One of the strongest drivers, according to Block, is the astronomical price that some tuna can command, and the lengths to which people will go to catch them. "It's a cocaine-of-the-sea type of problem where many people want it and no one's paying attention to the rules. Pirated tuna is a really big problem." Having used advanced technology in her own studies, she advocates its application to the problem of piracy. "What if we could barcode every tuna that's landed and keep track of them," she asks. "My dream is really to make a tag . . . that allows us to keep track of the fishery in a more accurate manner from point of landing to market, so we don't have any pirating." Soon, that dream may be realized: pilot programs in Mauritius attach a digital tag when a fish is caught, then store data in a blockchain so consumers can see the fish's entire history—from hook to plate—just by scanning a barcode.[68]

It may take more than just tags to save the ocean's large predators, but it has been done before. Wolves and mountain lions had been driven to the brink of extinction in North America before conservationists launched campaigns to restore these magnificent carnivores to the landscape. Despite boisterous but partly misinformed objections by hunters and ranchers, wolves have returned to many US states and robust populations thrive in Yellowstone and elsewhere. Quarrels with interest groups occasionally boil over, but earlier predictions of widespread disaster have failed to materialize. Instead, these wild predators inspire wonder in photo-snapping crowds who increasingly frequent national parks and hope merely to catch a glimpse. Tours in Alaska and Canada are dedicated solely to showing brown bears and polar bears to tourists, who bask in their august (but safely distant) presence. If a similar future can be created for tuna and billfish, by setting aside key protected areas and reducing fishery harvests, then these spectacular marine hunters will endure, roaming the seas, majestically slicing through the waves to seeking mates and snap up prey.

Like more familiar terrestrial predators, swordfish and tuna actually make their prey stronger, by picking off the weak, confused, and sick among them, thinning the herd and improving the gene pool. They are essential for the health of the ocean. And despite their power and speed, tuna can themselves be prey. Sea lions have been known to cleverly corner and capture them among rock pools in the Galapagos.[69] Pilot whales and killer whales feast on them in open waters.[70] But even more feared by all tuna are a group of silent ocean predators, formidable and relentless, the only other fishes swift enough to outrace them and massive enough to overpower them. These creatures of ancient origin are misunderstood, much maligned, and unjustly dreaded by humans. They are sharks, and they are the most awe-inspiring and formidable predators that rove the seas.

3 THE OLDEST FISHES IN THE SEA

Sharks and Rays

ELEVEN-YEAR-OLD RASHIDA TAYLOR had never been to an aquarium, but from the moment she spun through the entry door she felt as if she had stepped from a submarine into the sea. A meshwork of reflected light danced across the ceiling, splashing sounds reverberated off worn marble floors, and the salty humidity glazed her cheeks, still cold after a wintry dash from the school bus. Her eyes lit up with curiosity as she entered the great hall. Straight ahead, past a knot of gesticulating onlookers, was a gigantic tank of glass two inches thick. Rashida floated toward the pool as if drawn by an incoming tide, and suddenly a hammerhead shark slid silently and gracefully across her view. She was transfixed; it was the most beautiful creature she had ever seen.

After a few breathless seconds, the shark glided past again. The head reminded her of a spatula, or a snow shovel. Then came the five gill slits, and the long dorsal fin on top, the one that sticks out of the water. Her reading assignment said that hammerheads sometimes swim sideways, the dorsal pointing out to the side, to save energy. From the end sprouted a lopsided tail, the top way longer than the bottom. On the third lap, she paid attention to the little things. The eyes that bulged slightly from the tips of the snow shovel were not entirely black but showed some white around the edges, and a hint of expression. The shark's mouth was open slightly, revealing a few teeth, but they were skinnier and more claw-shaped than she had expected. And the shark was so muscular, she thought immediately of sprinters and wrestlers she had seen in the Olympics. The hammerhead had that same blend of grace, skill, and power.

A colorful sign next to the tank said shark ancestors swam the seas of planet Earth long, long ago, and it showed drawings of fossilized teeth frozen in time 400 million years in the past. Rashida had once learned in biology class that crocodiles may be as old as 200 million years; she had giggled about how they looked like her Pawpaw, who also seemed a few million years old. But sharks were even more ancient than crocodiles, and still they looked as fast and sleek as a race car. The sign said the earliest sharks appeared when only mosses and tiny bug-like animals lived on land. How had they survived all this time, nearly unchanged? Her eyes widened with respect as the hammerhead swung past on a third lap, elegantly.

According to the sign, proto-sharks appeared in the Devonian period, a chapter of planetary history that opened about 420 million years ago.[1] The interval was named after Devonshire, a rolling green and fossil-rich region in southwest England. Previous geological periods were named in honor of Wales, where even earlier animal fossils had been unearthed: the Cambrian period (540 mya) refers to the country's local name, *Cymry*, while later Silurian and Ordovician periods honor ancient Welsh tribes, the Silures and Ordovices. Rashida knew from class that the planet's continents were clustered together during the Devonian, like a jigsaw puzzle. The seas around this giant landmass were shallow, warm, and full of marine life. Her teacher had called it the Age of Fishes.[2] But beyond that welcoming continental shelf, most of the globe was covered by deep, dark, and cold seas. Sharks thrived in those depths—the sign showed that ancient sharks closely resemble today's deep-water frilled sharks—and the adaptations they developed to survive in the abyss held the key to their success to this day.[3]

Her school book had spoken in terrifying terms of a planet-wide extinction event that suddenly ended the Devonian's 50-million-year run, when three-quarters of all marine life had been wiped out. Hardest hit were the nearshore animals. Reef ecosystems, cradles of diversity and abundance built by colorful sponges and corals, were devastated, and many thousands of warm water species died. Sharks who favored shallow waters also disappeared, taken out by a marine environment changing so rapidly they could not respond, nor evolve. But those sharks who lived in the deep seas endured, against all odds. To Rashida, the hammerhead tirelessly circling the tank in front of her was a parade of survival: here was a creature who had cheated extinction and persisted into our modern world. She had

Scalloped hammerhead (*Sphyrna lewini*)

never seen a shark before today, but she would remember the encounter for the rest of her life.

THERE IS SAFETY IN THE DARK

My soul is full of longing
For the secret of the sea,
And the heart of the great ocean
Sends a thrilling pulse through me.
—Henry Wadsworth Longfellow,
"The Secret of the Sea"

Just a couple miles from the Pacific Ocean where his subjects swim, Chris Lowe directs the Shark Lab at California State University in Long Beach. There, he has led research into the lives of sharks and other fishes for decades. Drawn to sharks by their tenacious record of survival, he summarizes their ancient history as one of adaptation and good fortune. "The advantage that the chondrichthyans [sharks and rays] had was that their physiology enabled them to occupy deeper habitat. Those animals had to evolve all the adaptations necessary for surviving in an environment that was already lacking oxygen, is already really cold, and didn't have a lot of food. So many of their adaptations—the cartilaginous skeleton, the liver—probably all those features have been retained."[4] He underscores the resilience of sharks and rays, adding, "they survived these massive extinction events because those habitats weren't nearly as changed as the shallow ocean habitats. And then the fact is that maybe life radiated from the deep sea, and as the planet changed again, animals moved into those new habitats and then evolved adaptations for them."

A suite of successful adaptations has been fundamental to shark survival for several hundred million years. Sharks and rays are most readily distinguished from other fishes by their cartilaginous skeleton. Cartilage is the stiff but flexible substance responsible for the shape of our noses and the scrollwork of our ears. True bones begin as bendable rods of cartilage but are gradually stiffened by the deposition of calcium, a process known as ossification, that forms stony crystals replacing most of the cartilage. Inside bones resides the marrow, a complex factory floor for production of red and white blood cells, and plasma. By contrast, the skeleton of sharks and skates and rays is never fully ossified and possesses no marrow. In most species, the cartilaginous skeleton is suffused with mineral salts less rigid than calcium, creating more flexible structure. While the distinctions may seem subtle, the benefits are enormous. Fishes with cartilaginous skeletons weigh less and are faster than their bone-lugging cousins. More body mass can be dedicated to muscle, resulting in greater speed. The skeleton's flexibility allows impressive underwater acrobatics: compared with a rigid, torpedo-like tuna, a shark can spin and double back on itself almost like a snake, a boon in battle or when chasing prey through twisting reef canyons. And whereas all fishes must fight the implacable force of sinking, sharks are less dense thanks to their lightweight skeleton, and to their large and fat-rich livers, so they spend less energy staying up in the water column.

There are downsides, of course, as with any innovation. The flexible cartilage of rays and sharks is not as unbreachable as the rigid fortress of a bony fish's skeleton. Instead of relying on an internal safety cage, sharks evolved external armor not unlike the chain mail worn by knights of old. The skin of sharks, and rays too, is covered not by scales but rather by tiny and extraordinarily hard teeth called dermal denticles. Like the teeth in our own mouths, shark denticles are formed from dentine, one of the hardest compounds animals can manufacture, then capped with enamel, another substance with unparalleled toughness.[5] The result is a rock-hard spike, its tip pointing tailward, arrayed across the shark's body with thousands like it to form a nearly impenetrable armor. If you caress a shark, be sure to swipe your hand from head to tail, the skin will be as smooth as velvet; stroke from tail to head and the skin feels as rough as an iron file.

Shark skin also improves hydrodynamic efficiency, enabling sharks to swim faster, because the tiny teeth create micro-vortices that reduce drag.[6] Its unique form has been mimicked in a diverse range of human uses, from Olympic swimming suits to Navy battleships, all in search of greater speed through water.[7] The skin of a shark even has powerful anti-microbial properties, as the studded surface impairs bacteria from growing into dangerous mats. Similar manufactured textures are being applied to hospital doorknobs and even catheter tubes to mechanically defeat bacteria and save lives.[8]

One of the most entertaining, albeit grisly, uses of shark skin in history comes from knights of medieval battlefields. As it happens, stabbing Visigoths or Vikings with a broadsword spills a lot of blood, which makes the sword handle dangerously slippery. Some ingenious warrior, probably a fisherman on his day off, discovered that dermal denticles made shark skin one of the greatest anti-slip substances on earth. Thereafter, the best blades that doubloons could buy—in Europe and even Japan—had handles sheathed only with sharkskin leather, and never more did knights nor samurai find themselves embarrassingly empty-handed in battle.

Shark courtship closely resembles a battle, involving a great deal of passion, strenuous physicality, and a lot of biting. Males bite females, many of whom bear nearly ritual-looking scars testifying to their pre-mating ordeal; they withstand the toothy foreplay thanks to skin that can be twice as thick as their suitor's.[9] Males also slash at other males in an attempt to establish their reproductive dominance. All this sexual swordplay likely led to sharks evolving their armor of enameled denticles. Once a male has dissuaded his rivals and dentally signaled his intentions to a mate, he wrestles the female into a belly-to-belly position and holds her in place with a clamping bite. Unlike most fish that rely on external fertilization, with eggs and sperm released freely into the water column, sharks manually transfer sperm to the female. In males, the paired pelvic fins (located on the belly, just before the tail) each bear an elongate extension, deeply grooved and rolled like fleshy tubes, which can reliably be used to distinguish gender in the wild. Called claspers, these modified fins are inserted in the female's cloaca, her combined urogenital opening, and sperm packets are released into the groove. Gyrating like an earthworm on a hot sidewalk, the male generates currents of seawater that flush his sperm into the female; anywhere from six to sixteen months later, depending on the species, eggs are released or live pups are born. In the case of great whites, the newborns may be up to 4 or even 5 feet in length.[10]

After they enter the watery world, shark pups face a challenging future. They must avoid being eaten and find food for themselves, all without the slightest gesture of parental care. Studies of the energy used by swimming baby hammerheads, led by Dr. Lowe, calculated that they have only about three weeks to learn to hunt successfully; if they fail in that time, they will starve to death.[11] Even before birth, some shark pups must withstand a terrifying mortal challenge. Sand tiger sharks (*Carcharias taurus*), great whites, and a few other species experience a keen struggle for life within their very mother, a cage match so extreme that the winners even eat the losers.[12] Cannibalism in the womb is yet another example of the physical battles that sharks must endure, throughout their lives, in order to survive. But survive they do. Perhaps honed by their constant confrontations, sharks boast some of the most rapid injury recovery rates in the animal kingdom, and their immune system might just provide an avenue of safety through viral pandemics.

First, let us dispel a widespread myth: sharks are not immune to cancer.[13] This is a falsehood spread to bolster dangerously dishonest claims that shark cartilage can cure cancer. It does not: you cannot simply pop a capsule full of ground shark bones and beat cancer; to suggest otherwise is immoral, and criminal. But the myth persists and is responsible for gratuitously sending millions of sharks to the grinder every year. Scientists have, however, elucidated at least three compounds from shark immune systems that hold astonishing promise for modern medicine. Squalene, a building block of steroids that was isolated from dogfish sharks in the genus *Squalus*, is widely used to manufacture adjutants for vaccines. Adjutants are accompanying compounds that increase the body's ability to respond to a vaccine, ramping up production of the antibodies that constitute the front lines of protection against diseases such as influenza.[14] Fortunately, squalene can be extracted from olive oil and many other sources, which should limit the impact this discovery might otherwise place on wild shark populations.

Closely related is the chemical squalamine, also discovered in dogfish sharks, underappreciated fishes who right now should be lining up for a

Nobel Prize in medicine. Squalamine exhibits powerful antiviral activity against a broad range of human pathogens, by slowing the rate at which infected cells replicate the virus.[15] Laboratory research has shown squalamine to be effective against such killers as dengue, hepatitis B, and even yellow fever. The compound exhibits an entirely new class of antiviral effect, completely different from existing therapies, and thus represents a rich field for medical research to mine for new and effective treatments.

Surprisingly, sharks share one exceptional feature of their immune system with a very unlikely, and decidedly nonaquatic animal, the llama. Both carry in their blood a type of antibody so small it has been dubbed a nanobody, and it may be the key to tackling future viral pandemics. Human antibodies are large, complex molecular structures with two binding areas that seek out their target: a recognizable structure on the surface of a virus, for example. To directly manufacture antibodies is a Herculean task: the intricate folding of a human antibody is nearly impossible to replicate in the lab—like building a wooden ship in a bottle, at a microscopic scale—and you have to do it twice, once for each binding area. But nanobodies are short, much simpler to fold into the proper configurations, and possess just one binding area. Because of these advantages, shark and llama nanobodies now serve as a promising scaffolding on which to mount defenses against a gamut of diseases.[16]

Human monoclonal antibodies have already been engineered to treat rheumatoid arthritis and Crohn's disease. But shark nanobodies are the starting blocks for targeting Ebola, cholera, hepatitis B, botulism, and lupus.[17] Nanobodies can even be programmed to deliver minute radiation doses to tumors, killing them before the radiation is safely flushed from the body by our kidneys, another benefit of nanobodies' small size. Just before the COVID-19 pandemic paralyzed the planet, researchers in 2018 had already developed a nanobody programmed to block binding of the coronavirus MERS (Middle East respiratory syndrome) by targeting its spike protein.[18] This treatment holds promise for custom-designed cures for future coronaviruses. All these immunological discoveries were made possible by sharks' ability, honed over millions of years of evolution, to overcome bite wounds and viral infections suffered during hunting and courtship.

When sharks enter a pitched battle in the ocean, whether against fishes or sea lions or other prey, they are protected by their skin and armed to the teeth with row upon row of, well, teeth. Shark teeth vary enormously in shape and purpose, with only a few species brandishing the serrated perfect triangles stereotypical of great white sharks. Those teeth are adapted for cutting flesh, which the shark slices by fiercely slashing its head from side to side. The curved daggers of some hammerheads, however, are adapted for piercing and grasping swift and slippery squid. Other hammerheads, and many rays, have broad and flat-topped molars, suited to crushing shellfish and crustaceans. In lined cat sharks, males have daintier front teeth than females, the better with which to nip her in courtship.[19] Tiger sharks (*Galeocerdo cuvier*) also sport curled, claw-like teeth, which they wield to seize turtles and shatter their shells. Despite relying on jaws of cartilage, sharks can bite with a force that beggars the imagination, possibly a result of an evolutionary arms race with well-armored prey like turtles. The largest great whites ever recorded (more than 7000 muscular pounds) have jaws that can clamp with an estimated two tons of force, equivalent to the weight of a full-grown white rhino.[20]

Because shark teeth are not anchored to bone, they readily fall out: a typical shark loses about one tooth every week. That they fossilize so well, and litter the ocean floor, explains why we know so much about sharks from several hundred million years ago. These include the famous megalodon (*Otodus megalodon*), a shark so large it stalked giant whales as prey, measured an estimated 60 feet in length, and sported teeth the size of an axe blade.[21] But sharks have turned their habit of losing teeth from a dentist's nightmare into a dream scenario for hunting. When the outermost rows of teeth grow dull, they are replaced by new, sharp teeth from the rows behind, guaranteeing that sharks never enter battle with a dull sword. Shark teeth also are enervated, sending complex signals to the brain about the objects they touch, just like our fingertips. This may explain why sharks will mouth novel objects to find out more about them, just as a person entering a dark room will daintily brush the wall to locate a light switch.

To capture, slash, or crush prey, one first has to find it. Sharks and rays excel at detecting squid and fish and crabs and the like, no matter where they hide. Their eyesight is keen even in very low light, a holdover from life in the dark refuges of the ocean depths. Behind the eye's light-sensing retina lies a mirrored curtain, the *tapetum lucidum*, which reflects stray photons back through the retina. While this works brilliantly in the dark, sharks may face the problem of being blinded in well-lit waters by too many reflected photons. Port Jackson sharks, attractively banded denizens of the seafloor who frequent canyons and caves, have solved this problem by developing a type of internal sunglasses. When light levels get too high, pigmented cells packed with melanin spread over the upper layer of the tapetum, cutting down the light; when the shark dives into the depths the pigment cells retreat, in effect doffing the sunglasses.[22]

Beyond vision, sharks rely on other acute sensory systems to detect food at great distances, or when it is completely hidden. Sharks have an excellent olfactory system, capable of detecting schools of fish, or a single wounded animal, from miles away and following the scent toward the source. Great white sharks (*Carcharodon carcharias*), known formally as "white sharks"—one suspects the "great" has been added by movie directors and novelists—have been shown to detect blood at concentrations of just one part per million.[23] While this figure sounds impressive, it is worth pointing out that primates like spider monkeys can smell the odor of a predator at less than one part per billion.[24] Still, since sharks possess two pairs of nostrils, with water entering one and exiting the other, they can sniff a lot of water as they swim in search of food. Essentially, you can always find a needle in a haystack if you methodically pass the entire haystack through a needle detector. The pairs, located on opposite sides of the snout, permit detection of the direction from which an odor originates, just as paired ears allow location of sounds. Hammerheads, with their more widely separated nostrils possess even greater directional sensitivity.[25]

Like salmon, sharks also use their olfactory sense to assess the flavors rolling off nearby shorelines and pouring from river mouths. Leopard sharks (*Triakis semifasciata*) can navigate toward shore using scent alone, as was demonstrated by biologists from the Scripps Institute of Oceanography. In a simple yet ingenious experiment, the researchers caught leopard sharks in the shallows, plugged the nostrils of half of them, then released them all well offshore. Those sharks with unblocked nostrils swam directly back toward shore, as if following a compass, while the plugged sharks meandered aimlessly, unable to navigate, proving the navigational worth of a keen sense of smell.[26] That this research was led by scientist Peter Nosal is an amusing bonus.

Much of a shark's brain is dedicated to processing sensory inputs, more so than almost all bony fishes.[27] For an animal that roams the great, empty spaces of the ocean in search of infrequent food, prey detection is paramount. Chief among their sensory inputs are vision, smell, and the lateral line detection of pressure waves. But sharks and their cousins the skates and rays possess an extraordinary system that can detect a fish's beating heart, the muscular activity of a crab, and even the magnetic lines of force that encircle the globe. They can sense electricity, and this unique ability is one of the keys to their millions of years of success.

A SHOCKING DISCOVERY

I see no reason why intelligence may not be transmitted by electricity.
—Samuel Morse, *Samuel F. B. Morse, His Letters and Journals*

In 1678, the Italian anatomist Stefano Lorenzini announced a finding that would eventually electrify the world of sharks and rays. While dissecting

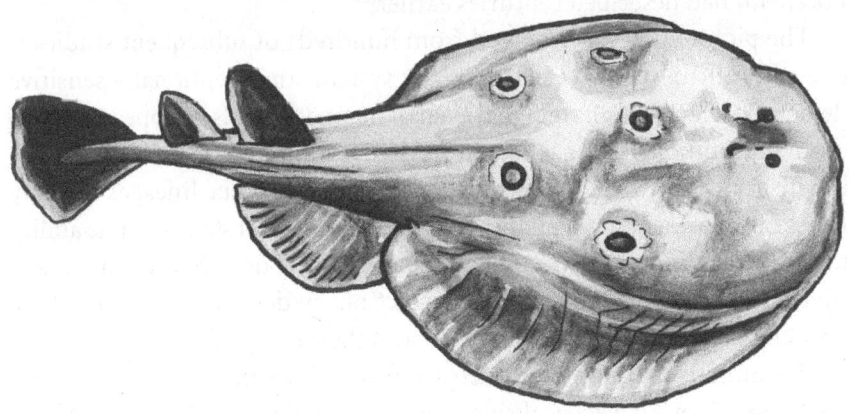

Torpedo ray (*Torpedo torpedo*)

torpedo rays, broad bottom-dwellers with attractive mottled patterns, he found a network of tiny organs clustered in the head, each filled with clear jelly and connected to nerve endings. These little sacs are shaped like a slender Greek urn, or ampulla, to which they owe their modern name: Ampullae of Lorenzini. Before Stefano could deduce their function, however, he became embroiled in the byzantine politics of the Medici family who ruled Tuscany. Just three years after publishing his work on rays, Stefano and his brother Lorenzo were arrested by order of Cosimo III de' Medici, Grand Duke of Tuscany, and tossed into prison for twenty years. The brothers were suspected of abetting a plot between Cosimo's wife Marguerite Louise, who had earlier fled the repressive and splenetic duke, and his worryingly popular son Ferdinando. It probably did not help matters that Stefano dedicated his book on torpedo rays to "the Most Serene Ferdinando III, Prince of Tuscany."[28]

What Lorenzini discovered before his imprisonment was a network of ampullae, each shaped like a miniature blind alley that opened to the outside world through a tiny pore. These pores dot the front and sides of the heads of rays and sharks and are visible to the naked eye on many species. Lorenzini was unable to deduce the function of the clear gel in the ampullae, however, and it would take almost 300 years before their true role was finally elucidated. In 1960, a zoologist in Birmingham, England proved that the system could detect faint electrical signals, and that these impulses were conducted into the ampullae by the pellucid jelly that Lorenzini had described centuries earlier.[29]

The picture that has emerged from hundreds of subsequent studies is one of a fully developed electrosensory system, an exceptionally sensitive detector of electrical charges and even of magnetic fields. It appeared more than 400 million years ago, again in the Devonian period, in freshwater lungfishes; since then, it has resurfaced in many other lineages of bony fishes but is present in virtually all rays and skates and sharks.[30] The ability to sense electric current would have been a tremendous boon to ancestral sharks who survived Earth's great extinctions in deep ocean waters where prey can be found only by their smell and their electrical signature.

All animals produce faint electrical fields, since all muscle movements are triggered by an electrical signal passed down a nerve.[31] This phenomenon was ghoulishly discovered in 1780 by Luigi Galvani, an Italian physicist who found he could make a dead frog's leg twitch by applying an electric current.[32] One imagines that dinner parties in eighteenth-century

Italy were quite the droll affair. A contemporary scientist, Alessandro Volta, helped clarify that the frog's body itself was generating current, the travel of electricity between two poles with different charges; as proof, he invented the world's first battery. Both men's names are preserved in our contemporary lexicon as the words "galvanize" and "volt." Muscular contractions for movement, breathing, and even heartbeats generate an electrical field around the animal, akin to a faint glow. Even at rest, individual cells harbor a slight charge differential, like Volta's battery, a charge that is detectable as an electric field.

In rays and sharks, electrical currents travel from the surrounding water into the ampullae through the skin pores and propagate through the clear jelly that conducts electricity better than any substance known in the animal kingdom.[33] At the end of ampullae are rows of sensory cells that detect minute voltage differentials. The system is so finely tuned that it can detect the equivalent current of a single AA battery passing through two poles separated by 1000 miles, the distance between Australia and New Zealand.[34] White sharks, some of the most sophisticated hunters on Earth, can respond to charges as infinitesimal as one millionth of a volt.[35] No matter how bafflingly a prey fish might be camouflaged, it cannot fool the shark's electrosensory system. Even tiny crabs buried in sand or silt can be detected with ease by rays and skates in search of a meal. Gliding over the bottom, a ray acts like a living metal detector, its wide snout homing in on the pings from hidden crustaceans, tube worms, clams, and other invertebrates which it unearths like treasure.

Rays and their various kite-shaped relatives—stingrays, skates, electric rays, and more—are uniquely suited to an existence near the ocean floor, where most make their living. They evolved from ancestral sharks around 200 million years ago, most likely from bottom-dwelling species similar to today's angel shark and wobbegongs, broad and flattened sharks we will shortly meet. Since the open ocean is often empty of prey, it is natural that some cartilaginous fishes would eventually specialize on the rich bounty found atop, and within, the sea's bottom sediments. Rays and their kin exhibit distinct forms of locomotion, hunting, defense, and even breathing, all honed over millions of years of cruising the seafloor.

Marine sediment accumulated on the bottom can be exceptionally fine, like corn starch, and is readily swirled into a murky cloud by the slightest flick of a tail or downward beat of fins. In such a fog of sediment, prey escape with ease, and predators can lurk dangerously. At some point in geological time, the ancestors of rays switched from powering their swimming with alternating thrusts of their tail to flaps of their pectoral fins. Tails concomitantly shrank, becoming a balancing organ rather than an oar. Over time, the broadened fins advanced from flapping up and down like bird wings to tracing a rippling motion that pushes water backward and the ray forward. Like a sidewinder in the desert, the undulating motion of their fins propels them ahead at a stately pace while barely disturbing the bottom. To avoid fouling their gills with inhaled sediments, rays possess a special pair of openings atop their head called spiracles that draw unsullied water from above. While most sharks have knife-like teeth specialized for capturing swift and slippery prey, many rays sport broad molars, the better to crush the armor of their favored crustaceans and mollusks. Rays find these buried creatures with their electrosensory system, but they differ from open-water sharks in sporting greater numbers of pores linked to their ampullae,[36] and an enhanced ability to divine their prey's exact location.[37]

Food detection is not the only capacity endowed by Lorenzini's ampullae. While rays, skates, and sharks rely on this system to find fish, squid, crabs, and other prey, their ancient ability to sense electric fields also served as an evolutionary launching pad for a host of novel abilities. Rays and sharks can detect Earth's magnetic field as they swim through it, a feat that humans could not replicate until lodestone compasses were invented in China just 2000 years ago.[38] They use this information to help navigate hundreds of miles of empty water and can even read the speed and the course of oceanic currents, since seawater moving through the planet's geomagnetic field generates a unique electrical signal of its own.[39]

It should come as no surprise that electrosensory systems can be used for communication. Skates are graceful creatures, closely related to stingrays, who prefer nearshore waters and are active mostly at night, when vision is limited. Skate tails are packed with twin sausage-shaped organs that can generate a weak electric current; moreover, the frequency of the current can be finely tuned, like the pitch of a trombone.[40] Their ampullae of Lorenzini permit them to discriminate specific frequencies, something sharks cannot do, allowing them to hear the full range of trombone tones.[41] Skates almost certainly use these tones to converse amongst themselves: the frequency ranges of different species do not overlap,[42] adult skates emit their signature frequency only when they are sexually mature,[43] and they emit more signals when they travel in pairs than when alone, as if they were conversing or courting.[44]

Returning to the torpedo rays studied by Lorenzini, one finds an evolutionary advance that takes the electrical signaling of skates one giant leap beyond mere communication. Torpedo rays carry electro-generating organs much like skates, but they are larger, more powerful, and located atop the head near their eyes. Each organ is composed of thousands of tiny plates, arranged like a honeycomb, that convert chemical energy obtained from the ray's food into electrical current. The organs closely resemble a battery and can accumulate the charge the ray produces until it is summoned in battle. When unleashed, the current can pack an impressive wallop. Some electric rays deliver jolts of more than 200 volts at 15 amps or more, the equivalent of tossing a toaster into a European bathtub (do not try this at home).[45] Chris Lowe, from whom we heard earlier, found that Pacific electric rays (*Tetronarce californica*) use moderate warning pulses to deter predators, but blasts of higher voltage to stun and even kill prey: those attacking bouts resembled a lightning drum roll of more than 1200 jolts each lasting just a few milliseconds.[46] Ancient Greeks and Romans were aware of the anesthetic effects of the lower voltage discharge from a torpedo ray: they prescribed light shocks to alleviate pains as diverse as headaches, toothaches, arthritis, gout, and even childbirth.[47] The very name "torpedo" is derived from a Greek word meaning to numb or stun. In modern times, torpedo rays are even being studied as a potential source of power, though the idea of hooking your cell phone charger to a tank full of electric rays seems more than a little preposterous.[48]

Most rays and skates cannot discharge the voltage of a torpedo ray and usually rely on camouflage for defense. Many species are speckled, spotted, or marbled, and some even change color to match the overall shade of the bottom. They have a habit of settling on the sediment, then shimmying until a light dusting of sand covers their body and obscures their outline. With eyes jutting only slightly from the top of the head, a ray buried in this fashion is almost impossible to detect, at least visually. But they can be unmasked by the electrosensory array of a hammerhead, and they must have another defense to survive the inevitable attack.

If you have ever skipped gleefully from the baking sands of a sunny beach into shallow ocean waters, you might have been taking a considerable risk. In certain locales, stingrays repose in those selfsame shallows, basking in the warm water while resting between nocturnal forays.

Were you to unwittingly plant your bare foot atop a dozing stingray, you might impale yourself on the ray's most fearsome defense: a cluster of spines as tough as teeth and tipped with an agonizing venom. Dagger-like in form, the spines arise from atop the tail and are modified dermal denticles, made from the same material that armors the skin. Each spine bears numerous backward-pointing serrations, making it all but impossible to remove from a wound, and each is wrapped in venom-producing cells that secrete complex toxins into a groove running down the spine's midline.[49]

If you are unlucky enough to be injected, the toxins immediately begin torturing your body, adding insult to the injury of being stabbed with a spine the size of a letter opener. Stingray venom is a complex cocktail of painful chemicals. Galectins attach themselves to tissues near the wound and induce cell death, triggering excruciating pain: tormented cells literally scream into your nerve endings as they perish one by one.[50] The stingray's chemistry set may include peroxiredoxin-6, an enzyme also found in jellyfish, snakes, and scorpions which prevents blood clotting and thus wound healing, provokes necrosis (tissue death, such as in gangrene), and can even inflict brain damage. Side effects include nausea, trembling, and a high temperature; the latter symptom explains why a gathering of these animals is collectively termed a fever of stingrays (like a murder of crows, or parliament of owls). While numerous myths circulate about treating stingray punctures, including the dubious recommendation of applying fresh urine, clinical research has shown that simply immersing the wound in hot water (not boiling), plus a couple aspirins, is sufficient to alleviate a victim's suffering in just half an hour.[51] The best treatment is prevention: when wading into shallows, use the "stingray shuffle" instead of lifting your feet and setting them down.

THE MERMAID'S PURSE

With more than 600 kinds of rays flapping through today's oceans, their diversity exceeds that of sharks by about 100 species. Their sprawling variety, however, can be clustered into a few distinct groups: sting-rays, skates, torpedo rays, pelagic (open water) rays, and those well-known giants the manta rays. Stingrays are largely, but not entirely, distinguished by their excruciating venom. A few other rays can deliver similar toxins, however, and most rays in the sea have at least one tail spine, whether toxic or not. Like skates and torpedo rays, the wing-like pectoral fins are so enlarged that they entirely engulf the head and fuse in front of it. A stingray's body resembles a diamond-shaped or circular pancake, like a broad sun hat, with head and body distinguished only as a thickened portion in the center. Stingrays also give birth to live young, perhaps up to a dozen in a brood. In truth, the offspring hatch from eggs while still inside their mother, then feed on the yolks plus protein-rich uterine fluid she provides, before being born.[52] This mode of reproduction, intermediate between egg-laying (ovipary) and true live-birth (vivipary), is known as ovovivipary, a word that would warm the tiles of any Scrabble lover's heart were there three rather than two Vs in the game.

Skates are superficially similar to stingrays. Their pectoral fins also envelop the head, although their tail is usually shorter and slightly thicker compared with the stingray's slender whip. Some, like the descriptively named little skate (*Leucoraja erinacea*), use the tips of their small, rearward pelvic fins to walk on the seafloor; researchers revealed that the same genes responsible for the skate's stroll also code for mammalian ambling, illuminating just how little genetic distance can separate unrelated animals.[53] They frequently have thorn-like scales down their back, and occasionally a single spine arises from their tail, but neither are charged with venom. Skates tend to live in cooler waters, while stingrays prefer warmer seas. But the most delightful distinction is in their reproduction.

If you have ever happened upon a mermaid's purse in the wrack line of an ocean's shore, you have found the egg case of a skate. The fancifully termed purse is leathery and rectangular, about the size of a deck of cards, with wiry tendrils or horns springing from the corners. The center often reveals a modest bulge, where the egg resided until a tiny skate hatched and slipped out of the purse to its seafaring future. Amazingly, a skate's electrosensory system is already switched on even in the purse-dwelling embryos. Clever lab experiments have shown that an embryonic skate only nine weeks old will instantly freeze when the electric field of a predator is applied, halting even its breathing until the threat has passed.[54]

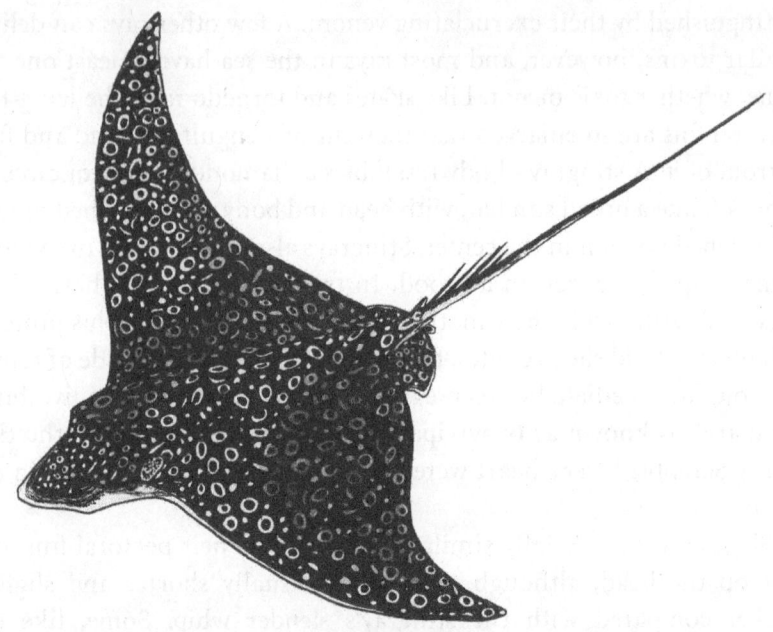

Whitespotted eagle ray (*Aetobatus narinari*)

Rising from the ocean bottom, some rays have retained the capacity of their shark ancestors to swim commandingly in open waters, flapping their way across great distances. Pelagic rays are more deep-bodied than skates and stingrays, and their pectoral fins are more pointed and wing-like. Diagnostically, a pelagic ray's pectoral fins do not meet in the front, leaving the head quite visible and distinct; the pectorals spread outward just at or behind the spiracle openings. A fine example is the spotted eagle ray (*Aetobatus narinari*), huge dark-bodied creatures decorated with a constellation of light speckles who swim through the water with all the fluidity and grace of their aquiline namesake. Their tail is long and whip-like and usually bears defensive but nonvenomous spines near its base. Their head terminates in a flattened snout, somewhat resembling a duck's bill, which they wave over the bottom when searching for food, in true metal detector fashion. These 500-pound creatures frequently catapult into the air from tropical seas, attaining heights that any pole vaulter would envy, before crashing back to the waves with a characteristic belly-flop slap. Some hypothesize the powerful slap is performed to dislodge parasites; others suspect the displays are a form of mating rituals, like college students cannonballing into a pool to impress one another.

Closely related to eagle rays are gentle creatures known pastorally as cownose rays for the snub muzzle that projects Holstein-like from the front of their head. A deep gold or mustard color, these rays congregate in shallow lagoons where they browse the sandy bottom in search of mollusks and other invertebrates. Cownose rays are notably smaller than eagle rays, and perhaps for this reason they carry venom in their single defensive spine. Like their larger relatives they frequently take to the open water and also engage in acrobatic, somersaulting leaps into the air. They have been known to undertake prolonged winter migrations, in massive schools of 10,000 rays or more, from Florida to the Yucatan peninsula or the coast of South America (this from no less an expert than University of Virginia scientist Dr. Ray).[55] Intriguingly, females from one tagged population departed Chesapeake Bay in July and swam to Florida, while the male rays headed northward, spending a few late summer months foraging off New England before ending the bachelor outing and veering south to rejoin the females.[56]

FROM SANDY BOTTOM TO SALTY BLUE

> If everyone were cast in the same mould, there would be no such thing as beauty.
> —Charles Darwin, *The Descent of Man*

Not all sharks fit the classic description of the shriek-inducing villains in blockbuster films: a muscular torpedo sporting triangular fins and a gaping maw bristling with razor-sharp teeth. Nearly 500 species of sharks span an impressive range of shapes and sizes, preferred prey, favored ocean zones, and behaviors. Some eschew open waters, preferring like stingrays a life near the bottom. These include the delightfully named carpet sharks, who earned their moniker thanks to striking patterns of alternating dark and light blotches that call to mind a Persian rug. Many carpet sharks also have a habit like their namesake of lying on the ocean floor. Other examples are the wobbegongs of Australia and Indonesia, moderate-sized sharks so decorated with frill and filigree and tendrils that some species have earned names like ornate (*Orectolobus ornatus*), floral banded (*O. floridus*), and tasselled (*Eucrossorhinus dasypogon*). The word wobbegong itself is believed to arise from an Australian Indigenous term meaning "shaggy beard." Their tasseled margins and complex spotting patterns endow these gorgeous animals with superb camouflage, invaluable as they lie in wait amid undulating tufts of seagrass and algae, poised to ambush unsuspecting prey.

Another carpet shark, though they have traded the wobbegong's flamboyant patterns for a matronly tan uniform, are the widespread and docile nurse sharks. Their downward-pointing mouths are adapted to preying on lobsters, stingrays, and other bottom-dwellers, whom they detect with two catfish-like fleshy barbels hanging from their upper lip. Like many sharks, they have a vigorously muscular throat which, when expanded rapidly, sucks in water and hapless prey. That sucking, combined with a soft-edged mouth that looks almost pursed, may be the origin of their name, as they can resemble a nursing baby. An intermediate relative is the diminutive epaulette shark, 3–4 feet in length, which combines the downward mouth and barbels of nurse sharks with the extravagant splotching of wobbegongs. Well adapted to life among the rocks and algae of the bottom, they frequently walk on their front fins, scampering between coral heads rather than swimming. Evolution has granted them longer and more flexible fins to improve their locomotion, and a higher tolerance to oxygen deprivation, probably an adaptation to life in tide pools that become anoxic at low tide. Incredibly, one epaulette species (*Hemiscyllium ocellatum*) can even walk on land: it scuttles on its belly and four fins, entirely out of the water, from one tide pool to the next in search of food.[57]

An environment as rich and diverse in prey and habitat complexity as the ocean bottom unsurprisingly yields an equally rich and diverse spectrum of shark species. Alongside the angel and carpet sharks are the bullhead sharks, named for the high forehead and sloping face that give them a slightly bovine head shape. These include several species of well-defended residents of rocky bottoms, including the Port Jackson shark (they of the retina-dimming sunglasses) and horn sharks. Both are handsomely marked: the back and head of a Port Jackson is draped with dark stripes and swirls, while horn sharks are irregularly spattered with small dark blotches. Bristling with thorny defenses that would make a stingray envious, Port Jackson and horn sharks bear spurs that jut menacingly from the leading edge of both dorsal fins. Like stingray spines, the spurs help defend these docile sharks against attack from above while they lie in wait for prey. Bottom-dwelling prey are topmost among these nocturnal species' diets, a preference manifested by their small and downward-facing mouths, and the predominance of molar-like teeth adapted for chewing tough crab shells and the like. Port Jackson sharks exhibit a novel solution to the challenge of breathing near the seafloor where silty water brought in through the mouth can foul delicate gills: they draw water in through the first gill slit instead, pumping it across the gills, and then out through the remaining four slits.[58] This elevates them into a select club of sharks who can eat and breathe at the same time.

Pushing off the ocean floor, you enter the open waters where the more widely known shark species are found. Pelagic predatory sharks, whether they patrol the rocky outcroppings of reef heads or cruise the wide open spaces of the ocean, fall into one of two types: requiems and mackerels. Requiems are a shark family that epitomizes our classic vision of a shark. Streamlined and long, head slightly pointed and fearsomely toothed, back drably colored but belly pale, they are built more than anything for hunting at great speed. The family includes tiger, blue, lemon, whitetip, blacktip, bull, and silky; the latter may be the most abundant sharks in the sea. They are among the most numerous of the open water sharks and include several species implicated in attacks on people, such as tiger and bull sharks (*Carcharhinus leucas*). Bull sharks have even been known to swim up freshwater rivers, and one was reported (in 1937) as far inland as Alton, Illinois, near St. Louis, a distance of some 1500 miles of Mississippi mud from the Gulf of Mexico.[59] All requiem sharks give birth to live young, with most species supplying nutrients to their developing embryos through a placental connection, just as in humans. As with many larger sharks, females require several years to reach reproductive age, and often a few more years will pass between successive pregnancies.

Tiger sharks, named for the tiger-like stripes that prominently mark juveniles, and for the attacking velocity of adults, take years to reach their full size of 12–15 feet. They are the largest of the requiems and may live to twenty-five years of age, reaching sexual maturity around the age of nine.[60] Females mate only once every three years and carry their young for fifteen or sixteen months before birth.[61] The math is simple: an average female tiger shark will reproduce just six times in her entire life, one of the reasons why this magnificent species has declined so precipitously. Tiger sharks are also famous for their eating habits which are anything but finicky: stomachs of captured adults have held everything from snappers, seals, seabirds, and turtles to license plates, tires, dead cows, bags of coal, a tom-tom drum, and even the head of a crocodile.[62]

Where sharks mate, and give birth, is critically important to the survival of their young, and thus the perpetuation of the species. Blue sharks (*Prionace glauca*), beautiful and willowy requiems stretching to 10 graceful feet in length, undertake monumental migrations between mating, birthing, and feeding areas. Tagging studies have revealed that Atlantic blues make routine trans-oceanic crossings between mating waters near Brazil, feeding areas off North America, and birthing wards in the warm waters west of Spain and Morocco.[63] To get the timing of birth just right, when a migration of 10,000 miles can separate courtship and birth, many females delay implantation, holding live sperm in their bodies until they trigger fertilization at precisely the right time to ensure their young will be born in just the right place.

In California, leopard sharks may have stumbled onto a thermal solution to the challenge of scheduling birthdays. Female leopards (not true requiems, but in a sister family) were observed by Chris Lowe and colleagues to gather in shallow waters during the day. In a narrow band of sun-warmed water along the shore, "females will aggregate in large numbers in this thermal envelope."[64] Reposing in the balmy shallows raises their body temperature and boosts their metabolism by as much as 17 percent. Ultrasound imaging revealed most of the lounging females were carrying young, and Lowe believes the temperature boost should accelerate embryonic development, leading to earlier births. Similar behavior has been seen in round stingrays (*Urobatis halleri*), who gather to enjoy warm estuaries in California and even bask in the heated water discharged from power plants. It is possible that both animals are speeding up gestation so that their young are born at the precise moment when food will be more abundant for the hungry infants.

The other distinctive group of large, oceanic sharks are the muscular hunters known as mackerel sharks. These powerful and athletic species include the famed white sharks, easily recognized by their brawny body and blunt nose, and the more streamlined, fleet-finned makos. Among the fastest fishes in the sea, shortfin makos (*Isurus oxyrinchus*) have been clocked at speeds over 30 miles per hour, with one estimate putting their top speed at an eye-popping 60 mph.[65] They reach such speeds by cranking up the thermostat on their swimming muscles by as much as 20 degrees; the effect is like engaging a futuristic warp drive that propels them into hyper speed.[66] White sharks also have this ability, though they cannot challenge a mako for pure speed. Instead, their thermal advantage allows white sharks to hunt at the thermocline—the upper limit of the ocean's deep and frigid waters—diving repeatedly below the boundary to capture unsuspecting or torpid fish.[67] Like marlin and swordfish, white sharks also possess networks of warm-blood arteries that wrap their eyes, enhancing their vision while hunting in the shadowy depths. In makos, a special vein passes through their red swimming muscles, picking up heat and transporting it to their head, where it warms the brain.[68] Thermal measurements of both mako and white sharks revealed their stomachs also can be 10 or 12 °F warmer than the surrounding water, supercharging gut enzymes and accelerating digestion.[69]

Blacktip reef sharks (in the requiem family) routinely linger in warm, shallow sites on coral reefs, basking before making hunting forays into cooler open waters. This behavioral thermoregulation—also practiced by lizards on sunlit rocks—allows them to maintain their body 2 or 3 degrees above the seawater temperature.[70] Blacktips retain this heat as they chase prey fishes down into deeper water: both predator and prey cool as they dive, but because the shark is slightly warmer when the battle commences, it maintains a decided advantage in speed and acceleration until its jaws close on the chilly and sluggish prey. Once again, science gifts us an amusing alignment of researcher and research, as the lead investigator on the blacktip study was none other than Dr. Conrad Speed.

One unique example of warm-blooded mackerel sharks is the uniquely sculpted thresher, whose extraordinary upper tail lobe resembles the curved blade of a scythe and can comprise half the shark's total length. The asymmetry of a thresher's tail generates lift, as the upper lobe pushes the tail down slightly, raising the head as the shark rotates around its center of mass. Tail-generated lift is particularly beneficial for slow swimmers, and all but the fastest sharks have the stereotypically longer upper tail lobe. As we have seen, threshers also wield their tails as hunting weapons. Hungry threshers will methodically circle schooling sardines, herding them into compact balls. When the sardines are perfectly packed, the leathery strap of a tail is whipped at the school so violently that dissolved gasses can bubble out of the water from the force of the blow.[71] Stunned sardines then make easy targets when the threshers, often working in pairs, slice into the school to snatch their dazed and disoriented prey.

TWENTY-THOUSAND TEETH UNDER THE SEA

> I could never write *Jaws* today. I could never demonize an
> animal . . . that we may, if we don't change our destructive
> behaviors, extinguish from the face of the earth.
> **—Peter Benchley, introduction to *Jaws***

People have a schizophrenic relationship with large, powerful predators. A grizzly bear spotted across Yellowstone Valley is thrilling, but one pawing at your sausage-filled cooler is terror incarnate. Safaris in the Serengeti are a roaring success if a lion or cheetah can be photographed, but few on the tour bus would volunteer to step down and stroll about the plains. Sharks are no different. We revel in their unchallenged mastery of the ocean realm and are electrified by their speed, their agility, even their savagery, but we harbor an unfounded terror cynically amplified by film and television. Our fear gives us license to turn a blind eye as more than 100 million sharks are annually slaughtered by people.[72] We justify our willful blindness by muttering that sharks kill four or five swimmers every year, but this justification rings hollow when we contemplate that dogs, our most beloved animal companions, will in that same year savage 400 or 500 people to death.[73] Other comparisons are less chilling, more amusing. For example, in a given year more people will die from teetering vending machines falling on them than from shark attack.

Despite our deep-seated dread of sharks, it is sharks who have more to fear from humans. Skates and rays, too, are swept up in a whirlwind of nets and baited long-lines that threaten their very existence. Studies tracking the precipitous declines are sobering. Assessment of twenty-one pelagic species of sharks and rays commonly targeted by fisheries found that all of them, every single species, is now under threat of extinction.[74] In a recent global investigation researchers placed blame on fisheries that directly target sharks, and those indirectly taking sharks and rays as bycatch, for rendering nearly one-third of all ray and shark species on Earth vulnerable to extinction.[75] A parallel review estimated that populations of sharks and rays have sunk by 71 percent in the last fifty years, a precipitous decline due to an eighteenfold rise in fishing pressure.[76]

Foremost among the threats to sharks is soaring demand for shark fins, an industry founded on the barbaric practice of catching sharks on the high seas, slicing off their dorsal and pectoral fins, even their tails, and dumping the dismembered body back into the water to die. A key ingredient in a popular Asian soup, the fins add neither taste nor color, but rather a peculiar gelatinous texture. As with any high value wildlife product, much of the trade is unreported, unregulated, and downright illegal. Estimates that valiantly incorporated not just reported fisheries catches but also data from the fin trade itself suggest that between 30 and 52 million sharks are abducted by the fin industry every year;[77] the Humane Society International puts that number at a ghastly 72 million.[78]

Making matters worse, for them, sharks are slow to reproduce, with some species waiting decades before mating. Ingenious new research—based on uptake of anomalous carbon atoms circulating the atmosphere since early atomic bomb tests—revealed that some female white sharks do not reach reproductive maturity until their thirty-third year.[79] The same study estimated that the largest male white sharks could live to an astonishing age of seventy-three years. As with many species that reproduce late in life, like gorillas and whales, lengthy delays before sexual maturity will effectively doom shark species to rapid extinction in the face of even moderate fishing pressure.

The only option is to mimic the approach applied to save rhinos, elephants, lions, and gorillas: establish safe havens where hunting is illegal and severely prosecuted, and ensure that long-term benefits like tourism revenue pass directly to local residents.[80] An uplifting example is offered by the Micronesian island of Palau, which in 2009 banned all shark fishing in its expansive territorial waters, inspired by the conviction that live sharks are worth far more to tourists than dead sharks are to commercial fishers. Soon after the ban, shark tourism was generating US$18 million annually for the island nation, about 8 percent of its gross domestic product.[81] Elsewhere, shark watchers contribute $25 million every year to Australia's economy, and a staggering $42 million to the island of Fiji;[82] globally, shark tourism generates a tidal wave of revenue reaching $314 million every year.[83] By 2017 the number of shark sanctuaries worldwide had swelled to fifteen, covering a combined 3 percent of global oceans, and dramatically reducing shark mortality within their bounds.[84] Fortunately, white sharks appear to be making a comeback in shark sanctuaries, and in the waters off California where their numbers have risen every year since 1994 when they were placed under state protection.[85] And in 2022, the US House of Representatives agreed to ban all shark fin trade within the nation.[86]

If shark sanctuaries can be vigorously defended and ambitiously expanded, these magnificent predators will endure. Generations of nature lovers will have the opportunity to view them in the wild, assuming they can survive pre-snorkeling encounters with ruthless vending machines. If weddings and other celebrations can find an alternative to shark fin soup, we can reel humanity back from the barbarism of finning. And if people can learn that few sharks resemble the implacable killers of movie screens and television spectacles, that most are docile, fascinating, and unique creatures like wobbegongs and whale sharks, they will embrace the wonder of sharing our planet with animals who have called the oceans home for 420 million years.

4 GREATS OF THE GREAT BLUE

Whale Sharks and Other Giants

SLIPPING FROM THE WOODEN PIER JUST AFTER SUNRISE, your tour boat skips over cerulean waters, a light chop slapping the hull like applause. Exmouth, a small port in northwestern Australia, recedes in the morning haze as you head out into the Indian Ocean bound for Ningaloo Reef. Odors of sea salt, windblown dust, and a hint of algae all tickle your nose. A sea turtle shows her head above the waves; soon after, a stingray leaps from the ocean and lands with a hearty wallop. Instinctively, you tighten your grasp on your mask and snorkel.

As you get farther from the rust-red coastline, you sense a change in the sea. Near shore, the water was so clear you could make out fine details on the bottom, a dozen feet below the keel. Now, a half-hour out, the sea has become cloudy, faintly greenish, like a light broth made with finely chopped herbs. And a tasty broth it is, for oceanic diners who traveled thousands of miles for their favored food: plankton. Those culinary voyagers, already gulping enormous mouthfuls of seawater for breakfast, are the stars you have come to see. Today you have a date to swim with whale sharks.

A cry rings out from the captain's deck, over your head. A sighting! Your guide clambers down the ladder, a wide grin on his face. Ladies and gentlemen, he says in a delightful Aussie accent, it's a great day for whale sharks! To your ears it sounds like "groyt doy for whyle shahks," and now you have a wide grin as well. After delivering a few firm rules about how to behave in the water, the guide invites everyone to get ready. Around you the group frenzies into action, pulling on wet suits, spitting into masks, checking camera settings. You don your own gear, waddle aft in cumbersome flippers, and giant-stride into the sparkling sea.

As you blow a snort of water from your snorkel, swarms of bubbles eddy in front of your mask. You see a fine snow of yellow-green particles dancing and swirling under the surface. You hear the shush of waves, and excited cries of nearby snorkelers, muffled by masks and water. An enthusiastic tap on your shoulder seizes your attention; a guide jabs the water with a gloved finger. You turn your head, and gasp involuntarily. The largest fish you have ever seen, indeed the largest fish on planet Earth, is only a few yards away. Your entire view is suddenly filled by a spotted and rippling wall of whale shark.

Swimming to your right, you try to keep up with this placid giant who moves surprisingly quickly through the water, seemingly without effort. You are momentarily mesmerized by the white spots, like a constellation of stars against an ink-black night. But they are more ordered and symmetrical than the night sky, calling to mind an oriental carpet or a quilted comforter. Atop the great head, itself larger than your entire body, the speckles are dense, almost hypnotically arrayed. Prominent ridges stretch from the gills to the base of the tail, with the white dots arranged in neat rows between them. Later, your guide explains that individual whale sharks can be identified by their spotting pattern, and that the long ridges improve their swimming efficiency.[1] Try as you might, the gentle behemoth easily outpaces you. Your last glimpse of this magnificent animal is its powerful tail, asymmetrical as in most sharks, fading into the hazy distance.

Another tap on your arm, and you spin underwater to see your guide hooking her thumb repeatedly over her shoulder. Looming like a train tunnel comes a gigantic open mouth. Your heart races as you try to remind your brain's amygdala that whale sharks are not looking to inhale a human being. You watch the approaching maw gulp methodically at swirls of greenish plankton twisting beneath the waves. The guide had explained that these leviathans filter about 1 million gallons of water a day.[2] You notice the seagoing titan is accompanied by a handful of diminutive escorts, like tugboats attending an ocean liner heading out of port. Hovering beneath the broad belly swim four or five slender fish, modest in size and painted with vertical black and white stripes like zebras.

You flinch as a clanging noise reverberates underwater: two guides point to their dive watches and signal a return to the boat. You've been in

Whale shark (*Rhincodon typus*)

the water 40 minutes and are only now feeling a twinge of cold. The whale shark recedes silently into the greenish blue as you try to commit the sight to deep memory, then you turn and swim for the ship. Climbing the ladder, you barely feel the rungs; you remove your gear in a daze. It's only once the guide shouts, "yah, good?" that a smile spreads and it all sinks in. You're sipping a cold fruit juice, the sun is warming your forehead, and you just swam side by side with two immense whale sharks. Yah, good.

THE LARGEST FISH ON EARTH

> He doth bestride the narrow world like a Colossus
> —William Shakespeare, *Julius Caesar*

The oceans harbor some of the largest, longest, and most colossal creatures to call our planet home. The blue whale is Earth's most massive mammal, tipping the proverbial scales at just under 200 tons. Lobsters are the heaviest arthropods, saltwater crocodiles the largest reptiles; the longest animal on the planet is a siphonophore, distant relative to corals and jellyfish, that stretches to more than 150 ribbon-like feet. And whale sharks are undoubtedly the planet's biggest fish. These titanic creatures can reach 60 feet in length and weigh over 20 tons, three times that of the largest African bull elephant.[3] Such superlative creatures thrive in the sea partly because of its salty buoyancy. On land, if an elephant were enlarged to the size of a whale shark, its bones would shatter when it stood: bone strength grows only with the square of an animal's length, but weight increases as the cube of length, so in goliaths the former is overwhelmed by the latter. Not even evolution can overcome this fundamental limitation of physics.

Giant animals face another fundamental problem the oceans have helped solve, one of thermodynamics. All living bodies generate heat from the workings of every single cell, each humming along like a tiny furnace, and that heat is shed to the outside world only across the skin. Again, cubes battle squares: as animals get larger, the number of cells in their body rises as the cube of length, but the amount of skin only expands as the square. At extreme sizes, an immense animal's cellular furnaces would smolder with more heat than its skin could dissipate, cooking the poor creature in its own juices. This inescapable constraint of thermodynamics may have set the maximum size of the dinosaurs: any larger and they would have been the evolutionary equivalent of a hot dish. Oceans crack this conundrum because water strips heat away four times faster than air, making you shiver after an hour-long swim in the sea, but saving a blue whale or a whale shark from overheating.

Once the evolutionary doorway to gigantism was nudged open by the sea, a bevy of animals charged right through, since being oversized yields many rewards (if your weak bones and multitudinous furnaces do not get in the way). The giants who call the ocean home—whale sharks and basking sharks, true whales, manta rays, and the peculiar mola molas— take full advantage of the benefits of bulk and are surprisingly similar in behavior, metabolism, and diet.

Most of the total weight of any fish is muscle, although marine giants also pack a lot of fat on their frames, insulation that buffers them from cold but highly productive waters. Their mouths are bigger, so they can swallow enormous swigs of tiny food like plankton. And their metabolism is lower than that of a smaller animal, requiring less nourishment relative to their body size. On land, for example, a diminutive shrew must eat nearly its entire body weight in insects every day to stay alive; a massive elephant, however, would take more than a month to ingest an equivalent amount of food. The shrew, if deprived of sustenance, will starve to death in a matter of days as its frenzied metabolism blazes through meager fat stores. In contrast, elephants, and whale sharks, can travel for months without eating as the slow boil of their metabolism languidly works through fatty energy reserves.

An unhurried metabolism provides other, more internal advantages. Animals with low metabolic rates tend to suffer less tissue damage, and fewer harmful genetic mutations.[4] These protections might even save you from extinction, if you are a mollusk: paleontological data show that clams and snails that disappeared rapidly from the fossil record had higher metabolic rates than those species that survived.[5] Of course, having a mellow metabolism also makes you slow-moving, which means you cannot readily escape predators; few predators, however, are willing to tackle a titan ten times their size.

It is plain to see why whale sharks and the other giants of the sea evolved to be so huge. They feed on explosions of plankton which are intermittent and far-flung, their locomotive stamina propels them across the planetary distances that separate those eruptions, they are nearly unassailable by smaller predators, and their titanic bodies offer fat stores and the slow metabolism needed to fuel the long journeys. And yet there is one final benefit of being a giant, one which lasts a very, very long time.

Beneath the arctic waters of the north Atlantic another ocean giant drowsily cruises. Charcoal-grey and cylindrical, she slips forward like a submarine running silent. When prey is encountered, though, she can pick up speed, engulfing fishes like cod and haddock in her cavernous maw. Reaching nearly 20 feet in length, and weighing almost two tons, this Greenland shark (*Somniosus microcephalus*) belongs to the expressively named family of sleeper sharks. In the summer she favors cold waters down to 6000 feet and rises to the surface only during winter months.[6] A few of these somnolent leviathans were captured accidentally in 2010 surveys of Greenland's commercial fisheries. Researchers from Denmark and Norway seized the unprecedented opportunity to determine the age of these enigmatic creatures. By peeling back the lens layers on each shark's eye, they were able to date the oldest, central layers using radiocarbon techniques. What they found astonished them. The largest Greenland shark's age was an extraordinary 392 years, making it literally the oldest fish in the sea.[7] It had been born sometime around 1620, the same year a few score of intrepid (if slightly underprepared) Pilgrims set sail on the *Mayflower*.

A MOUTHFUL OF MOTES

Whale sharks and nearly all the ocean giants (apart from Greenland sharks) sustain themselves with helping after helping of the richest broth in the world: plankton. It is astounding, at first blush, that the largest animals on the globe can thrive on such tiny food particles suspended almost invisibly in the ocean. But plankton can be mind-bogglingly abundant, and they replace themselves swiftly. In highly productive oceanic zones, this brisk growth yields tons upon tons of plankton, enough to support pods of great whales and parades of whale sharks. In the fertile waters off Yucatán, Mexico, researchers observed more than 400 of these speckled behemoths feeding in a patch of ocean just 7 square miles in size.[8] The short-lived gathering of whale sharks revealed something critical about plankton: despite its omnipresent nature, explosions of plankton rich enough to sustain whale sharks erupt in only a couple dozen places on the planet, and there for just brief periods of time.

Algae flourish where nutrient-rich and well-oxygenated waters are brought to the surface, and join with sunlight to supercharge phytoplankton photosynthesis, growth, and reproduction. The Humboldt current drives a powerful upwelling off Peru, which almost never flags; the Benguela current propels another alongside the southern tip of Africa. But elsewhere, upwellings are influenced by seasonal winds, and plankton growth booms only when conditions are ideal. In the warm tropical waters surrounding Indonesia and the Philippines, and in the Timor Sea which separates those island nations from Australia, the appearance of rich phytoplankton broth is heralded by the departing monsoons. Cloudy skies give way to warm sunlight in the early months of each year, and by spring the soup is ready. Soon whale sharks, manta rays, and other giants flock to the buffet.

To feed on miniscule phytoplankton and miniature zooplankton, petite jellyfish, diminutive fish larvae, and other tiny tidbits, these oceanic titans must ingest, filter, and expel massive quantities of water. A whale shark's cavernous mouth can inhale 2600 gallons of water every minute.[9] At that rate it could process an entire Olympic swimming pool—660,000 gallons in all—in just four hours. Each hour the whale shark strains up to 6 pounds of plankton from all that seawater, a prodigious harvest considering the miniature size of each morsel. For a human equivalent of this feat, try walking through a snowstorm with your mouth open until you have gulped enough snowflakes to fill a glass of water.

Whale sharks, basking sharks, and manta rays use filter pads to separate suspended particles from the water, before it flows over the delicate gills. The filtered food is then directed to the esophagus and swallowed. Filter pads are long, slender ribs of cartilage, packed into dense ranks, each rib covered by fine whiskers. Like the filaments of a dust broom, every whisker traps a few particles: the more hairs, the more particles caught. In whale sharks, the slender ribs are cross-linked, the whole apparatus arranged like a basket deep within the funnel-shaped mouth. This reticulated basket traps particles larger than about one millimeter, which then are swallowed. Small particles eventually become stuck, however, clogging the sieve. Whale sharks and other plankton eaters routinely "cough" to back-flush the filter basket and expel the jammed particles. Filter-feeding manta rays (whom we will shortly meet) have evolved an even more advanced technique that never requires back-flushing, so elegant that engineers have been inspired to apply it to intractable industrial problems.

Filter feeding allows ocean creatures to subsist on the smallest of foods, but it does have its drawbacks. First, it is inherently nonselective. Anything large enough to be trapped by the filter basket will be eaten. For millions of years on planet Earth this was not a risky proposition: the occasional gulp of stinging jellyfish might have caused a stomach ache, nothing more. In the last century, however, people have been dumping plastic waste into the environment, and one 2010 study estimated that some 14 million tons of plastic reached the ocean in that year.[10] By 2016, the United States was producing more plastic waste than any other nation, some 280 indigestible pounds per person every year.[11] Thanks to this disappointing facet of humanity's so-called progress, whale sharks feeding today on a soup of plankton will inevitably consume obscene quantities of plastic. In 2018, a whale shark was found stranded on a beach in the Philippines, and it died shortly thereafter. The autopsy revealed plastic particles wedged in its gills and shreds of plastic crowding its stomach.[12]

A second drawback is that filter feeding is a drag—a hydrodynamic drag, that is. Fishes who rely on filter feeding cannot travel rapidly when forcing thousands of gallons of water through a tiny mesh; the effect is like hauling an open parachute through the water. Fortunately, whale sharks can alter the shape and orientation of the filters to allow water to pass through them more easily, a trick they employ when swimming to cross great distances. Once they reach an area with bountiful plankton, they fan open the filters and slow to a speed that maximizes their sieving efficiency, about 2 miles per hour.[13] At this speed they expend less energy swimming but must still battle drag. Ultimately, if the calories they gain from feeding on plankton fall below the calories burned by swimming, they face an energetic deficit that sets the minimum density of plankton on which they can afford to feed. When spasmodic plankton blooms subside, whale sharks and their ilk are energetically forced to abandon feeding, lest they starve, and embark on a long journey to another plankton hot spot. In some cases, those journeys can lead them halfway around the world.

CRUISE CONTROL

I am tormented with an everlasting itch for things remote.
I love to sail forbidden seas and land on barbarous coasts.
—Herman Melville, *Moby-Dick*

In September of 2011, off the Pacific coast of Panama, biologists from the Smithsonian Tropical Research Institute spotted the fins of three whale sharks.[14] The team leapt into action, and using a modified spear they attached satellite tracking tags to the back of each shark, just behind the dorsal fin. All three were females, and one, nicknamed Anne, would set a world record. Over the next 841 days, Anne's tag delivered positional fixes that revealed a shark on a mission. After deep dives near the Panamanian coast, she swam to the Costa Rican island of Cocos, then to the famed shark gathering grounds in the northern Galápagos Islands. Moving on, she cruised west through deep blue waters to Hawaii, halfway across the world's largest ocean. When the transmitter's batteries finally expired two years later, she had made it all the way to the western rim of the Pacific amid the Mariana Islands. All told, Anne had completed a journey of 12,500 miles, smashing all previous records.[15] While some have questioned the validity of this particular voyage, more recent satellite studies have confirmed tracks of 4500[16] and 4000 miles over considerably shorter stretches of time.[17]

The travel logs of Anne and other adult whale sharks suggest they have discovered the marine equivalent of a moving sidewalk across the vast Pacific. The North Equatorial Current, a stream of surface water gushing westward a few hundred miles north of the equator, offers a highway linking productive feeding grounds off Central and South America to the warm, soupy waters of Indonesia and the Philippines. Because the current slackens during winter and early spring, however, the one-way trip can take as long as two years. And that, precisely, is why plankton feeders like whales and mantas and whale sharks are united in being so massive: they need a gargantuan gas tank to fuel the journey on this long aquatic highway. If you were on a road trip with a whale shark, she could drive tirelessly all day and all night and never stop for a snack.

Charles Darwin, in his earthshaking study of Galápagos mockingbirds and finches, discovered that isolation of one island's birds from another's induced the birds to evolve into distinct species. The same principle applies to many fishes, especially small ones. Species endemic to the Atlantic Ocean differ significantly from their relatives in the Pacific because of the isolation imposed by distance, their weak swimming, and a modest obstacle called Central America. Whale sharks, however, present quite a different picture. Thanks to their marathon transoceanic cruises, a whale shark can visit all the oceans on Earth in its lifetime, estimated to be as long as 130 years.[18] Their unparalleled mobility means whale sharks are among the least isolated organisms on the planet, polar opposites of Galápagos finches. For this reason, whale sharks belong to a single species (*Rhincodon typus*), and because of their genetic mixing they form a single population, with only faint distinctions between sharks plying Atlantic waters and those favoring the Indian and Pacific Oceans.[19] But from a conservation perspective, their planetary voyages convert isolated dangers in a single region, like an illicit fishery, into jeopardy for all whale sharks everywhere. Protection of the planet's largest fish requires cooperation across all oceans, and all continents.

Jennifer Schmidt has studied whale sharks at the Shark Research Institute for more than twenty years. From her very first encounter in the waters of the Philippines, she was captivated: "This massive thing, it looked like it was the size of an aircraft carrier, just passed right beneath me."[20] Her research applies genetic techniques to shed light on whale shark movement and reproduction dynamics, and to steer their conservation. She laments that the threats faced by whale sharks are manifold. "We believe that fishing, or poaching—depending on the rules of the country—is a huge player, both intentional and accidental. Commercial factory fishing probably takes a great toll; much of that is unregulated in international waters, so it's hard to say how much." Whale sharks regrettably command high prices in Asian seafood markets, traded covertly because of their endangered status, where the gentle giants are dubbed tofu shark because of the soft texture of their meat.[21] Majestic whale shark fins are also sold, after the owner has been dispatched, as trophies to be exhibited like macabre ornaments in emporiums touting shark fin soup on their menus. Hunting of this sensitive species, easily captured and slow to reach reproductive maturity (at 20–25 years of age), has decimated their numbers. Estimates for population declines range from 40 percent around Ningaloo reef[22] to 80 percent in the waters off Mozambique.[23]

Dr. Schmidt continues to tick off threats: "We think boat strikes probably take a big toll, particularly from large container ships. I mean, we tend to forget how many ships cross virtually every part of the ocean all of the time, and whale sharks are on the surface." To serve a swelling global economy, the number of oceanic shipping vessels has quadrupled since the 1950s;[24] by 2015 they were transporting more than 10 billion tons of freight per year.[25] "And I'm sure habitat change takes a toll. The ocean is warming; patterns of food—plankton—are almost certainly changing." Earth's seas are absorbing more carbon from our changing atmosphere, making their waters more acidic, which impacts shell-forming plankton, among other life. As evidence, researchers surveying such creatures on routes first sampled in the late 1800s found that shell thickness had plummeted by three-quarters in every single specimen.[26] Last but not least, Schmidt cites the impact of marine pollution, such as the oil spilled by the *Deepwater Horizon* platform into the Gulf of Mexico in 2010.[27] "We have a lot of whale sharks off that coast, and the track, the oil plume, went straight through their habitat." Her summation was stark: "It's bad to be a surface-feeding animal in an oil spill."

The collective impact of all these threats is maddeningly difficult to assess. Most of the time, whale sharks are solitary creatures, and the ocean is indeed a very large place. Apart from individuals who wash up on beaches or are reported as bycatch (such honesty is distressingly scarce) or happen to bump into a scientific exploration, we know very little about how these giants live in the deep blue waters of the open ocean. To illustrate just how rare such firsthand encounters have been, consider that by 1985—a century and a half after this enormous animal was first described by science—whale sharks had been observed in the wild only 320 times. But all that changed with an amazing discovery: in a handful of tropical locations, juvenile whale sharks gather in dense aggregations, shedding their solitary ways and joining crowds that can number in the dozens or hundreds.

CONGREGATION STATIONS

Since congregations of whale sharks were first found at Australia's Ningaloo reef and a few remote scuba locations, the number of known aggregation sites has swelled to more than two dozen worldwide. Like a scene from a city park in summer, throngs of juveniles and a smattering of adults gather to socialize and gorge themselves on eruptions from the marine equivalent of a picnic basket: a nearshore upwelling. Whale sharks are remarkably faithful to such locations—one individual nicknamed Stumpy has returned to Ningaloo for over twenty years—a fidelity that finally allows marine scientists to study elusive creatures whose biology is shrouded in mystery as murky as the waters they favor.[28] And those locations are generating more than just data. Their popularity has spawned a whole new industry, one that might just save these gentle giants from the dangers our civilization has been throwing at them.

Whale shark tourism has boomed in the past two decades. Many aggregation sites are found in shallow waters near tropical shores, where paddling in turquoise waters with the largest fish in the world has swelled in popularity. In the Maldives, some 75,000 whale shark tourists visiting a single atoll contributed more than US$9 million to the local economy in 2013.[29] Four years later, a similar study in Indonesia's Cenderawasih Bay National Park showed the industry netted over $10 million; based on these local figures, one economic analysis calculated whale shark tourism injects some $2.6 billion annually to the nation. As with predatory sharks, the value of a single whale shark over its entire life span has been estimated at ten times its worth as a one-off fisheries catch.[30] Tourism is not a panacea, however, and under-regulated sites risk disturbing or injuring whale sharks when too many swimmers, boats, and motors clog the waters. Conservationists today are focused on enforcing codes of conduct that permit rewarding experiences for visitors but safeguard the star attractions.

Much of what scientists like Jennifer Schmidt have discovered about whale sharks comes from observations at these popular aggregations. Even citizens can contribute; conservation-minded tourists upload underwater photos to identification databases, from which more than 12,000 separate individuals have been identified. Schmidt counsels caution, however, when generalizing the discoveries made in these aggregations, where adults are less often seen, and some two-thirds of congregants are males. She notes wryly, "I compare it to an alien landing on Earth and ending up at an all-boys high school, and taking that to be the sum of what humanity is." Many mysteries remain: where adults travel after leaving aggregations, where they mate, or even give birth, are still completely unknown.

Whale sharks seen at aggregation sites sometimes bear the unmistakable signs of boat collisions, the result of oceanic voyages that regularly cross major shipping lanes. Researchers working in the Maldives, however, discovered something amazing. When reviewing multiple photographs of the same whale shark, its identity confirmed by the fingerprint-like pattern of its spots, they were able to trace wound healing rates in individuals with visible injuries. They estimated that even severe gashes achieved nearly complete closure after just one or two months, with one animal showing 50 percent recovery in just four days.[31]

How have whale sharks evolved their swift healing rates? Partly thanks to their toothy ancestors who evolved to withstand biting and clawing by flailing prey. But a filter-feeder's habit of eating near the surface may have also contributed. In areas where currents bring productive waters together, these creatures collide with all sorts of marine flotsam: logs, sticks, dead fish, jellyfish, coconuts, and more. Whale shark eyes, for example, show an adaptation unique among fishes: they are covered by the same tooth-hard denticles that armor the skin of predatory sharks. These scales armor the entire eye, even the iris, and shield it from damage by floating objects. Further protection is granted by the shark's ability to retract its eye almost fully into the head, using an exaggerated eye-roll reminiscent of a flamboyantly exasperated teenager.

The sites where whale sharks aggregate are remarkably similar: warm and shallow seas, adjacent to deep water, with steeply sloping bottoms leading to an abrupt drop-off. This trifecta of conditions virtually guarantees upwellings of nutrient-rich water that fertilize plankton in the balmy and sunlit shallows. Whale sharks cruise happily and open-mouthed through the surface waters to feed, but during the day plankton dive to escape visual predators and hide in the shadowy depths, taking an escape elevator to the basement bunker. Fortunately, whale sharks enjoy a suite of adaptations allowing them to follow their food wherever it may go.

Near the surface, a whale shark's armored eyes can readily see dense clouds of soupy plankton. Their retinas are packed with both rods and cones, like our eyes, but this duplex design is uncommon among fishes. Rods work well in dim light; cones provide color vision in brighter light. In whale sharks the color receptors are tuned to match the broad spectrum available near the surface, their favored depth, whereas in most fishes they are calibrated for the heavily blue-shifted colors characteristic of deeper waters.[32] But when the sun arcs high in the sky, and whale sharks dive after plummeting plankton, they take advantage of their unique duplex retina by switching off cones and allowing rods to take over. Like donning a pair of night-vision goggles, they can now target swirls of plankton even in deep, dark waters.

To reach the tasty depths, a hungry whale shark takes advantage of his negative buoyancy and enters a glide descent.[33] Sinking in a controlled fashion, he steers himself forward with the pectoral fins but holds his tail motionless to save energy. Whale sharks may glide to depths of 6000 feet or more, where the temperature hovers just above freezing.[34] Once at the nadir of the dive, our aquanaut opens his mouth to feed and gently uses his tail to propel himself slowly forward. The bracing cold penetrates his thick skin, gnawing remorselessly at his warmth. Icy water flushes over gills rich with blood vessels that carry the chill deep within his body. But whale sharks show an unusual arrangement of musculature: a huge mass of white muscle, poorly enervated by blood vessels, lies in the core of their body, surrounded by highly vascularized red muscle.[35] Chilled blood mostly reaches only the red muscle, while the white muscle core remains warm and isolated, functioning like a battery for storing heat. This arrangement inverts the blueprint of warm-blooded tuna, in which the core is composed of red muscle that generates metabolic heat through vigorous swimming.

Despite the appeal of abundant plankton, cold soon saps muscle power and compels our deep-water diner to return to the sunny surface and replenish his lost body heat. On the ascent, his swimming angle is twice as steep as was the descending glide path, and he travels more slowly, at about half the speed of the dive.[36] Thus the round trip is highly asymmetrical, pairing a long and swift glide to depth with a short but deliberate swim back to the balmy shallows, like trudging up the stairs at a water park after a gleeful slide to the splash pool below. Compared with horizontal swimming, this asymmetrical glide and climb technique saves swimming energy, maximizes the time spent feeding, and yields a welcome savings of nearly 30 percent in overall foraging efficiency.[37]

When our whale shark reaches the surface, he triggers another special adaptation: a surge of blood flows to his gills, speeding absorption of much-needed warmth. This phenomenon, also seen in deep-diving blue sharks, supercharges the reheating process and recharges his thermal battery of white muscle. Researchers also discovered that whale sharks spend more time surface-basking after colder and deeper dives, showing that behavioral and physiological adaptations unite to meet the challenges of foraging in deep, cold waters.[38]

A similar strategy is employed by the appropriately named basking sharks (*Cetorhinus maximus*). These 50-foot cousins of whale sharks strain zooplankton from boreal surface waters where great oceanic fronts meet and boost productivity.[39] They visit such regions only during the summer, when they can simultaneously feed and bask in the summer sunshine. In winter, they undertake great migrations toward equatorial seas, off the coast of Brazil and elsewhere. Upon arrival they eschew the hot surface layer and instead spend their days foraging in comfortably cool waters 2000 or 3000 feet below the waves. Incredibly, they may remain at these depths, without surfacing, for up to five months at a time.[40] Basking sharks most likely have slower metabolisms than their tropical cousins the whale sharks, permitting them to forage for such lengths in the cool depths. Or perhaps they are just escaping the frenetic jangle of samba music pulsing from Brazilian beaches.

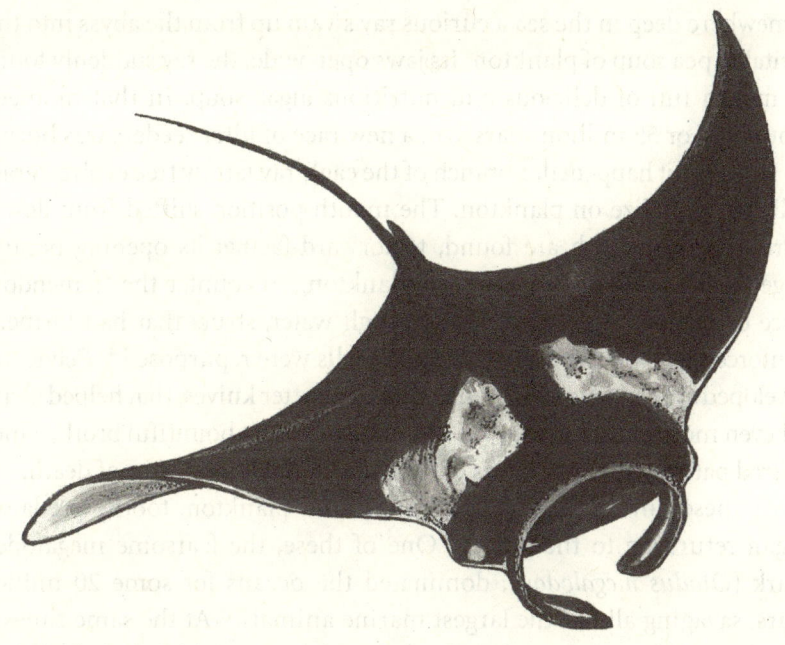

Giant manta ray (*Mobula birostris*)

A WHOPPING WET BLANKET

Round about 66 million years ago, a cataclysm on Earth wiped clean the slate of life. But its aftermath heralded the arrival of most modern fishes. Evidence suggests a colossal meteor strike off the Yucatán Peninsula blasted enough debris into the atmosphere to trigger a fatal shift in planetary climate. Dust storms dimmed the sun for centuries, plant life collapsed, dinosaurs went extinct, and oceans turned to acid. Fossil records show that more than three-quarters of Earth's plants and animals perished, altering marine ecosystems so dramatically that a million years would pass before they could recover.[41]

During the planetary calamity, the deep sea provided a refuge where bottom-dwelling rays persisted, munching away on the few shellfish that survived. Above, a race of huge plankton-eating fishes known as pachycormids disappeared in the paleontological blink of an eye.[42] As the choking dust settled, continents and oceans embarked on recoveries, and marine productivity surged. Plankton turned the seas emerald green, thanks to nutrient runoff from now-verdant lands, and the extinction of pachycormid planktivores. On one fateful day, it is tempting to imagine, somewhere deep in the sea a curious ray swam up from the abyss into this veritable pea soup of plankton. Its jaws open wide, the ray suddenly found its mouth full of delicious and nutritious algae soup. In that moment, around 50 or 55 million years ago, a new race of filter-feeders was born.[43]

However it happened, a branch of the eagle ray family tree evolved gradually to specialize on plankton. The mouth position shifted from downward, where shellfish are found, to forward-facing; its opening became larger, welcoming more water and plankton. To counter the tremendous force of pushing open-mouthed through water, struts that had formerly reinforced jawbones for chewing hard shells were repurposed.[44] Pelvic fins developed stubby outgrowths, like rubbery butter knives, that helped channel even more plankton into the mouth. But in that bountiful broth, which offered pathways to new kinds of life, there lurked new forms of death.

As these novel rays developed a taste for plankton, toothy predators began returning to the oceans. One of these, the fearsome megalodon shark (*Otodus megalodon*), dominated the oceans for some 20 million years, savaging all but the largest marine animals.[45] At the same time an ancestor of the modern sperm whale, nicknamed Melville's leviathan (*Livyatan melvillei*), was attacking squid and other large prey with foot-long teeth.[46] These animals co-occur in fossil deposits in Peru, testament perhaps to an evolutionary race to gigantism in which each relied on its enormous size for defense against the other. In the face of these terrifying and massive titans, placid plankton-eating rays had only one option, evolutionarily speaking: join the race to enormity.

Today's manta rays are the culmination of this evolutionary saga. Their newfound bulk afforded them some measure of protection from giant predators, filled an ecological niche vacated by extinct planktivores, and granted the diverse benefits of gigantism. What were once modest-sized, shell-eating, bottom-dwelling rays evolved after the meteor strike into the filter-feeding colossi of today. Giant manta rays (*Mobula birostris*) are built like flying carpets, and broad enough to transport your entire family, and a couple camels for good measure. "Manta" derives from the Spanish word for shawl or blanket, and what a whopping blanket they are. The largest mantas can weigh 4400 pounds and boast a 23-foot wingspan, capable of blanketing six king-sized beds.[47]

Manta rays and whale sharks have much in common and often frequent similar locations, like gourmands favoring the same obscure restaurants. Both are supremely efficient feeders, cruising slowly with enormous mouths agape, filtering a meal of plankton with highly modified sieves.

Mantas, however, swim with bird-like flapping of their expansive pectoral fins, the largest animal on earth to use this form of locomotion. Reinforced by an interwoven skeletal mesh, those powerful yet rubbery fins propel the animal with a hydrodynamic efficiency that rivals tuna and mackerel, albeit at lower speeds. Each aquatic flap is asymmetrical, the wings reaching much higher on the upstroke, pausing briefly while the ray glides effortlessly, then driving shallowly downward to complete the cycle. Superimposed on this stroke is an oblique undulation that twists the fin, further increasing power output. Their novel swimming motion is so economical that a team of scientists studying manta kinetics applied their findings to a unique invention.[48] They built a wholly new type of remotely operated underwater vehicle, the MantaBot, whose silicon wings beat in a convincing mimicry of the real creature. The team leader's name, Dr. Frank Fish, is a droll bonus; one presumes he also speaks most forthrightly about his chosen subject.

Manta rays and their close relatives are instantly distinctive, thanks to their catamaran-shaped head. Projecting forward from either side of the mouth are twin paddles called cephalic (head) fins. These arise embryonically as lobes of the pectoral fins, a developmental process spurred by genes (like *Hoxa13* and *Hand2*) that also promote the appearance of claspers in male sharks, another set of modified fins.[49] Those two genes even played a key role in the evolution of terrestrial animals like ourselves, spurring the fin-to-limb transformation that allowed our ancestors to crawl from the sea.[50] In the gigantic oceanic mantas and reef mantas (*Mobula alfredi*), cephalic fins are robust and oar-shaped, while in eight modest-sized kin the smaller and more pointed protuberances resemble horns, earning this group the nickname of devil rays. Their purpose, however, is more dietetic than diabolical: the paddles stretch forward and curl downward, forming a funnel that channels vast volumes of plankton-rich water into the mouth. Once engulfed, the green chowder is passed through gill rakers, cartilaginous arches lying atop the large gills, that sieve a meal of plankton from the water using an innovative approach which is today inspiring novel engineering solutions to an age-old problem.

In a typical filter—imagine a kitchen sieve straining a pot of watery rice—the filter pores eventually get clogged, and water can no longer run through. Most filter-feeding animals will backflush their filters with a forcible cough of water, rinsing the pores clear but losing a plateful of food in the process. Whale sharks cough once every 7 or 8 minutes when feeding.[51]

But mantas never do, a fact that caught the attention of marine scientists and mechanical engineers. In sieve-filtering fishes like whale sharks, water passes perpendicularly into the filter, and through tiny pores that trap all larger particles. But scientists discovered that particles smaller than the pores were still being filtered and swallowed by mantas, yet no caking or clogging of the filter occurred. How was this possible? No filtration system invented by engineers could achieve such results.

It turns out that plankton particles in mantas are skittering across the top of a filter that resembles a washboard: tightly packed rows of miniature speedbumps, with pores at the bottom of each valley. As edible particles stream along, they collide with the first bump and are bounced upward into the flow; at the same time, a little water drains through the pore. As the process repeats, particles ricochet off successive bumps, eddying in the moving water until a dense cloud of particles has accumulated. Amazed engineers dubbed this innovation "ricochet separation" and are hotly pursing its application to industrial filtration applications, where cleaning of filters is a tedious and expensive chore.[52] For mantas, the innovation means that more plankton are filtered more efficiently, since smaller particles can be captured, and there are no delays for coughing. Pushing water over a washboard rather than through tiny pores also imposes less drag, so manta rays burn fewer calories while swimming through their soup. Unfortunately for giant rays, their filters have caught the eye of more than just engineers. Gill rakers are prized in traditional Chinese medicine as an alleged cure for ailments from acne to cancer (no evidence whatsoever supports these claims), and surging demand has devastated ray populations around the world.[53] Their decline is heartbreaking in part because of a growing understanding that mantas are highly intelligent and remarkably social.

Nestled between their cephalic fins, manta rays have one of the largest brains ever measured in a fish.[54] This discovery was somewhat surprising as filter feeding would not appear, superficially, to be intellectually challenging: just open your mouth and swim. But manta rays assail clouds of plankton in much more intricate performances, engaging in choreography befitting a ballet company. Feeding mantas will swim head-to-tail like a conga line, or perform repeated somersaults to corral plankton, or swirl into a coordinated cyclone, compacting helpless plankton in the center before diving through the storm's eye to feed. Giant manta cyclones may even draw cold, nutrient-rich water up from the depths to fertilize surface plankton. Engineers (who really are starting to owe a professional debt to manta rays) have applied these social feeding techniques to optimization algorithms, mathematically guided searches for solutions to convoluted problems like aiming solar panel arrays on partly cloudy days.[55] Impressively, the manta-inspired formulas outperform existing optimizers in nearly all cases.

Despite being most commonly seen at the surface, mantas are known to feed at considerable depths. In New Caledonia, reef mantas equipped with satellite tags regularly dove to below 1000 feet (occasionally to 2200 feet), almost always at night, and lingered there for up to 10 minutes in a U-shaped dive profile that suggests a foraging pattern.[56] Off the coast of Ecuador, isotope analysis revealed that giant mantas habitually feed in deeper waters, rather than on surface plankton.[57] This biochemistry technique compares the relative proportion of carbon and nitrogen isotopes—molecular variants with slightly different weights—in manta tissues against their preponderance in plankton at various depths. The isotopic signature gradually accumulated in muscle and fat can reveal lifelong foraging patterns. After a feeding plunge, mantas warm themselves at the surface, basking like whale sharks. Their large brains benefit from having heating blankets like those in marlin and swordfish, dense networks of blood vessels that keep the brain warm and functioning acutely to plan and execute the next dive.[58]

In the shallows where divers can observe them, mantas routinely socialize with each other. Individual rays can recognize one another by variations in their distinctive patterning theme: the black backs of reef and giant manta rays are marked with broad strokes of white chevron-shaped clouds behind the eyes and stripes pointing to the wingtips, and their white bellies usually display black speckles. These patterns can change, often dramatically, during feeding or intense social exchanges.[59] Extreme variations occasionally occur, like the bright pink reef manta ray discovered in Australian waters and nicknamed Inspector Clouseau after the bumbling detective of Pink Panther films. On the reef, mantas gather into distinct groups, respond to one another by wagging and curling their cephalic fins, and show complex, lasting bonds.[60] Females prefer the company of other females, and teenagers hang around their own age group (as any parent can confirm); males, in contrast, show few lasting associations and congregate indiscriminately at the richest of feeding sites.[61] In the flamboyant coral reefs of Raja Ampat, Indonesia, rays spend hours at cleaning stations where small fishes scour the bodies of larger animals, plucking parasites and exfoliating algae. Reef mantas gather in large numbers and remain faithful to a particular spa, jostling for space even when alternate sites are available quite close by.

Sociality is particularly inflamed when mantas reach the age of reproduction. In this, however, they are less like fish and more like humans. Mantas are long-lived creatures, delay mating until they are nearly full sized, bear live young but infrequently, and give birth to only a single offspring at a time. Reef mantas, for example, may live for forty years and reach sexual maturity only around their tenth year.[62] Gestation can require twelve months or more, and females will usually delay up to three additional years before mating again. Courtship is intensely social, with receptive females leading multiple suitors on a "mating train," a vigorous dash through open water punctuated by somersaults that are mimicked by the males. This underwater steeplechase helps the female select the most vigorous male. Once a worthy partner has outlasted all rivals, he swims atop the female, seizes her pectoral fin in his mouth (oddly, it is almost always the left fin), jostles her body until his belly presses against hers, then uses his clasper fins to manually inseminate her.[63] During gestation the developing baby ray is bathed by uterine milk that provides nutrition and oxygen. By the time she or he is born, a manta ray pup is enormous, up to half the size of the mother.

Across the marine giants, a full range of reproductive strategies are on display, revealing how many pathways there are to evolutionary success. Manta rays emphasize the survival of a single offspring, investing as much energy as they can in the pup. Whale shark females carry 200 or even 300 eggs and give birth to tiny youngsters. Incredibly, fertilization of these embryos can be staggered in the mother, with some offspring born early while later-fertilized embryos remain unborn for weeks or months.[64] Females may achieve this assembly-line approach by storing sperm, then releasing it for intermittent fertilizations, something seen in blue sharks over periods as long as twelve months.[65] Staggered birthing is literally the opposite of placing all your eggs in one basket: by spreading their young across the ocean like a necklace of pearls, female whale sharks increase the chances that at least one or two find themselves in a perfect nursery, predator poor and nutrient rich. If winning the lottery of reproduction is your aim, then playing a lot of tickets can be the best plan, precisely the strategy pursued by superlative ocean sunfishes.

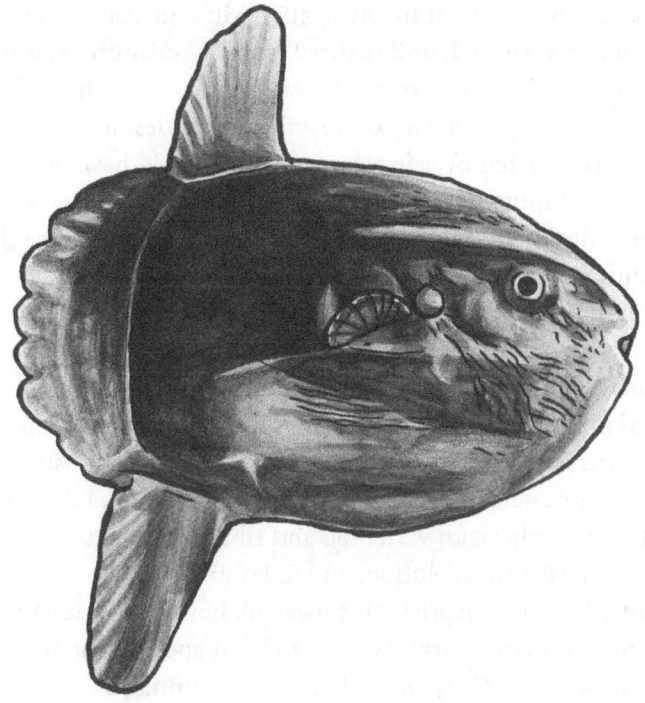

Ocean sunfish (*Mola mola*)

THE GREATEST AND THE SLIGHTEST

An egg is always an adventure: it may be different.
—Oscar Wilde, *The Wit and Humor of Oscar Wilde*

When a prickly little ball was sieved from Australian waters in 2017, a fish larva resembling a tiny balloon studded with transparent pyramids, its identity remained a mystery. Unlike mantas and whale sharks, most fishes release their eggs to hatch in open water rather than developing inside the mother. For each miniscule larva it must be quite a shock to emerge from a cozy egg, set abruptly adrift in an empty ocean to survive all by itself. As soon as their yolk is exhausted, they must rely on nascent eyes and barely formed fins to find and pursue food, until they reach their adult form. Identifying these underdeveloped larvae, which look maddeningly different from their parents, can be an equally intimidating task. But after three diligent years, researchers at the Australian Museum solved the enigma of the "conspicuous, rotund, and distinctly spiky" creature who measured just one-tenth of an inch across.[66] It was the newly hatched offspring of a bump-head sunfish (*Mola alexandrini*), the heaviest fish in the world.[67] Had it not been netted by science, its destiny would have been to attain an adult mass of more than 6000 pounds, a flabbergasting 60 millionfold increase over its birth weight.[68] Even more astonishing was the discovery that an adult female can carry up to 300 million eggs, making it the most fecund fish in the world, yet another sunfish superlative.[69]

Sunfish are peculiar in just about every way imaginable. Their scientific name, *Mola*, is Latin for millstone, an admirable description of a weighty fish shaped improbably like a giant disc. They have only two fins, one projecting toward the surface, the other toward the seafloor. Instead of a tail, they sport a truncated, fleshy lobe that serves as a rudder. Like a millstone, their body is entirely rigid: vertebrae and ribs have been severely reduced over 10 million years of evolution, and today all propulsion is provided by the two fins. They are surprisingly powerful, however, permitting sunfish to swim against ocean currents and travel at speeds that rival cruising marlins and sharks.[70] Only one other kind of animal swims as they do, and it is not even fish, but a penguin. Both sunfish and penguins flap their paired fins in unison, generating lift that is converted to forward momentum by rotating the fin angle, more like a bird winging through the sky than a rowboat thrust across a lake.

Closely related to the bump-headed sunfish is the equally spectacular ocean sunfish (*Mola mola*; the fish so nice they named it twice); a third species, the hoodwinker sunfish (*Mola tecta*) was described only in 2014.[71] All three molas are massive, with ocean sunfishes reaching 9 (or possibly 13) feet across. Their skin, covered in multi-cusped, jagged scales, is gritty and rough. Beneath the skin, however, lies a thick layer of rubbery gelatin that can account for more than 40 percent of its total body weight.[72] Researchers surmise the layer helps insulate the fish during deep dives for food, a habit it shares with mantas and whale sharks. It also helps them reach and return from these depths, as its low density renders the fish neutrally buoyant. The gelatin may be acquired from one of their favored foods, jellyfish, a preference that has earned molas the title of gelativore, or jelly-eater.

After sunfishes dive, they bask on the surface, turning sideways until they lie completely flat just inches below the waves, catching rays until they warm again for another plunge. The longer the dive, the longer they sun-bathe, suggesting that basking is a behavioral mechanism for thermoreg-ulation after long forays into deep, cold waters.[73] All the ocean giants rely on behavior changes to maintain their body thermostats, save one. Opah (*Lampris guttatus*), a large disc-shaped species superficially similar to the molas, was discovered in 2015 to be the world's first truly endothermic fish.[74] These open-water giants, tipping the scales at some 600 pounds, can sustain internal temperatures 8 degrees above that of the water surround-ing them, the heat generated by dark red pectoral swimming muscles.[75]

Drifting back to the basking molas, several researchers suggest their surface loafing may be an attempt to encourage cleaning of parasites that grow copiously on their large frames. Seagulls have commonly been observed visiting lounging sunfish to pluck crustaceans and other pests from their upturned flank, as if from a crumb-strewn picnic table. The common name of sunfish stems from this basking behavior, as most early encounters occurred when seafarers stumbled upon them during tanning sessions (one imagines a startled fish hastily reaching for a towel to cover her modesty). When lolling at the surface, an enormous fin may occasion-ally flop into view, provoking anxious but deceived beachgoers to report an imminent shark attack. In southern Massachusetts a flood of such calls to 911 prompted emergency services to respond forcefully via text: "The sunfish is doing normal sunfish activities. It's swimming. It is not stranded or suffering. The sunfish is FINE . . . PLEASE STOP CALLING THE POLICE DEPARTMENT ABOUT THIS SUNFISH!!"[76]

Unfortunately for sunfishes, their habits of luxuriating at the surface and feeding (like basking sharks) at oceanic fronts where productivity is rich, put them on a collision course with the world's fisheries. Though rarely the explicit target, they are trapped and killed by indiscriminate gillnets with shocking frequency. In the Mediterranean during the early 1990s, ocean sunfish composed between 70 and 93 percent of the total catch by Spanish drifting gillnets.[77] A decade later, more than 30,000 sun-fish were still captured every year by the Moroccan gillnetting fleet.[78] Even in well-managed waters like California, where swordfish fisheries also rely on gillnets, nearly a third of all bycatch (nontarget captures) hauled on board were ocean sunfishes, their discarded bodies preposterously out-weighing the total catch of the swordfishes.[79]

The decline of ocean giants like sunfishes, mantas, and whale sharks may have cascading consequences for their oceans. Filter feeders and jellyfish specialists have long been thought to control, through predation, the abundance of jellies and their ilk. With the decline of these giants, the swift reproduction of their prey threatens to overwhelm the sea, with disastrous costs. Jellyfish are notorious predators on fish larvae, and the overabundance of the former can smother the latter's survival. Numerous studies around the world have signaled increases in jellyfish abundance.[80] Some evidence suggests we are merely on the upswing end of a twenty-year cycle driven by regular oscillations in ocean conditions.[81] But too many human-driven factors have been implicated in the surge to ignore our own impact: global warming, increasing nutrient loads, ocean acidification, and overfishing of filter-feeders are all proven to provoke blooms of jellyfish.[82] In one assessment of combined human impact on oceans, jellyfish outbreaks had plagued six of the top ten impacted coastal regions.[83] In crisis, however, opportunity can be found. People have been eating jellies for centuries, and today nearly a half-million tons are consumed annually in Southeast Asia alone.[84] Japanese food scientists have even engineered a way to squeeze the water from these gelatinous creatures, rendering them into a rubbery and edible patty. So you might double-check the fine print next time someone offers you a peanut butter and jelly sandwich in Tokyo.

Just as the absence of ocean giants can lead to an unraveling of the marine realm, their protection can yield tremendous benefits, beyond the massive value of their tourism appeal. We now know that large marine creatures like whales and seals, but also sharks and mantas and whale sharks, transport nutrients from the open ocean to nearshore reefs. After feeding in the depths, all these animals return to the surface where they defecate and release valuable nutrients that sustain healthy reef communities. Fancifully termed "the whale pump," one estimate suggests that whales alone recycle more nitrogen to the surface waters of the Gulf of Maine than the combined input of all the region's rivers.[85] In Pacific coral reefs, more than three-quarters of the nitrogen released by grey sharks was brought in from open-ocean feeding forays.[86] And the fidelity shown by manta rays to specific coral sites, where they return for routine spa treatments, underscores just how much investing in local conservation efforts can provide lasting returns for coral reef ecosystems, and for the people whose livelihoods depend on them.[87]

Part II
ROCK, SAND, AND REEF

JULES VERNE PROVIDED THE WORLD ITS FIRST GLIMPSE of life under the waves in his brilliant and lovingly detailed novel *20,000 Leagues Under the Sea*. Few people on Earth had the slightest conception, when the watery tale was published in 1870, of what lay beneath the surface of the sea. Verne himself had never pierced the briny depths, but his formidable imagination was kindled when he beheld the world's first mechanical submarine, the *Plongeur*, unveiled at the 1867 International Exposition in Paris. The 140-foot steel vessel, cylindrical in shape with a rapier-like bow built for ramming enemy ships, inspired him to concoct the now-famous story of Captain Nemo and his own craft, the *Nautilus*. Verne was an avid amateur naturalist, and his meticulous research of sea life provided rich detail for the spectacular chapter modestly titled "A Walk On The Bottom Of The Sea."[1]

Having been captured by Nemo, the protagonist Professor Pierre Aronnax is introduced to the wonders of undersea life on an excursion across the ocean floor led by the captain himself. The party don diving suits and bell helmets, sip air recycled by rebreathers, and set their weighted shoes on the bottom. "And now, how can I retrace the impression left upon me by that walk under the waters," begins the Professor. "Words are impotent to relate such wonders!" But relate them Verne does, majestically and reverentially, as his underwater party explores a diversity of seafloor habitats in breathtaking detail. First they stroll on "fine sand . . . sown with the impalpable dust of shells." Approaching some rocky outcrops, "hung with a tapestry of zoöphytes of the most beautiful kind," the colors and range of forms arrest the Professor's gaze. "It was then ten in the morning; the rays of the sun struck the surface of the waves at rather an oblique angle, and at the touch of their light, decomposed by refraction as through a prism, flowers, rocks, plants, shells, and polypi were shaded at the edges by the seven solar colors. It was marvellous, a feast for the eyes, this complication of colored tints, a perfect kaleidoscope of green, yellow, orange, violet, indigo, and blue; in one word, the whole palette of an enthusiastic colorist!"

Verne's enthusiasm was fanned by painstaking zoological research, and his potent imagination made organisms that he could have known only as sallow and shrunken specimens in jars come brilliantly alive. Coral, anemones, and sea stars are gorgeously depicted, as are giant basket stars that remind the author of "fine lace embroidered by the hands of naiads, whose festoons were waved by the gentle undulations caused by our walk." In the space of ten pages, the expedition visits a half-dozen distinct marine habitats, including kelp forests and underwater canyons cleaved through stone; they even take a leisurely nap on the seafloor, to recover from their exertions. Verne flashes his genius by staging this first marine exploration on foot, a means of travel immediately familiar to anyone who has ever strolled through a forest or prairie. It is an invitation to the reader to open the eyes of their imagination and see the undersea world where it is most similar to land: on the ocean floor.

When a submarine dives beneath the surface, it enters a vast void, a three-dimensional volume filled only by water. It is the equivalent of stepping outside a space station into the great empty vacuum of the universe. Of course, the aquatic void is not empty, it is freckled with life, from phytoplankton to whale sharks that dot the sea like stars in the cosmos. But when that submarine reaches the seafloor, it encounters something new: physical structure. The bottom is decorated by a panoply of monuments and edifices. Some are the product of geology, like the great stone blocks around which Nemo's party trudged, while others are engineered painstakingly by living organisms. Towers and pedestals soar over archways and tunnels; crevices and canyons are crenellated with filigree scrollwork that would satisfy the most flamboyant baroque architect. It is this structure that sets the seafloor apart, making it so fundamentally different from the open ocean.

All manner of marine organisms collaborate to build these structures, in convoluted and dynamic processes that unfold over millennia. Sponges decorate rocks, form small blobs, or blossom into enormous towers and barrels that can dwarf a person. Corals of a thousand varieties cohabitate with photosynthetic algae and put their combined energies to work building palaces, ramparts, and cathedrals of calcium which link to form immense reefs. Leafy fronds of algae festoon underwater surfaces, but algae can also be rock-hard and encrusting, coralline varieties that build their own structures and glue together loose rocks and boulders. Barnacles, mussels, oysters, and other shellfish also secure and accrete loose materials in a constant struggle against the destructive energy of waves, tides, and cyclones. Collectively, these living organisms build on the seafloor an intricate architecture that is wholly lacking in the open seas.

Undersea structure offers a nearly endless variety of benefits to marine fishes. Hiding places abound in rocks and corals, into which a fish can slip to escape predators. Conversely, hunters conceal themselves in holes and crevices, then burst out to attack a passerby. Some fishes sleep in caverns or beneath shadowy overhangs, or loiter in those dim spaces until night falls on the reef. When the siren song of mating beckons, fishes use undersea structure as courtship parlors, then bury eggs amid pebbles or glue them to rock faces. Juveniles seek safety amid seagrasses and mangrove roots, intricately woven refuges from predators. Whereas nearly all open ocean fishes are built to chase and attack smaller fish swimming in front of them, a host of adaptations have sprung up in reef fishes enabling them to feast on the diverse life encrusting the bottom's structures. Parrotfish chew on coral, butterflyfish pluck invertebrates from stone walls, and triggerfish blast divots in sandy bottoms to expose buried worms and crustaceans. Industrious damselfish even tend tiny, leafy crops of algae on rock outcrops, then pugnaciously defend their harvest from marauding schools of herbivores looking to steal their lunch, like schoolyard bullies.

Out in the high seas, plankton and tiny larvae floating near the surface are the foundation of the food webs that sustain sardine, tuna, whale sharks, and the other denizens of the great blue. But down on the ocean floor, that foundation is much more expansive. Where light reaches the bottom, coral and algae and even photosynthetic sponges convert solar energy into food, like a vegetable garden. Near land, nutrients flush from rivers into the sea, providing an endless source of food. And everywhere, even in the ocean's deepest places, edible material falls like snow from above. Referred to unromantically as particulate organic matter, everything from fish scales to feces, expired plankton to jellyfish tentacles, even dead whales, all this edible material tumbles from the surface as a veritable rain of food morsels. In most of the abyssal depths of the ocean, where light never penetrates, this organic snow represents the only source of food for a multitude of deep-dwelling organisms.

In those great depths, the waters are inky black and icy cold. Some 500 feet below the surface, all light fades and a transition is passed beyond which the temperature hovers just a few degrees above freezing: you have crossed beneath the thermocline. In these depths, photosynthesis is impossible, oxygen is scarce, muscles are frigid, and the cost of movement is exorbitant. The animals who live below this threshold have adapted to a life so different from ours that they might as well inhabit the surface of the moon. Indeed, the seafloor in these great depths often resembles the moon, a featureless landscape of fine silt, only infrequently interrupted by rocky outcrops. But great challenges breed evolutionary innovation, and among the strangest and most entrancing of creatures to inhabit the planet are the fishes of the abyss.

Between the surface and the abyss, one finds a cold water world in the so-called twilight zone, where light filters only dimly but nutrients and structures sustain surprising productivity. Fishes in this watery gloom rely on built-in flashlights to find prey and mates, but they also sniff the water for chemical messages of danger and opportunity. In the twilight zone, the seafloor supports little algae or coral but is rich with crustaceans, sponges, anemones, and other invertebrates that thrive on organic snow. Daily vertical movements of plankton add to this plethora of food. These middle depths may be home to more fishes (by weight) than in all the rest of the oceans,[2] and the diverse banquet supports massive schools of some of the most important commercial fish on Earth, like pollock and haddock and cod.

But let us begin like Professor Aronnax and embark on a tour of the sunlit seafloor to visit one of the most biodiverse and ancient places on the globe: a coral reef. In this oasis of abundance, bright tropical sunlight boosts coral and algae into levels of productivity that rival a rainforest. Plentiful food blended with complex physical structure is a world-class recipe for biodiversity, and evidence suggests that coral reefs have been hotbeds of diversity for millions of years. They have even survived planetary shifts in climate, like the one that confronts Earth today, and served as refuges for fish diversity through those challenging times.[3] The result, today, is a magical location teeming with life, festooned with color and texture, and boasting some of the most unique, extraordinary, and downright bizarre fishes in the sea.

5 AN OASIS OF ABUNDANCE

Life on a Coral Reef

LABORING IN HIS UNDERSEA GARDEN, a little damselfish pauses, rolls slightly with a wave of his fin, and listens intently: the thrum of a motor tells him the divers have returned to the reef. A few minutes later, booming splashes overhead herald their entry into the water. Wrapped in dark rubber, bristling with hoses and belts and dangling accoutrements, they descend. Soon raucous bubbles are all around, flippers stir sand onto his carefully tended algae, cameras snap photos of his brightly marked face and fins. Fortunately, he knows they will remain for only an hour or so, then ascend away from his world, and he can go back to weeding his crops. There is much work to be done, no time for fooling around. He is tempted to make a few playful lunges toward the visitors, they do seem to enjoy that. But what grabs his attention, as it does on every visit by the glass-faced tourists from above, is just how bloody noisy they are.

For the farmer damselfish, the reef is never silent. Sounds carry information, and his ears are always tuned to the messages reverberating through the water. Like the stations of an FM radio, each frequency broadcasts a different tune by a different artist. Clicks and pops are ever-present, as neighboring damselfish chitchat about their farms and argue over boundaries. A few minutes are spent in weary complaint over the latest rampages by thieving surgeonfish. Why don't they just grow their own food, instead of stealing ours? Lower on the scale, a hollow burping sound signals a toadfish objecting to some intruder; on other days the same toadfish will toot a higher-pitched whistle, wooing females with his sonorous charms. Over at the edge of the damsel's patch of reef, a mechanical crunching sound signals a parrotfish munching on coral, her beak-like teeth making short work of the cement edifice. The radio spectrum is full of song, the sea full of singers.

When divers arrive, the undersea symphony is obscured, as if a marching band high-stepped into the neighborhood, horns blasting and drums booming. Their equipment bangs and clatters, so loudly the damselfish winces. Every clang from the boat overhead makes him want to dart to safety, even though he has heard it a hundred times before. Divers hoot and moo into their breathing tubes, pointing wildly at some new attraction. Black gloves grasp rocky ledges as they steady themselves, cameras brandished. He wonders why they don't just float calmly, as he can; it's easy to turn yourself with just a few light strokes of a fin. They are so ungainly, and are forever sinking or rising, something he can control intuitively. Well, he imagines, I suppose I would be uncomfortable and awkward in their world, too.

One of the visitors clutches a flat sheet, like a blade of algae but square and white, covered with some sort of pattern. Curious, the damselfish hazards a quick swim up from his farm and catches a closer glimpse. He recognizes the shapes on the shiny card, they are crude drawings of some of his reef-mates. Here are images of two or three parrotfishes—a lot of detail is missing, but he concedes their coarse resemblance—also a couple of pufferfishes, a bright yellow trumpetfish, and even a rough approximation of himself. Entranced, he barely notices the diver reaching for the card; startled, he bolts for his farmstead, dodging the outstretched arm. From the safety of a crevice, he watches the diver examine the card, then peer at his neighbor, a little goby who is a friendly sort of fellow. With a gloved finger the diver points at one image, then a thumbs-up is brandished to a masked companion, along with loud murmur. They are learning, thinks the damsel, to identify us. The surprising realization makes his heart swell with pride. Those noisy, clumsy visitors actually appreciate what I see: the diverse carnival of colors and textures, shapes and patterns that decorate and populate my inviting little reef.

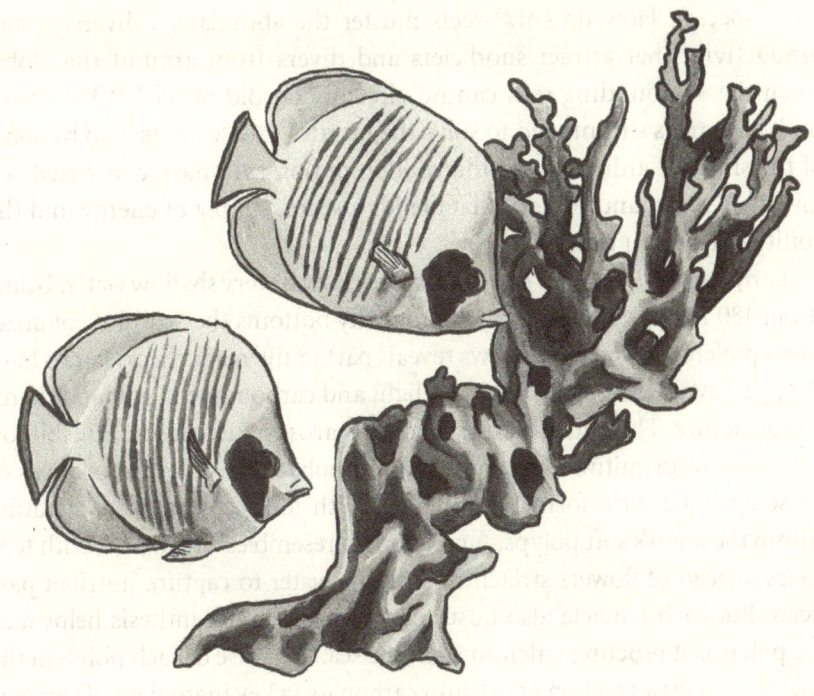

Bluecheek butterflyfish (*Chaetodon semilarvatus*)

AN OASIS IN THE DESERT

"What makes the desert beautiful," said the little prince,
"is that somewhere it hides a well."
—**Antoine de Saint-Exupéry, *The Little Prince***

The sea can be a lonely place, featureless, and nearly devoid of life. Ask those pitiable sailors shipwrecked by Moby Dick and tormented by famine in the central Pacific; ask whale sharks who journey thousands of miles to find a robust meal of plankton. Far from tropical shores, equatorial waters can be pretty as a postcard but deadly as a dustbowl. Their transparency is at once beautiful and menacing, a measure of just how little nourishment can be found. Yet a few outposts of richness dot this expansive, watery desert. Here and there, tiny islands rise to the surface, ornamented by palm trees and surrounded by a narrow fringe bursting with life. Despite the desperately poor water all around them, coral reefs are among the most diverse and productive ecosystems on the planet. Only tropical rainforests, growing on soils similarly bereft of nutrients, boast more species. How do coral reefs muster the abundance, diversity, and productivity that attract snorkelers and divers from around the globe, when the surrounding seas can be virtually devoid of life? What magic trick have reefs summoned to solve the puzzle? The secret is held by some of the smallest animals, and the voodoo potion they have concocted is a mix of alchemy and sorcery that blends a secret supply of energy and the ability to travel in time.

Early naturalists noted that corals grow only in very shallow water. Below about 180 feet, coral reefs give way to rocky bottoms they cannot colonize. Their preference for the shallows reveals part of the magic trick: corals have the photosynthetic capacity to turn light and carbon into sugar, and therefore structure. This life-sustaining alchemy arose on Earth some 3.5 billion years ago, in primitive algae known as cyanobacteria. Modern relatives of those cyanobacteria form a mutualism with today's corals, cohabitating within the coral's soft polyps. A coral polyp resembles a tiny vase, with tentacles instead of flowers stretching into the water to capture nutrient particles. But each tentacle also hosts algae, whose photosynthesis helps feed the polyp as it procures calcium from the sea. The base of each polyp in the colony secretes a skeleton of calcium carbonate (a key ingredient of cement) that builds over time into great spires and parapets and atoll-fringing reefs. Without the symbiotic algae the coral would have to subsist only on nutrients floating in the water, which provide a meager diet indeed.

There is one group of animals, albeit very humble ones, that can survive on that lean diet. Simplest of all marine animals, sponges are colonies of cells working in coordination to pump vast amounts of water through hundreds of tiny pores. In a single day, a hard-working sponge can process as much as 40,000 times its own volume in water. Along the way, filter cells lining the sponge's internal plumbing harvest nutrients suspended in the water. Dramatic forms emerge, as colonies build themselves into barrels the size of a bear, cylinders like a pipe organ, and undulating sheets like parade bunting. Their growth occurs at a glacial pace, but sponges can live for many years, steadily gathering nutrients all the while. The sponge you see today is a time traveler, having compressed years of painstaking construction into a towering and majestic organism, like a time-lapse film of a skyscraper's construction.

While a sponge is pumping water and sieving it for dissolved organic matter, it continually jettisons filter-specialist cells that have reached the end of their shelf life. Organic matter, sparse in the ocean, is concentrated into sponge detritus and dumped onto the reef where hungry denizens descend on the tasty tidbits. Called the sponge loop, this process of accumulation and repackaging of nutrients delivers a previously invisible source of food to reef communities.[1] Like corals, some sponges also enjoy an energy subsidy from photosynthetic algae that take up residence in their outermost tissues. There is even a direct link between corals and the sponge loop: some sponges gather mucus produced by corals, rich in carbon and nitrogen, and later release more than a third of it to the reef as loose food particles.[2]

A reef's complex food web, the feeding interactions of all its members, is laced between the foundation of corals and sponges and algae. A healthy reef is a dynamic and shifting patchwork of algal meadows, boulder fields of hard corals, the fingers and fans of soft corals, massive sponges, and hundreds of other cohabitating species. But like narcissists on stage, these actors all compete for space. Algae are held in check by herbivorous grazers, from damselfishes and surgeonfishes to limpets and sea urchins, who hungrily browse on the rich pasture and keep it from overgrowing the corals. Branching corals with ungulate-inspired names like elkhorn and staghorn grow upward toward brighter sunlight but suffer most when battered by waves and storms. Mound-shaped colonies such as brain coral and boulder star coral are better protected from the ocean's assault but get overtopped and shaded by their faster-growing competitors. Sponges employ chemical weaponry to attack and usurp mound corals, but they too can be engulfed by the creeping carpet of algae if there are too many nutrients in the water, or too few grazers to manage the mowing.[3]

While the pillars and slabs of the reef's foundation are fighting among themselves, they are also sustaining scores of species of fishes, each of whom tilts the battle this way and that. Hard corals feed parrotfishes, who take noisy bites with beak-like teeth adapted to crush the calcium-hardened skeleton and extract the nutritive polyps. Sponges, seemingly soft from the outside, are themselves defended against predation by glass-like spicules and a variety of toxins in their tissues. Nevertheless, angelfishes specialize on sponges, feasting on a smorgasbord of species to avoid overexposure to the particular defenses of each sponge type.[4] Diminutive bottom-dwellers like blennies and gobies gorge themselves on the detritus continually shed by sponges, and they themselves provide prey to the rest of the food web.

And a host of fishes, from massive manta rays to tiny cardinalfish, feast on plankton that gushes from the reef, including clouds of eggs and larvae launched hopefully into the blue sea.

Should any of these fishes be surprised by a rapacious snapper or barracuda, they have but to spring into the cracks, crevices, and holes that abound on a coral reef. To better their odds, most reef fishes live in compact territories they know well, with safe hideouts just a tail flick away. The intricate structure of a reef provides food as well, because the walls and pillars are lined with barnacles, mussels, tube worms, and anemones, sessile animals that use the pillars and pedestals to hoist themselves into rich currents of edible plankton. Once they have colonized a spot, however, they are stuck in place like statues and are fair game for dozens of reef fishes. Wrasses, butterflyfishes, triggerfishes, and many more scour the crenellated surfaces of the reef, probing holes and slipping into fissures to pluck the invertebrates from their plinths.

Reef fishes can effectively exploit the crevices and cracks and candelabras of their undersea cafeteria because they are compact, compressed from side-to-side like a pancake on end, and because they can manipulate their body position with incredible precision. No tuna, as agile as they are in open water, can stand on its head, hover an inch above a rock, then roll onto its side. But angelfishes, tangs, butterflyfishes, and more perform this circus trick all the time. Like gymnasts in slow motion, reef fishes leisurely advance and retreat, pivot and spin; they can even swim backward. Rivaling a NASA spacecraft in their degree of control, they effortlessly adjust pitch and yaw and roll with precise rippling and fanning of fins. Fish are not floating in zero-G, though; their gymnastics are performed in an ocean where gravity pulls at them inexorably. But thanks to a unique organ, they have conquered this pull and maintain perfect neutral buoyancy. This organ set the stage for a sprawling diversity of shapes, fins, and behaviors that gave fishes an all-access pass to the backstage crannies of a reef.

A few million years after sharks and rays branched from the fish family tree, a few primitive mouth-breathers evolved. Mostly denizens of estuaries, those fishes could gulp air and absorb a little oxygen into dense networks of blood vessels in their mouth. Later, out-pocketings of the throat and gut developed, permitting longer air storage and more surface area for gas exchange. And a little over 100 million years ago, a remarkable invention emerged: the swim bladder. This self-contained pouch deep inside the fish can be inflated or deflated like a balloon and provides enough buoyancy to counteract the grasp of gravity. Fishes with swim bladders are perfectly neutrally buoyant, no matter the depth, and their fins are no longer conscripted into a continuous fight to stay off the bottom. Early fishes (and a few still living today) filled the swim bladder by swallowing air at the surface and emptied it by belching. But modern fishes took one more evolutionary stride and invented a sophisticated pump known, inelegantly, as the gas gland.

In advanced fishes, the swim bladder is inflated using dissolved gases in the bloodstream; no gulping and belching for them. The gland itself encases a rich bed of capillaries adhered to the side of the bladder. Inside the gland, a remarkable feat of blood chemistry occurs, relying on the Bohr effect. Christian Bohr studied human physiology in the late 1800s and discovered that when blood becomes more acidic, its hemoglobin has less affinity for oxygen. He also discovered the best way to get a Nobel prize into your house is to raise a brilliant son and grandson, Niels and Aage (both physicists), who each won the coveted medal. Inside the gas gland, the Bohr effect creates a tiny pump. Glucose is converted into lactic acid within the capillaries, and the acidity forces hemoglobin to dump oxygen.[5] The only place for the gas to go is into the bladder, and voilà, the gland pumps up the pouch.[6] To shrink the bladder, the lactic acid is withdrawn from the capillaries, boosting hemoglobin affinity and sucking oxygen back into the bloodstream. Molecular evidence reveals that key elements of this ingenious gas-release system first evolved in fish eyes, where rapid delivery of oxygen helps retinas process images faster, the better to see prey and predators.[7] Later, the same system is repurposed to inflating the swim bladder, an evolutionary leap that unlocks an explosion of reef fish diversity.

Paroled by their swim bladders from the law of gravity, the evolution of reef fishes was unfenced by the need to swim ceaselessly in open waters. Natural selection explored a cornucopia of bizarre shapes, like wafer-thin butterflyfish, tubular trumpetfish, and lovably lumpy frogfish. Freedom from relentless swimming also heralded the emergence of peculiar behaviors, brilliant colors, and novel lifestyles, each a response to the ecological opportunities and physical structures provided by the reef. That oasis of productivity, paired with possibilities bursting from newfound evolutionary elasticity, has yielded a community brimming with creatures so gorgeous and diverse as to rival any ecosystem on our planet.

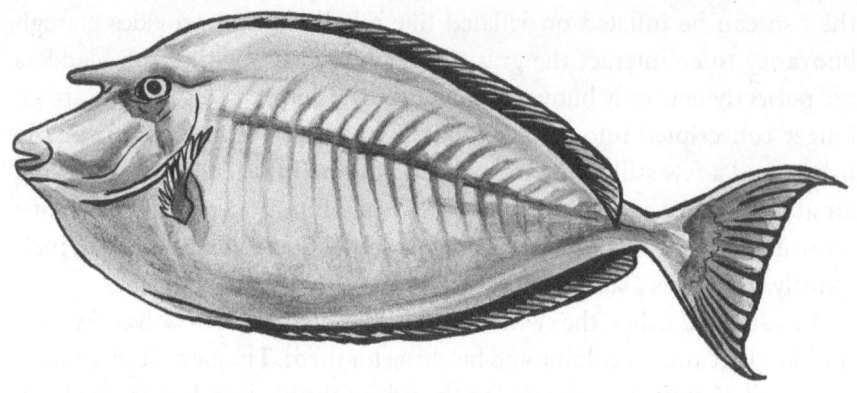

Bluespine unicornfish (*Naso unicornis*)

SEAWEED SALAD

A person who is growing a garden, if he is growing it
organically, is improving a piece of the world.
—**Wendell Berry,** *From a Continuous Harmony*

On dry land, whether in the Amazon rainforest or on the sunbaked plains
of the Serengeti, life begins with plants. Small herbivores such as mice or
rabbits eat tender green leaves and juicy fruits but become targets for carnivores one link up the food chain. Those carnivores—foxes and weasels
and their ilk—are also prey for larger, faster, and more ferocious predators.
Under the waves, the role of plants is played by algae and seagrass (though
only the latter is a true plant). This aquatic greenery is the foundation of
the food chain, its anchor link. Nibbling on this foundation are a guild
of grazing fishes, small to medium-sized vegetarians who themselves are
targets of sea basses and jacks and other predators. Like rabbits feeding
foxes, they represent a critical link between the reef's primary productivity below and the predators above.

Crescent-tailed surgeonfish, brilliantly marked tangs, and vigilant
damselfish are among the grazers who frequent shallow, sunlit reefs where
algae thrive. Reef vegetarians resemble one another, with characteris-
tically compressed bodies and pointed snouts, lips pursed as if to deliver
a kiss. Their fins are broad and blunt, like ping-pong paddles next to a
tuna's long scimitars, and set further forward. This arrangement gives

them ultra-precise control over their movements and body position, permitting them to meticulously inspect and pluck the tastiest pieces of algae from the intricate jigsaw puzzle that is a coral reef. Incisor-like teeth are adapted to clip the algae, which is passed to a muscular throat studded with mounds of enamel and dentine: these pharyngeal teeth grind tough algae like a pepper mill. Once swallowed, acids and enzymes break down the algae's complex polysaccharides, but the fish also gets a little help from some tiny, and not-so-tiny, friends.

Much like cows, which rely on bacteria and protozoans in their gut to digest grass, grazing fishes host a diverse microbial community that specializes in breaking down algal cells. These microbes produce enzymes that chop up the algae's toughest compounds, which the fish could not digest on its own. Without the microbes, each bite would yield far fewer nutrients, and the fish would starve like a cow pastured on a parking lot. Researchers found something extraordinary, however, when studying bacterial digestion in surgeonfishes (named for scalpel-like spines that protrude defensively from their tails). The fishes all harbor a highly diverse microbial community, but each is dominated by a clan of titanic bacteria 1000 times larger than the others.[8] Known as *Epulopiscium*, these mega-microbes would be visible to the naked eye were they not living deep within the fishes' intestines. There, they churn out digestive enzymes matched to the diet of each fish: bluespine unicornfish (*Naso unicornis*), named for the fleshy spike poking from their foreheads, favor brown algae and harbor one species of *Epulopiscium*, while sohal surgeonfish (*Acanthurus sohal*) teem with a different species better-suited to its preferred red algae. The mega-bacteria even adopt their hosts' daily schedules, ramping up enzyme production early in the morning, then shutting down the enzyme assembly line as night begins to fall.

Surgeonfishes, who can reach forty years of age if predators are evaded, follow a pattern typical of reef grazers.[9] They spend their nights in tight crevices, their compressed bodies allowing them to squeeze into the narrowest of sanctuaries. As the sun rises, they emerge to browse on delectable turfs of algae, meticulously maneuvering their snouts to clip preferred flavors. Many surgeonfishes defend algal territories, and smaller species even pair up to guard a patch against larger competitors.[10] They know their territories well, having mapped the best hiding places, shortcut tunnels, and algae-rich plateaus, but will occasionally switch territories, perhaps in search of greener pastures of algae.[11]

Algae, more commonly known as seaweed, may not be particularly fond of their role as food for grazers. They gird their fronds with biochemical armor that makes their tissues unpalatable or downright toxic. Some are simply mechanical defenses, like calcium carbonate, which renders coralline and encrusting algae nearly as rigid as cement. Other chemicals have more potent effects, interfering with digestion or even working as a poison. Red tide is the most well-known example, an infamous algal bloom provoked by overabundant nutrients—often from agricultural run-off—and excessively warm waters. Staining the sea surface a characteristic brick-red, the algae release toxins that can trigger fish die-offs, shellfish poisoning, brain damage in sea lions, and severe illness in people who eat affected seafood, or even simply inhale the fumes of algal blooms tossed in nearshore waves.[12] The inspiration for Alfred Hitchcock's film *The Birds* was a true story about California seagulls disoriented by red tide toxins in their fishy diet that dive-bombed homes in Monterey Bay; driven psycho, they notoriously hurled themselves at rear windows in a frenzy of vertigo.[13]

Back to the surgeonfishes, captive trials revealed they turn up their snouts at unpalatable and dangerous seaweed, and concentrate on the most delectable and least-well defended varieties.[14] From a buffet of fifty seaweed varieties, surgeonfish selected only ten for lunch: none of the favored algae were calcified and none produced toxins, while nearly all species with robust chemical defenses were left untouched. Other scientists working in the Red Sea discovered that seaweed preferences shift as the calendar unfolds. Favoring red and brown algae for much of the year, surgeonfish switched to green algae in the winter, apparently to fatten up prior to the spring and summer mating season.[15] And who could be better qualified to study the reproduction of surgeonfishes than lead researcher Dr. Lev Fishelson?

If grazers choose from their menu of seaweed based on palatability, then a simple solution emerges. Instead of foraging in the wild, plucking what one can find and warily avoiding toxic fronds, why not cultivate your favored food crops on a well-tended field? That very approach has been adopted by damselfishes, the farmers of the coral reef, and it has been so successful that the group has diversified into more than 400 species around the world.[16] It seems that the farming life, in tropical seas, can be a mighty fine life.

Damselfishes are compact creatures, moderately compressed but giving a robust and muscular impression when seen next to a slender and delicate butterflyfish. A bantamweight boxer, with eyes slightly bugged and nose blunted, scowling at the ballerinas drifting lightly by. In adults, the markings are unremarkable: dull browns and blacks, in some species embellished with yellow on the tail or fin tips, or turquoise speckles along the back. Juvenile damsels, in contrast, can be among the most brilliantly painted compositions in the galleries of any reef. Decked out in blazing orange or searing yellow, washed with contrasting azure, ornamented with sparkling cobalt or adorned with peacock-like spots, their flamboyant colors may advertise to pugnacious farmers that they are not a threat to steal from the homestead. More often than not, eye-popping childhood colorations account for adult names: Neon, Sulphur, Azure, Golden, Yellowtail, and Black-and-gold are just a few dazzling damselfish monikers. Sadly, in all species the youthful flamboyance fades as youngsters adopt the business-like dress of their parents.

Adult damselfishes vigorously defend agricultural estates, rocky outcrops with good sunlight and firm substrate on which they find a patch of algae growing. They diligently tend the patch, encouraging the growth of seaweeds they find delicious and nutritious. Depending on where the farm is located—warmly sunlit reef flat, boisterously wavy reef crest, or the cool and dim reef slope—the ideal crop selection will vary.[17] But the end result is a diverse crop of delectable greens surrounded by fallow, rocky tracts dominated by just a few tough and nearly inedible species.[18] Much work goes into tending the homestead. Undesirable species of algae with too many defensive compounds are ripped out. Drifts of sediment are cleared by energetic fanning with their fins. Small pebbles are grasped with strong teeth and dragged to the border. The resulting lustrous algal meadows attract admirers, some of whom come to plunder but others that actually benefit the damsels.

Tender and leafy fronds invite tiny and even microscopic organisms to colonize the rows; like bacon bits, they are eaten with the salad of seaweed, adding protein to the meal.[19] Some farmers have added shepherding to their résumé, by domesticating tiny mysid shrimp whose excrement fertilizes the farm like so many cow patties: damsels with shrimp flocks have richer fields of algae and are healthier thanks to the fertilized produce.[20] Though the majority of damsels source most of their food from the farm, recent research using stable isotope techniques (described in the previous chapter) has revealed they select items from a broader take-out menu. Nearly all species will opportunistically grab tiny crustaceans or mollusk larvae swaying in the waters around the farm; for a few species, these pelagic prey make up a whopping three-quarters of their diet.[21] It appears that some farmers have given up the vegetarian lifestyle, pulled back to carnivory by the savory delights of meat.

A school of yellowtail surgeonfish (*Prionurus punctatus*) meandering across a coral reef cannot help but be drawn to the appealing greenery of a well-tended damselfish farm. Amid all the stiff, unpalatable, and downright inedible algae the soft and nutritious fronds must be irresistible. And resist they do not. Swooping onto the field like a gluttonous flock of crows (a murder of crows, if you prefer) drawn to fresh seedlings, they descend on the poor damsel's farm. As the lead intruder approaches, the farmer charges. Though only half the weight of his rival, this damsel is pugnacious by nature, and skilled in the art of defense. He feints an uppercut, then lands a blow. A thumping jab of his snout staggers the surgeon, who turns aside. But another takes its place, and another. Soon, the damsel is overwhelmed and must retreat as the shoal descends. Surgeons clip and gnaw at his precious algae, ravaging his field until they deign to depart, leaving the farmer to assess the damage. Perhaps one day damsels will invent the undersea equivalent of a scarecrow.

Other trespassers include close relatives like the Achilles tang (*Acanthurus achilles*), a fiery orange spot marking its heel (actually the tail base), who rove in farm-raiding packs. Sea urchins, who lumber onto the fields like implacable porcupines, would utterly denude his farm if not ushered off: damselfishes will seize a spine in their teeth and bodily drag any urchin until it falls off the plateau. But there are others who slink onto the farm, and they come not for the crops but for the landowner himself, including one wolf who dons farmer's clothing. Brown dottybacks (*Pseudochromis fuscus*) are elongate fish, shaped like bratwurst, and not much larger than the damselfish they stalk. Like many reef fishes, they come in a few different color morphs—in northeast Australia, a dusky brown or saffron yellow—and can change between them. In the same area, there are yellow and brown species of damselfish. When the dottyback spends a couple weeks around a damsel's homestead, it gradually adopts the color of the landowner, until it is disguised perhaps as a jolly neighbor coming over for a cup of sugar. Before the murderer can be unmasked, it pounces: the damsel is taken by surprise, overpowered, and eaten. Dottybacks who mimic their prey's coloration are three times more likely to consummate the kill, a triumph for their mastery of masquerade, but a distressing state of affairs for a damsel.[22]

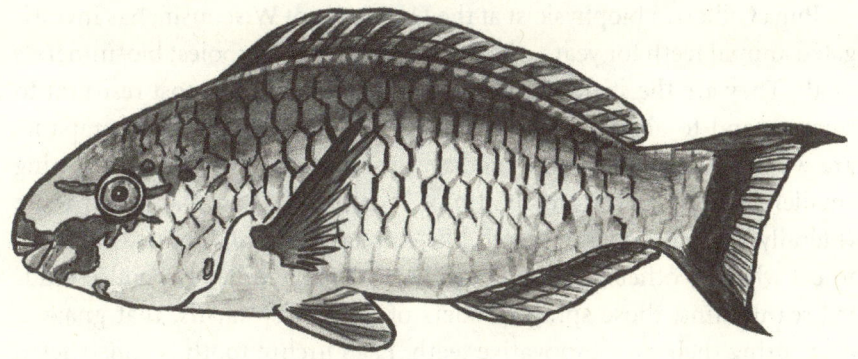

Quoy's parrotfish (*Scarus quoyi*)

FISH OF A FEATHER FLAP TOGETHER

It is tough to survive on salad alone. Other than damsels who cultivate a carefully selected menu of nutritious seaweeds, most grazers augment their diet with plankton, detritus, and a host of tiny organisms—crustaceans, worms, mollusks—that live amidst the algae and are swallowed with every mouthful. Some 30 million years ago, however, a suite of adaptations appeared in a few of those grazing fishes that allowed them to gnaw a chunk of coral, crush the cement exoskeleton, and feast on the tender contents.[23] It was like taking a bite from a cinder block and finding candy inside. With every bite, a distinctive crunching sound boomed across the reef: the coral's defense had been breached. Gaining access to this piñata of soft and nourishing polyps was like a starter's pistol, setting in motion the diversification of some of the most numerous and flamboyant creatures on any reef.

Parrotfishes earned their name because they swim by flapping pectoral fins like a bird's wings, and because their large front teeth are fused together into a decidedly beak-like shape. It also helps that most species are brilliantly colorful, dominated by tropical hues of green and blue, and dazzlingly marked with stripes and splashes of red and yellow. Names of many species sound like a 1960s festival roll-call: would the parents of Rainbow, Tricolour, Festive, and Azure please report to the stage? But the oversized, white, and slightly comical buck teeth are what first grab your attention. These represent an incredible feat of evolution, as they are enameled with one of the hardest biological substances on planet Earth.

Pupa Gilbert, a biophysicist at the University of Wisconsin, has investigated animal teeth for years. "Parrotfish teeth are the coolest bio-minerals of all. They are the stiffest, among the hardest, and the most resistant to fracture and to abrasion ever measured."[24] Tiny crystals of fluorapatite are woven into bundles, like medieval chain mail, with bundles getting smaller and harder toward the tip of each tooth. Parrotfish wield these (literally) ground-breaking teeth like chisels into coral, permitting them to eat what no other fish can chew. Dr. Gilbert and her team also studied sea urchins, those spiny invaders of damselfish farms, that gnaw at algae using their own innovative teeth. Each urchin tooth is constructed of offset layers, like a deck of cards fanned for a magic trick, each layer composed of rock-hard polycrystals of calcium and as thin as a scalpel. As the tooth tip is worn down, the topmost layer is shed, exposing the next razor-sharp card in the deck.[25] Applied to home cutlery, this advance could put a self-sharpening knife in every kitchen drawer. "The sky's the limit at this point," enthuses Gilbert.

Once a parrotfish has bitten a chunk of coral with its formidable teeth, it passes the cement to a specialized grinder in its throat, a separate chewing apparatus known as the pharyngeal mill. These "throat jaws" are set deep within the mouth of the fish, chewing swallowed food even while the front teeth are taking another bite. The mill evolved in parrotfishes (and their cousins, the wrasses) a few million years before the famous fused teeth and was the key adaptation that kicked off their explosive diversification. It is made up of grinding, molar-like teeth attached to a complex set of bones, themselves derived from repurposed gill arches, and a muscular sling that saws the teeth back and forth. Chewing heartily, the pharyngeal mill pulverizes the exoskeleton of coral into a slurry of ground cement mixed with polyps, algae, and all manner of micro-animals living atop the coral. Parrotfishes have no true stomach, instead passing this slurry directly to their intestines, which extract the copious nutrients.[26] Combined, the teeth and mill allow parrotfishes to eat what no other fishes can, unlocking access to an immensely abundant food source on the reef.

In a single year, reef parrotfishes are estimated to consume more than 5000 pounds of coral cement per acre.[27] All that food has to end up somewhere, as the intestine is a one-way trip, and like a goat browsing on fine fodder, parrotfish release excrement continuously. Parrotfish feces are clouds of almost pure sand, the milled remnants of the calcium-rich coral skeleton. So abundant is parrotfish guano that it forms the principal component of nearly all the blazingly white beaches of the tropics. Something to ponder the next time you are luxuriating on the shores of Fiji: you are rolling and lounging in parrotfish compost.

Because coral is abundant and parrotfishes face little competition while eating it, they can feed themselves with ease, leaving time in the rest of the day to attend to other affairs. Such as affairs of the heart, and the gonads. As in many animal species that enjoy an abundant food source, parrotfish form polygynous mating systems: females greatly outnumber males and choose the largest and fittest male with whom to mate. The same system arises in birds like manakins, who engage in elaborate dances punctuated by wing snapping, and in bowerbirds whose males entice females with elaborately constructed display courts replete with romantic archway. In parrotfishes, the females select males who are brightly colored, which reflects the quality of their diet, and who are swift and vigorous swimmers. As in peacocks, favored characteristics can become extravagantly enhanced, and parrotfishes today are among the most brilliantly colorful species on any reef. Sexual selection within parrotfish arose around 10 million years ago, and the groups in which it dominates (*Scarus* and *Chlorurus*) have shown the most rapid diversification since. Of the nearly 100 recognized species of parrotfishes, more than half are in those groups with female choice.[28]

Polygynous mating systems, known colloquially as harems, have been the subject of intense study, both behavioral and genetic. In birds, males are flamboyantly colored while females are drab, the better to camouflage the nest. The same is true in parrotfishes, where most females are pale or dark reddish or striped; however, there is an added twist. In parrotfishes and closely related wrasses, every single male was born female. These fishes are known as serial hermaphrodites: they emerge from eggs with the capability to express either gender, and to produce either eggs or sperm, but they begin life in a female phase (known as "initial") and later, if the chance arises, transmogrify into a male phase ("terminal").

To understand why this happens, consider the sex life of a stoplight parrotfish (*Sparisoma viridae*). Initial phase adults are modest in size, perhaps a foot long, with a bright crimson belly and the rest of the body checkered with a regular pattern of pale greenish blotches. Nearly all are females (though a tiny fraction are already males, bearing functional testes). Far less common are the terminal phase males, brilliantly painted in a seafoam green, their heads decorated with three stripes of orange-pink, their gill covers and tails adorned with bursts of yellow. Each male dominates a large territory, within which a dozen or more initial phase stoplights may reside. Terminal males initiate a courtship dance involving short bursts of speed, quartering turns, and lunges toward the surface. Disco moves are paired with distinctive popping calls, bursts of sound pulses that entice females and menace rival males.[29] The syncopated song-and-dance invites females to join the male in an upward surge above the reef, where eggs and milt (fish sperm) are released in a choreographed cloud. Occasionally, an initial phase male will sneak into the dance and release his own milt, a sly maneuver known to scientists as streaking (and charmingly called bedswerving in ye olde English). Fertilized eggs are then swept from the reef by currents, and juveniles hatch in deeper waters where, hopefully, fewer planktivores are to be found.

From a genetic standpoint, the male's DNA will appear in every offspring—setting aside the contribution of an occasional streaker—while each female's genes will be passed on only to her own eggs. Evolutionarily, there is a great advantage in being the terminal male. But what happens when that gloriously colorful male is eaten by a hungry shark? His disappearance will incite the initial stoplights into a frenetic race, physical and hormonal, to become the next terminal male. The female who succeeds in replacing him will seize his evolutionary advantage and dominate the harem's gene pool. To win the race, initial stoplights ramp up their levels of blood testosterone as estrogen levels decline.[30] Initials gradually assume male-like characteristics, including an enlarged forehead and greater body weight. If they outgrow and outduel the other proto-males, they ascend the throne of dominance. Soon they are draped in the colors of a mature male, a process itself triggered by soaring keto-testosterone levels.[31] Pretenders to the throne, those who fail to surpass the competition, will be evicted from the harem by ruthless attacks from the dominant male. Such bachelor males would find no eggs to fertilize, unless their bedswerving skills were particularly acute, and almost none of their DNA would make it into the future. Thus, most of the contenders revert to the female phase and remain members of the harem where they gradually but steadily toss eggs into the lottery of evolution, biding their time for another shot at the throne.

After a long and exhausting day of courtship, mating, feeding, and streaking, stoplight parrotfish knock off for the night. When evening falls, the risk of being nabbed by a predator soars, so they take a few precautions. Parrotfishes may lounge beneath a coral ledge, or tuck themselves into a cave, or simply recline against a rock on the bottom. But wherever they sleep, they are protected by a unique cloak of invisibility. Special glands in their skin secrete a clear mucus that forms a sleeping bag, wrapped entirely around them like a bubble. The sleeping bag wards off blood-sucking parasites like a mosquito net.[32] But the cloak also prevents keen smelling predators from picking up the parrotfish's scent, safeguarding them through the long night when sharks and moray eels and snappers emerge to patrol the reef for a meal.[33]

NATURE RED IN TOOTH AND CLAW

I bring to life, I bring to death
—Alfred Tennyson, *In Memoriam*

Among fishes who frequent coral reefs, the overwhelming majority are predators. Over half of all species specialize on eating other fishes, a diet known as piscivory. Much of the predation occurs at night, when killers like barracuda, many snappers, some groupers, jacks, and others emerge to hunt their finned neighbors. Predatory fishes are the largest in any reef community because they can eat only what fits in their mouth. Unlike sharks, who can thrash and tear chunks from large victims like seals and whales, nearly all piscivores must gulp their prey whole. Some, like snappers and jacks, are built for speed. These are the pursuit predators. Accelerating to the speed of an Olympic sprinter or more (sleek wahoos jog at 20 mph but can surge to 48 mph), pursuit predators will chase down slower fish and devour them, aided by their own pharyngeal teeth that crush bone and flesh as a parrotfish mills coral.[34] Other predators adopt a sit-and-wait strategy. These artists of ambush, like groupers and stonefish, are often slow moving, large-bodied, and so sumptuously camouflaged as to bamboozle even the most keen-eyed reef fish. Just after sunrise, you might find a grouper waiting patiently, silently, for an unsuspecting mark to swim too close to her enormous mouth. Throwing her jaws open wide, she inhales an enormous volume of water. The hapless victim is sucked into the gaping maw like driftwood into a whirlpool. Noiselessly, her jaws snap shut, water is expelled from gills, and she swallows her first morsel of the evening.

Barracudas, cylindrical and dead-eyed predators who inspire terror in slow-swimming fishes, and more than a dollop of dread in snorkelers, have a different approach. While most piscivores are restricted to swallowing only those fishes that fit in their mouths, a barracuda's definition of bite-sized prey is far more broad. When a great barracuda (*Sphyraena barracuda*) attacks, it rushes its victims at speeds in excess of 40 feet per second and seizes them with stiletto-like front canines.[35] But if the prey fish is too large to be swallowed, the barracuda grips it firmly in the corner of the mouth, and bites it repeatedly while violently shaking its head in short back-and-forth strokes. All the force of the barracuda's muscular body, which can weigh over 100 pounds, is employed in this fearsome thrashing.[36] Upper and lower teeth in the rear of the mouth work together like scissors, slicing through flesh and bone until the prey is no longer a fish, but rather one mouthful and a couple of bloody chunks. During the butchering, opportunistic fishes often hang around in hopes of grabbing a chunk, large or small, for themselves. Throughout the day, a great barracuda may swim as far as seven miles in search of a meal and can migrate over longer periods to reefs as distant as sixty miles or more.[37]

Although many predators must patrol large territories to feed themselves, a few piscivores are more stay-at-home types. Moray eels, a diverse group that hunts mostly by dark of night, are far less mobile than barracudas but are highly successful nonetheless. More than 200 species of these snake-like fish slide and slither through the Swiss cheese caves and tunnels of coral and rocky reefs, for which they are finely evolved. Their skin lacks scales, and evolution has jettisoned nearly all fins: both might otherwise hamper their cave-dwelling habits. Only the dorsal and anal fins remain, fused into a continuous ribbon along the top and bottom of their bodies. Some species snatch small fishes from the water, grasping them in dog-like canine teeth, while others specialize on crustaceans and clams that they chew with broad-topped molars. The piscivorous morays specialize in chasing fishes into their sanctuaries, hurtling down narrow tunnels like a kingsnake after a fleeing mouse. Deep in a moray's mouth are a second pair of jaws that thrust forward to snap at prey and drag them into its throat, a terrifying sight mimicked years ago to startle moviegoers in the science fiction thriller *Alien*.

Complex hunting techniques have evolved in these wily predators. Morays have been spotted leaping from the water to seize crabs on the shore,[38] and even tying their bodies in knots around large prey before tearing the immobilized victim into bite-sized pieces.[39] Amazingly, yellow moray eels (*Gymnothorax prasinus*) tie as many as five distinct knots, including an overhand bend and a figure eight; such dexterity would qualify them for a merit badge at any summer camp. But the most unique hunting method relies on a partner, the roving coralgrouper (*Plectropomus pessuliferus*). In Red Sea reefs, these huge groupers ambush smaller fishes, sucking unwary prey into their colossal mouths. Those fishes fortunate enough to escape the vacuum attack flee for the safety of the reef, ducking into the first narrow crack they can find. The grouper, unable to wedge its hefty body into the crevice, remains undeterred and instead enlists some help. It cruises to the resting chamber of a giant moray (*Gymnothorax javanicus*) and solicits assistance by shaking its head and fluttering its dorsal fin.[40] Aroused from its slumber, the eel dutifully accompanies the grouper to the unfortunate fish's hiding place, which the grouper points out by performing a headstand. In slips the eel and gulps the absconding fish whole; should the twice-pursued fish manage to evade this slithery attack and bolt from the crevice, it emerges only to find the grouper hungrily waiting to finish the hunt.

While piscivores are busy snapping up unwary fishes, most nonpiscivorous predators on coral reefs hunt invertebrates. They choose from a diverse menu of crabs, clams, urchins, marine worms, and more, plucked from rocks and corals or unearthed from sand and silt. To access these morsels, invertebrate specialists boast an impressive toolkit of highly evolved forms and behaviors. Copperband butterflyfish (*Chelmon rostratus*) and others use extended snouts like needle-nosed pliers to reach into cavities inaccessible to other fishes. All butterflyfishes are highly compressed and gorgeously colorful, spectacularly barred with yellow or orange or blue, and occasionally stamped with a false eyespot to dupe a predator into chasing the wrong end. Masters of slow, pivoting acrobatics, they undulate pectorals and curtain-like dorsal and anal fins to maneuver into the precise orientation for feeding. Within the snout, brush-like teeth are specialized to glean coral polyps from their holes and pluck invertebrates clinging tenaciously to hard surfaces.

Triggerfishes and filefishes, some of the most fantastically marked fish on any reef (with names like Picasso triggerfish to prove it), hunt invertebrates in a wholly different manner. A triggerfish will hover in a headstand just inches above the bottom, undulating its long dorsal and anal fins like curtains rippling in front of an open window, its mouth pointing at a promising patch of sand. It sucks in a large gulp of water, takes aim, and blasts a powerful jet straight down. Spurts of water stir up the sand, over and over again, excavating a broad hole. Hidden crabs and sand dollars and mollusks are unearthed, which the triggerfish swoops down to capture. Often a dozen opportunistic freeloaders like wrasses and hogfishes hang close by, waiting to slip in and grab an overlooked morsel. But triggerfishes are uniquely specialized for hard-shelled prey, using teeth and jaw muscles that rival those of a parrotfish for toughness. They will occasionally spit out the mouthful, then snatch it again in their jaws to reposition the clam or crab like shifting a walnut in a nutcracker.[41] After a day of digging in the sandbox, the triggerfish will retire by night to a restful cavern, where it wedges itself safely in place using its namesake trigger. Two stout spines in the front of the dorsal fin can be erected, the first sticking straight up to the cave's ceiling like a deadbolt, the second (the trigger) locking the first in place: only if the trigger spine is released can the deadbolt be folded down. Pinned securely in place, the triggerfish is almost impossible to drag from its cave.

Closely related, and resembling triggerfishes in their oddly rhomboid silhouette and forward-jutting mouth, are the filefishes. These large-bodied reef fish also feed on hard-shelled invertebrates, though they prefer to pluck prey from rocks and corals rather than use the sandblasting technique. Like triggerfish, filefishes swim with symmetrical undulations of their anal and dorsal fins, a highly efficient way of moving their large and heavy bodies through the water, albeit slowly.[42] They differ in having only a single dorsal fin spine, thus no trigger, and in their unusual scales. Where most fish have smooth and overlapping scales, filefish are cloaked with scales that lie adjacent to one another like bathroom tiles, each tipped with numerous little spikes.[43] To the touch, these fishes are rough like sandpaper. Fishermen of yore would keep a dried filefish or two lying about for whenever a rough patch or splinter on their wooden boat needed smoothing. Wielding the dead and hardened fish like an eponymous file, the fisherman swiftly sanded down the offending bit of wood. Today those same spiky scales are being studied by engineers, as their property of encouraging oil droplets to slide only in the head-to-tail direction can be applied to oil pipelines where it would increase efficiency and reduce transport costs.[44]

When not serving as household tools or pipe linings, filefishes show a charming dedication to what can be called parental care. They share these habits with triggerfishes, a complex set of behaviors that distance them from parrotfishes and improve their chances of successfully reproducing. When a male triggerfish or filefish is ready to mate, he prepares a bassinet on the seafloor, a shallow hollow made by blowing jets of water. In some triggerfishes, males and females blow sand together and touch abdomens in a wooing dance ritual. In others, the male may dig the hollow alone, but amid a crowd of males similarly engaged in amorous domesticity.[45] Resembling a showroom floor of do-it-yourself cribs set out for maternity-curious females, the crowded mating territory is often called a lek. When a female filefish or triggerfish has made her choice, she deposits eggs into the shallow bowl, where they stick fast to sand grains. After fertilization, the male studiously guards the eggs against hungry fishes who would be overjoyed by the prospect of a free caviar lunch.

FISH THAT GO GRUNT IN THE NIGHT (AND DAY)

To greet one who is belching or breaking wind is carrying politeness too far.
—Desiderius Erasmus, *Collected Works*

Karen Maruska listens to fishes for a living. As a marine biologist at Louisiana State University, she spends her days (and a few nights) exploring how and why fishes communicate. Although Jacques Cousteau titled his first book about marine life *The Silent World*, Dr. Maruska knows the sea is far from silent. "Anyone who's put their head underwater in the ocean and gone snorkeling, especially on a coral reef, one of the first things you hear is a lot of snapping and popping."[46] Her early work was on the diminutive farmers we met earlier. "There's a lot of species of damselfishes that we now know make sounds. They have a huge diversity in the kinds of sounds they produce and the behavioral context that they use them in." In her dissertation research, she found male damsels sing an ample repertoire, particularly when enticing females to mate. "There was definitely a sound when he was leading the female back to the nest, that was probably the most characteristic sound, a long series of pulses that kind of sounded like grunts. Males would make shorter pulse trains toward any rival males that were nearby. They also made sounds when they were clearing out the nest . . . so the female would feel good depositing her eggs there."

Variations in repertoires can even keep species separate. Bicolor and threespot damselfishes (*Stegastes partitus* and *S. planifrons*) live in close proximity, look almost alike, and have very similar courtship displays known as signal jumps. Male rush upward in the water, then dive back down while simultaneously warbling a chirping call. But threespots trill a chirp with three pulses of sound, while bicolors add a crucial fourth pulse. To a discerning female, the difference is like night and day, sparing her the embarrassment of choosing a male of another species.

In well-lit coral reefs, fish rely on their excellent vision to find prey, detect predators, and communicate with neighbors. Bright colors and bold patterns are often found in species like triggerfishes and filefishes who compete over territories and for selection by females. Their virility and fitness are directly on display, since a fish who is a poor hunter or forager cannot engage in vigorous courtship nor match the brilliant colors of better-fed competitors. But as Maruska has found, fish also possess keen ears. Standouts in the chorus of carolers are fishes known as grunts, for the burping rumbles they intone (though the family name, Haemulidae, refers to the blood-red color found inside the mouths of some species). Bearing a close resemblance to their relatives the perches, most family members are nocturnal. Grunts spend their days hanging passively in shaded parts of the reef but venture in groups over open sands once night falls to search for tasty invertebrates. If startled, they will emit a series of their namesake grunts, which sound for all the world like a pigsty full of talkative (albeit aquatic) hogs.

Grunts, and many fishes who use sound to communicate, rely on the swim bladder to amplify their message, like the resonating throat pouch of a frog. In fish, the actual origin of the sound can be quite diverse. Many species have special sonic muscles, attached to the swim bladder wall, which contract and rub the bladder like a balloon.[47] The result can be a single click, or a sequence of pops, or a drawn-out grunt. This technique is shared by cod and their relatives, who sing in cooler, more temperate waters. Other fishes generate sound with their pectoral fins, which are connected to the body by tightly strung tendons that twang like guitar strings as the fins beat.[48] Sunfishes, whom we have met before, and clownfishes (those adorable anemone-dwellers made famous in animated films) depend on their pharyngeal teeth to make sound, grinding their molars and amplifying the vibrations with the swim bladder.

In grunts, the vocalizations appear to warn of threats, as first noticed by fishers when the unfortunate grunt was hooked or netted. In others, like damsels and parrotfishes, a variety of sounds are broadcast during courtship. Parrotfish strenuously slap their opercula (the gills' hatch-covers) like a tambourine. As with color, those fishes who are large and well-fed produce more vigorous calls, so choosy females can use sound quality to estimate the calling male's fitness. Sea robins, curious creatures who scuttle over the seafloor with spines projecting like crab legs from their dorsal fins, make conspicuous drumming sounds during spawning. This vocal habit was known as far back as Aristotle's time and was later likened to the tireless crooning of their namesake bird.[49] As sea robins grow larger, their underwater voice gradually changes from soprano to bass, thus communicating the singer's size.[50] Closely related gurnards, who resemble sea robins in using fin spines to comb the seafloor for small crustaceans, vocalize strenuously while competing over food.[51] Like hounds vying for food from a single bowl, a gurnard growls and makes aggressive rushes toward challengers as it circles a feeding site. Once it seizes a crab, the lucky gurnard trumpets a series of intimidating knocking sounds while scampering off with its prize.

Among the diverse undersea chorus, the most expressive vocalists must surely be the toadfishes. They owe their amphibian nickname to their broad heads and wide gaping mouths, and to their tendency to crouch, camouflaged, on sandy or silty bottoms. But mostly for the delightful croaking noises they produce, honks readily audible even to snorkelers. Short pulses resembling burps from a French horn are produced by both sexes, while a longer tone like a boat whistle is tooted by males alone.[52] This sonorous boat-whistle call is aimed by a swim bladder specially evolved to amplify the sound and focus it forward. Toadfish possess special sonic muscles, attached to the swim bladder, that contract much faster than typical white and red muscle fibers, generating sound frequencies greater than 100 hertz (about two octaves below middle C).[53] After building a nest of shells and stones, the male's song entices females to lay eggs there.[54] Maruska noted that the nocturnal habits of these fish influence their use of sound. "Toadfish and midshipmen rely heavily on acoustic communication. They do a lot of their spawning at night, so visual cues aren't really important. They make these really loud, long-duration hums that males make in the nest . . . trying to attract females. Females will home in on male sounds, and choose a male based on his sounds."[55]

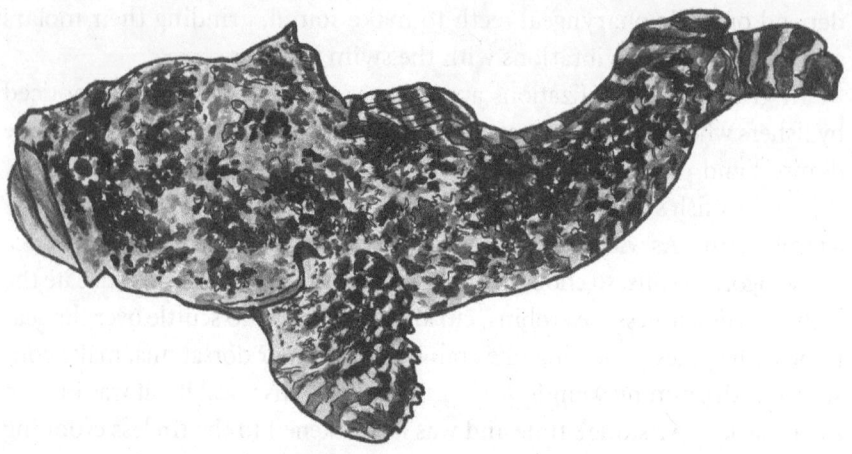

Gulf toadfish (*Opsanus beta*)

Unfortunately for toadfishes, tooting your own horn can get you in trouble. Dolphin trouble that is. It turns out that bottlenose dolphins, who sing complex underwater songs of their own and boast exceptionally keen hearing, have learned to home in on toadfish whistling. In the waters off Florida, toadfishes make up an eighth of all fishes eaten by these dolphins, a staggering haul considering each dolphin can polish off more than 12 pounds of fish in a day.[56] Playback experiments show that dolphins can track their prey by sound alone and distinguish delectable menu items from inedible ones. As the evolutionary war of water words heated up, toadfishes adopted a prudent course of action: shut up while danger is near. When the clicks and pops made by dolphins are audible, toadfish cut their boat-whistling in half (while stress hormones flood their bloodstream) as they cower until the threat has passed.[57]

Neither toadfish nor dolphin can tune in to the other's song, however, if there is noise blaring in the background. Like a safecracker trying to open a vault, they need silence to hear the fine details. Regrettably, today's oceans are far more noisy than they were a hundred years ago. World boat traffic increased fourfold in two decades (1992–2012), and all those engines are filling the seas with a loud rumble that threatens to drown out fishy communication.[58] Motor noises cut in half the distance at which the Lusitanian toadfish (*Halobatrachus didactylus*) can hear boat-whistle calls from neighbors,[59] and oyster toadfish (*Opsanus tau*) are known to call more loudly when boats are near, a wasteful expense of much-needed energy.[60] In Brazilian damselfishes (*Stegastes fuscus*), who communicate volubly with underwater pops and clicks, even the noise of people partying on shore can be an annoyance. During the peak days of that country's famous carnival, roving trucks can pump the sound of samba at well over 100 decibels. As the racket mounts, damsels were observed to abandon grazing more readily, and even overlook predators until they were only inches away.[61] What might be a party for people sounds like a jangling, distressing nuisance for fishes.

We are only just beginning to pull back the curtain and understand the pageantry of coral reefs. Thanks to Karen Maruska, we know more than ever about fish communication, but countless mysteries remain. "I've always just been fascinated by fishes and their diversity. And just understanding how they do things, like reproduce. They're just so weird, and it's fun. Everything you do is learning something new. As a scientist, that curiosity always gets you up in the morning."[62] That same curiosity has inspired millions of people to swim, scuba, and snorkel through coral reefs, appreciating and documenting the diversity they find. That diversity includes fish species, but also their colors, patterns, shapes, sounds, and performances. What they continue to discover is staggering. On the stage set by coral reefs the greatest show in the sea is playing, and the cast of characters includes some of the weirdest, wildest, and most extraordinary creatures the planet has ever seen.

6 WEIRD AND WONDERFUL

Where Horses Swim and Bats Walk

ON THE SECOND MONDAY OF JULY 1772, Captain James Cook set sail from Portsmouth, England to circumnavigate the globe's southernmost reaches. His mission? To discover a massive continent rumored to lie deep in the southern Pacific Ocean, across the most violent seas on Earth. For two years he and his crew tolerated petrifying conditions: ice froze their rigging, frostbite was an ever-present peril, and starvation loomed during the long months below the Antarctic circle. But they prevailed, and Cook eventually laid eyes on his prize, the frozen landmass then called *Terra Australis Incognita* but now known as Antarctica. While icebergs and freezing seas denied him the triumph of setting foot on the continent, the HMS *Resolution* had sailed farther south than anyone before. When he could progress no further, he turned his ship toward warmer tropical waters; little did Cook know, however, that some of the most perilous seas he would ever sail lay before him.

First, his companion ship the HMS *Adventure* lost several men to cannibals in Australia. But a more insidious threat awaited him on September 8, 1774, when he and his crew purchased a few innocuous-looking fish. The ship's naturalist made a couple of sketches, then shared some of the liver and roe with his captain. The decision nearly cost both men their lives. In Cook's own words: "At three to four o'clock in the morning we were seized with the most extraordinary weakness in all our limbs attended with numbness of sensation like that caused by exposing one's hands and feet to the fire after having been pinched much by the frost. I had almost lost the sense of feeling nor could I distinguish between light and heavy objects, a quart pot full of water and a feather was the same in my hand. We each took a vomit and after that a sweat which gave great relief. In the morning one of the pigs which had eaten the entrails was found dead."[1]

Four score years later, two sailors on the Dutch ship *Postilion* inadvertently repeated the gastronomic experiment and fared no better than Cook's pig. Arriving just 10 minutes later, the boat's physician described how he encountered the boatswain. "The patient was extremely uneasy and in great distress, but was still conscious. [He] quickly assumed a paralytic form; his eyes became fixed in one direction, his breathing became difficult, and was accompanied with dilation of the nostrils; his face became pale and covered with cold perspiration; his lips livid; his consciousness and pulse failed; his rattling respiration finally ceased. The patient died scarcely 17 minutes after partaking of the liver of the fish."[2] Clearly, this was not a fish meant to be eaten.

In the case of Captain Cook, one wonders whether he might not have been sold the fish with malicious intent. Following his own poisoning, the captain wryly noted the distinct absence of a timely warning. "In the morning when the natives came aboard and saw the fish hanging up they immediately gave us to understand it was by no means to be eaten, expressing the utmost abhorrence of it, and yet no one was observed to do this when it was sold or even after it was bought."[3] Was this simply a case of unwitting merchants, or perhaps clever agitators who intentionally kept mum about the fatal risks? If the latter, it would be just the first of several attempts on Cook's life by Indigenous peoples. His storied career came crashing to an end when he was stabbed to death and torn asunder by a throng of enraged Hawaiians. To be fair, he had pilfered their cemetery, wrongly accused a chieftain of theft, and kidnapped a nobleman out of spite, so they had good reason to be infuriated.

Front and center in these tragedies of toxicity are pufferfishes, who rank among the most poisonous animals on the planet. Even the tiniest portion of the liver or ovaries of the most toxic species is enough to kill a person. Mitsugoro Bando, a renowned Kabuki actor, famously boasted in 1975 that his resilient constitution would permit him to eat not one, but four entire livers of the famous Japanese *fugu* puffer.[4] He died a few hours later, a dramatic exit if ever there was one. In China, the poison was once used to commit murder, when two hired assassins stabbed a twenty-nine-year-old man in the buttocks with a syringe full of deadly puffer toxin.[5] A couple in Minneapolis fell ill in 2014 after eating dried pufferfish they had purchased unwittingly. One described the earliest symptom as a strange numbness of the mouth, saying "my teeth can't feel themselves."[6] Fortunately, the family recovered after a few days.

All the victims, from Cook to the couple with anaesthetized teeth, had ingested (or been poked with) a nerve agent called tetrodotoxin, one of the deadliest poisons on earth. Named after the family of pufferfishes, Tetraodontidae (a reference to their four large front teeth), tetrodotoxin is twenty-five times more potent than cyanide. Once ingested it blocks the sodium channels of nerve cells, silencing their electrical impulses like a piece of paper slipped between a battery and its terminal. While moderate amounts may cause only numbness, severe poisonings result in paralysis of large muscle groups, then death by asphyxiation or cardiac arrest in as little as 20 minutes.[7] All too often, the victim remains fully aware of what is happening to them, though the paralysis renders them powerless to speak. Tetrodotoxin victims have even been buried alive, after physicians could detect neither pulse nor respiration and prematurely issued a death certificate. In Japan, some townships responded to such horrifying tales by requiring that casualties of *fugu* poisoning be laid in state for several

White-spotted puffer (*Arothron hispidus*)

days alongside their coffins, giving them one last chance to reanimate and escape the grave.

Puffers from which *fugu* is prepared are lethargic and gawky fishes. Tiny fins and a blunt tail are ill-equipped to propel their rotund and large-headed bodies with any speed or agility. Their fused teeth are adapted for scraping invertebrates rock faces, not for biting a foe. They are unable to slip into narrow fissures to hide. Instead, for defense they evolved an ability to inflate their bodies with a gulp of water, and to erect sharp spines like a porcupine. Only the largest predators can swallow a puffer-fish that has doubled in size, and none are keen on a mouthful of spines. But enough predators breached these two defenses, and puffers came to rely on tetrodotoxin, the ultimate bodyguard. There is only one problem with this evolutionary story. Pufferfish cannot manufacture tetrodotoxin, indeed no vertebrate can.

TOXIC RELATIONSHIPS

> I could not love except where Death
> Was mingling his with Beauty's breath
> —Edgar Allen Poe, "Romance"

Coral reef fishes live and evolve not in isolation, but enmeshed in an intricate ballet with thousands of creatures around them. In this company, however, many of the performers are trying to eat you, and some of the tiniest stage hands can save you. Like all higher animals, puffer-fishes lack the biochemical pathways to assemble tetrodotoxin. But they have drafted into service a few lower organisms that can brew the deadly potion. Puffers acquire tetrodotoxin thanks to several groups of bacteria, some hosted and some eaten. Strains of toxin-making *Pseudomonas* will colonize a puffer's skin and render it poisonous,[8] while a few types of *Vibrio* bacteria that also manufacture the poison are routinely absorbed from gobbled snails, shrimp, and some sea stars.[9] During the spawning season, when predation risk soars, pufferfishes will even stock up on defensive poison by eating egg plates of flatworms (*Planocera multitentaculata*) that are packed with tetrodotoxin. The fishes themselves show resistance to the toxin's effects, and they are able to re-direct the poison to their skin and internal organs, particularly the aforementioned liver. If reared on a tetrodotoxin-free aquarium diet, puffers quickly lose their toxicity, elegant proof that diet plays a key role.[10] Predators learn to avoid pufferfishes since the smallest mouthful of their tetrodotoxin defense (if they even survive) teaches a nauseating and numbing lesson. Certainly Captain Cook would never again have eaten the fish that nearly killed him. So powerful is this association that even toy models only faintly resembling pufferfishes are unfailingly avoided by piscivores in research trials.[11]

Puffers are not the only ones who have developed a degree of resistance: predators have fought back, through evolution, against this powerful defense. The best example comes from California, albeit on land, where common garter snakes hunt rough-skinned newts, the most toxic salamander in North America. Tetrodotoxin protects the newt, sequestered in skin glands; the same *Pseudomonas* bacteria (among others) live on its skin and are responsible for initially producing the toxin.[12] The newts evolved the ability to marshal this poison as a defense against predation, but garter snakes have had some time to develop a counter-strategy. Garters in the region show varying degrees of resistance to the toxin, and this resistance peaks in sites where newts are routinely encountered.[13] In those same areas, the newts have fought back with higher levels of tetrodotoxin. An evolutionary arms race seems to be underway, with each species ramping up its defense and counter-defense in response to the other.

Back on the coral reef, predators and prey continue the tango of coevolution. Hunters become faster, their senses more acute. In response, the hunted have adopted a wide range of defensive strategies, from camouflage to sharp spines, armor plating to toxicity, and the interplay of offense and defense kindles the diversity of reef fishes. Like puffers, most fishes possess skin glands which secrete mucus, a protein-rich slime that reduces friction while swimming, shields against external parasites, and makes fish slippery when attacked by predators or grasped by anglers. Soapfishes, close relatives to groupers, produce so much mucus when threatened that their skin becomes covered in foam as the fish literally works itself into a lather. That soapy foam is toxic, capable of sickening or killing other fishes that come in close contact;[14] it also has antibiotic properties that shield the soapfish from infection.[15] Its active compound is a poison called grammistin, whose effects were discovered inadvertently by researcher John Randall. After spearing a 9-inch soapfish, he made an unwise decision: "rather than carry the fish all the way to the boat at the surface, it was stored temporarily inside [his] bathing trunks. Very soon it became apparent that a secretion from this fish was a powerful urethral irritant, and it was promptly removed from the bathing suit."[16] The defense appears successful against predators as well as foolhardy scientists, since soapfishes are rarely found in the stomachs (or trunks) of piscivores.

Many reef fishes use poison as but one deterrent in their defensive arsenal. Boxfishes, square-bodied and slow-swimming creatures with a tiny tail for propulsion, are heavily armored by a bony carapace of skeletal elements interlocking just beneath the skin. Despite an eponymously boxy outline their shape is remarkably hydrodynamic and has even been applied to experimental vehicles designed by Mercedes Benz. Yellow boxfish (*Ostracion cubicum*) augment their protective armor with toxic skin secretions of unique protein boxin, a hemoglobin-destroying poison three times more powerful than cyanide.[17] Meanwhile, the finless sole (*Pardachirus marmoratus*) has been dubbed the shark-proof fish thanks to paradaxin, its own brand of toxin that serves as a powerful deterrent to sharks.[18] An unhurried bottom-dweller related to flounders, the sole also relies on camouflage, hiding in plain sight on sandy or pebbled bottoms. Flounders and soles match the pattern and color of the bottom by transforming pigment skin cells into splotches of different hue and size, a disappearing act that happens in as little as 6 seconds.[19] If you are small, or slow, or both, you must have formidable defenses if you are to survive. So it is not surprising that the tiniest fishes on the reef are among the most well-protected. Gobies, a group that includes the smallest vertebrate animals on Earth, rely on an impressive range of defenses—from toxins to camouflage to tunneling—and the result has been an astronomical success: the family of gobies contains more fish species than any other.[20]

Gobies resemble miniature cylinders, tapering toward the tail, with a dorsal fin distinctively split in two parts, and pelvic fins modified into a disc-like suckers. Apart from these unifying features, roughly 2000 species present a diverse mix of habitat and diet preferences, courtship and mating ceremonies, and defenses against predation. A couple dozen species live among branching and stony corals, but most gobies have taken up residence on the bottom. In this they are aided by the lack of a swim bladder, their negative buoyancy allowing them to rest comfortably on the sand or stone. There they scuttle about capably on bent pectoral fins, looking for all the world like a fish doing push-ups. So diverse and abundant are gobies that on tropical reefs they make up a fifth of all fish species, a third of all individuals, and can account for as much as 50 percent of all the energy flow. Most measure between 2 and 4 inches in length, although the record-setting coral reef pygmy goby (*Eviota sigillata*) reaches only three-quarters of an inch. Being small usually correlates with a short lifespan, and here gobies are superlative as well. At least two species can expect to live no more than eight or ten weeks, the briefest blink of an eye that is another world record among marine vertebrates. They become sexually active at age five (weeks!), reproducing early to offset swift and sweeping losses to predation:[21] annual mortality in gobies can exceed 97 percent.[22] To fuel their rapid growth (granted, to a tiny finished product), gobies eat copious amounts of detritus, organic matter falling from above or cellular ejecta from sponges; many of the smallest species, however, lean toward carnivory.[23] Hyper-abundant and quick to reproduce, gobies are critical to reef food webs, assimilating nutrients and conveying them to predators (in the form of their own bodies).

Perhaps nobody informed gobies of their chief role, for they certainly try to avoid that fate. Many species have toxic skin secretions, a few even loaded with tetrodotoxin. They warn predators, with bright bands of turquoise blue or spots of maroon, that a nasty mouthful awaits. Coral-dwelling gobies (genus *Gobiodon*) live inside the safe refuge offered by staghorn, elkhorn, and other stony corals. As if in payment for protection, gobies fastidiously groom the coral, and will charge hungry butterflyfishes nosing about in hopes of a mouthful or two of their home. The butterflyfish soon relents, preferring to avoid castles defended by toxic palace guards.[24] Those guards do face one serious challenge imposed by their palatial host. Coral respiration peaks at night and uses up a lot of the available oxygen near its branches. In response, coral-dwelling gobies have a fantastic tolerance for low-oxygen conditions and will even mouth-breathe when the tide is low.[25] Their skin lacks scales, helping their tiny bodies absorb oxygen across the skin, but leaving them unarmored. In exchange these gobies are flush with toxic glands that slather their skin in noxious chemicals, bolstering their defenses.

Some gobies choose not to live among the parapets and ramparts of coral castles, opting instead for the sandy plains beyond the battlements. There, they thrive among reduced competition for falling food, sifting sand for tiny invertebrates, but they also face a much greater risk of predation. In response, they have learned to dig tunnels, or better yet, to hire a contractor. On tropical reefs, more than a hundred goby species have struck up a professional relationship with more than twenty species of snapping shrimp that specialize in excavation.[26] Known also as pistol shrimp, these crustaceans possess one oversized claw that pinches so rapidly it cavitates the water, forming bubbles and discharging an ear-splitting pop. Captive pistol shrimps held in jars have even been known to shatter the glass, so powerful is the bubble wave. The odd-couple flatmates are linked by tactile communication. While the shrimps make excellent diggers, they are completely blind. Instead, they rely on the keen eyes of the goby and use long filamentous antennae to stay in perpetual, tickling contact with the fish. Watchfully, the goby sits at the tunnel mouth, and when predators approach it flicks its tail to warn the shrimp.[27] It helps that some goby roommates are colored like the sandy seafloor, speckled or wrapped in pale stripes. Those simple costumes, however, are only crude examples of true camouflage, a disappearing act that our next fishes elevate to a mesmerizing art form.

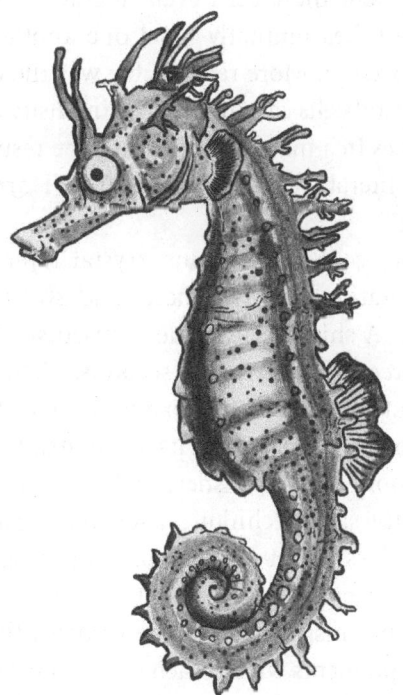

Long-snouted seahorse (*Hippocampus guttulatus*)

NOW YOU SEE ME, NOW YOU DON'T

Flamboyant portraits of tropical fishes are painted from two entirely different palettes, each daubed with distinct types of paint. Reds, yellows, blacks, and whites are known as pigment colors and are displayed by skin cells known as chromatophores. Cells loaded with melanin, called melanophores, appear black, while those packed with uric acid are white. Xanthophores and erythrophores accumulate carotenoid pigments and are yellow, red, or intermediate shades of orange. The other palette is loaded with structural colors, reflective sheens that share more with a prism than a can of paint. Cells known as iridophores contain plate-like crystals of guanine—the much-valued fertilizer ingredient—that reflect light off the top and bottom of the plate simultaneously. Interference between the two reflections creates shimmery colors of silver, green, and blue. Most fishes have several of these cell types, often working in concert. A reflective iridophore of blue overlaying a pigment cell of yellow, for example, will combine to produce a shiny metallic green. Sometimes during a fish's embryonic development these cells even interact. White and black pigment cells in some fishes mutually repel one another, producing alternating stripes like a zebra. More remarkable was the discovery that both pigment and structural cells can modify the intensity and even color they broadcast, sometimes in a matter of seconds. The result is a magical ability made famous in literature from H. G. Wells to Harry Potter: the power to become invisible.

Inside iridophore cells the guanine crystal layers can be distorted, altering their orientation and thickness, and shifting both color and reflective brilliance. A shimmering blue damselfish (*Chrysiptera cyanea*) can become dull green in a matter of seconds.[28] This "now you see me, now you don't" strategy allows a fish to appear attractive to choosy partners, but swiftly throw a cloak over its vivid attire if a predator passes through the neighborhood. Pipefishes, endearing and elongate relatives of seahorses, share the same technique: a female broadnose pipefish (*Syngnathus typhle*) displays bright ornamentation during courtship but flips the switch to dim her embellishments when hungry cod enter the mating arena.[29] More impressive still are the changes that chromatophores can express. When pigments within each cell are clustered into thick dots the skin appears dark, but when they are dispersed into tiny pinpricks, the skin looks white or even translucent.[30] If chromatophores of different color classes alter their pigments in harmony, a fish can transform from red to yellow or from white to blue, in a matter of seconds. Localized modifications will trigger brilliant stripes to appear, flaming spots to vanish, or appealing bars of black and white to fade. Wielded together, chromatophores and iridophores can conjure a mesmerizing array of patterns and colors, and some fishes have such finely tuned control that they flawlessly mimic the backdrop and literally vanish before your very eyes.

Among reef fishes, the champions of camouflage must be flounders. These oddly shaped animals are actually compressed, their bodies flattened side-to-side, but they swim with one side always toward the bottom, making them appear squeezed from top-to-bottom (a form known technically, but forlornly, as depressed). They acquire this tilted-over orientation during larval development, a metamorphosis that involves some shocking alterations. The eye destined to look at the bottom slowly migrates all the way across the head, until it comes to rest next to the other; thus both eyes lie improbably on the fish's uppermost side. Clearly, one can see approaching predators better this way than with one peeper buried in the mud. Flounders are separated into left-eye and right-eye species depending on which one remained fixed in place. Pectoral and pelvic fins on the lower side are much reduced, and even absent in some species. Flounders feed on invertebrates and small fishes, spending most of their time scouring sandy and silty seafloors for buried crustaceans and worms. In these open plains, they are at great risk of predation, a danger that has honed their ability to disappear. When a threat appears, they glide to an immediate halt and adopt the precise color and pattern of the bottom beneath them. They can mimic finely speckled yellow sand, smooth reddish mud, and even the coarse grey and green splotches of seafloor pebbles. One tropical species, the eyed flounder (*Bothus ocellatus*) can marshal the entire transformation in just 2–8 seconds, thanks to its nimble chromatophores.[31] To simplify their disappearing act, flounders favor substrates they can readily mimic and avoid hard-to-match patterns like live coral and sponges.[32] The advanced skill of copying living backgrounds is what puts our next magicians in a class all their own.

Frogfishes are possibly the weirdest looking, most improbable creatures on a coral reef. They barely resemble a fish at all. Instead of the sleek disc of a butterflyfish or streamlined torpedo of a tuna, they are lumpy and irregular, clumsy and slow, yet utterly and whimsically lovable. Fins are mounted on thick lobes, the blunt face and rounded body appear lopsided, and a downturned mouth lends them a charmingly grumpy appearance. Distinct from the similarly named but more musical toadfishes, frogfishes thrive in places that flounders avoid: they specialize in mimicry of live corals and sponges. So perfect is their camouflage that they are rarely seen, despite being quite common. One could spend an hour examining a few blobs of red sponge and never realize that one blob was actually a fish nestled against the sponge and cloaked in perfectly matching tones. Frogfishes do not adjust their coloration as rapidly as flounders, preferring instead to settle on a color pattern and stick to the background it matches. If forced to migrate, it takes a few hours to change their costume. They can match corals and sponges of vivid yellow and red, or display multicolored mottling typical of coral walls dotted with algae and invertebrates.[33] Some species are even adorned with fleshy tubercles and flaps to further break up their outline, none more so than the sargassumfish (*Histrio histrio*). This little magician specializes in mimicking the speckled fronds of sargassum weed, a tropical seaweed that floats on the surface in tousled masses. Concealed in this tangle of yellow and brown algae, the sargassumfish is undetectable even to its prey, tiny shrimp and larval fishes whom it lures by jiggling a worm-like extension of its dorsal fin. So theatrical is the performance that the fish's scientific name references (twice!) an overly dramatic stage actor. Or perhaps it was the discovery of their histrionic habit of gobbling juveniles of their own species, nothing being more melodramatic than a cannibal in camouflage.

Intermediate between frogfishes and flounders, and earning honorable mention in the category of weirdest fish, are batfishes. Unlike bats they do not fly, but unlike fishes they rarely swim; instead, batfishes prefer to walk. Around eighty species inhabit tropical reefs, where they rest their flattened, triangular bodies on the bottom, perched on stout pectorals that resemble arms more than fins. Patiently they wait, tempting prey to approach by dangling a retractable lure in front of their jaws. The two-part lure, formed of a dorsal fin spine tipped with a fleshy knob called an esca, is similar to those dangled by deep sea anglerfishes, close relatives to both bat and frog fishes. Unlike their abyssal cousins, however, the batfish's lure is not illuminated, although the polka-dot batfish (*Ogcocephalus cubifrons*) does employ a chemical bait to attract its lunch.[34] Secretory cells in the esca, not unlike those in the skin of soapfishes, exude a compound that attracts marine snails. Large numbers of their shells can be found in batfish stomachs, where they make up nearly half the contents, a testament to the bait's appeal.[35] Batfishes even take advantage of underwater currents, facing down-current so the chemical temptation streams away from them like a fishing line, reeling snails to a toothy end.[36]

Camouflage is a defense found most commonly among slow-swimming fishes who risk murder whenever they move, even within their tiny territories. None are more effective at this strategy of hiding in place than the most famous parental care iconoclasts in the ocean, seahorses. Some 25 million years ago, tectonic uplift created vast basins of shallow, warm water where seagrasses proliferated. And in those marine meadows, seahorses evolved their curiously equine shape.[37] Their ancestors the pipefishes swim in a standard fashion, head first like a tiny barracuda. But in the dense blades of seagrass, a new opportunity for camouflage arose. Taking full advantage, seahorses evolved an upright posture, the better to mimic the vertical blades, and a grasping tail, the better to anchor themselves to the stalks. Pipefishes were already good camouflage artists, and their ability to match background colors and patterns was supercharged in seahorses. Today, nearly fifty species thrive in the world's oceans, pastured amid seagrasses, seaweeds, corals, or sponges against which their odd shape and superb color matching render them undetectable.

Seahorses settled into their new habitat and posture, but continued to adapt rapidly thanks to higher rates of protein and gene evolution than occur in most fishes.[38] They have evolved unique armor, a novel tail design, and a one-of-a-kind technique for catching their prey. Instead of scales, seahorses are wrapped with overlapping bony plates beneath their skin; so stiff is this armor that the fishes evolutionarily jettisoned their ribs as unnecessary structural supports. The tail, lacking a terminal fin, is one of the few square tails in the animal kingdom. Inside are a stack of bony squares like picture frames, each linked to the other by three distinct kinds of joints: ball-and-socket, peg-and-socket, and gliding. These joints permit a remarkable degree of flexibility, and the tail can grasp seagrasses and seaweeds as effectively as a monkey's tail clutches a branch, curling through more than two full revolutions. Three-dimensional models of the picture frame arrangement show square seahorse tails also offer greater resistance to crushing than a comparable round tail, very useful against the jaws of fishes and bills of wading birds.[39] Engineers responsible for that study recommended adapting the design to medical or robotics applications, and researchers in Japan galloped with the suggestion, manufacturing a meter-long wearable tail whose wagging movements can help people counteract balance disorders.[40]

While clasping seagrasses or corals, seahorses are on the lookout for planktonic invertebrates, like tiny shrimp and jumpy copepods. When it spies a suitable morsel, the seahorse moves its long snout ever so slowly into position. Research by fluid dynamics engineers revealed the seahorse's head shape minimizes disturbance of the water around it, allowing the fish to gingerly advance its snout toward the copepod without setting off any alarms, to as close as four one-hundredths of an inch.[41] Then the seahorse lunges, using a pivot-and-suction motion unique among fishes. Were it not for the stealthy approach, the copepod would escape, but because the ambush is launched from exceptionally close range, the seahorse is able to slurp prey at will, all from the security of its seagrass perch. So effective is this feeding skill that seahorses lost their teeth several million years ago, and like a high schooler after a wisdom tooth removal, they simply eat their meals through a straw.

After kangaroos, seahorses probably have the most famous pouches in the animal kingdom. Unlike Australia's hopping marsupials, however, it is the doting fathers who bear the young. Seahorses reside in compact territories, often as little as one foot in diameter. They are highly monogamous, thought to be an adaptation to increasing reproductive success for low mobility fishes who live far from one another: better to pair with a single partner than risk wooing new mates again and again.[42] During courtship a female (call her Helen) will travel to visit her mate every day at dawn (call him Paris, for those who remember Greek history).[43] On one magical morning, both seahorses begin to brighten in color and engage in reciprocal quivering, trembling and twisting their bodies, a dance that indicates their readiness. The next day is the big one. While Helen points her body toward Paris, he repeatedly opens his pouch and pumps water in and out. This action helps oxygenate the space where eggs will soon be placed. In some species, the male also makes a clicking sound while pumping, serenading his partner.[44] After a few rounds of pointing, brightening, and proud pouch pumping, Paris and Helen let go of their shared perch and rise in the water column. While drifting upward, Helen transfers her eggs into Paris' pouch; these copulation rises are repeated several times, until anywhere from a hundred to a thousand eggs are delivered.

After receiving Helen's eggs, Paris fertilizes them, his sperm slipping through the pouch opening before it is drawn shut. Inside, the eggs benefit from oxygen and nutrition supplied by his pouch. Genetic studies have revealed pregnancies are astonishingly similar between seahorses and humans, with dozens of shared genes that regulate nutrient transport, gas exchange, salt balance, and immunological protection.[45] Throughout gestation, which lasts from ten to forty days (depending on the species), Helen faithfully visits Paris every morning until the eggs hatch and seahorse colts trot into the open water.[46] His camouflage protects him while the couple await the glorious day, and holding eggs internally safeguards them from predators until they hatch.

Sadly, adult seahorses face heavy predation themselves, from a voracious species known as *Homo sapiens*. Seahorses are swept up as accidental bycatch, particularly by bottom trawlers, an indiscriminate practice from which camouflage cannot save them. One study estimated seahorse bycatch at 37 million animals per year, in just twenty-two countries with large fishing fleets.[47] But millions more are actively collected and sold in a dodgy Chinese medicine trade, where they are reputed (with no substantiating evidence) to cure everything from asthma to incontinence to impotence. Asian markets are not the only culprits, however, as some 65,000 dried seahorses are sold annually in the United States as "curios" for seaside tourists;[48] similar numbers have been recorded in Portugal and elsewhere.[49] One hopes that people will put a halt to these callous practices; exquisite and gentle seahorses deserve better.

IT TAKES TWO TO TANGO

> And hand in hand, on the edge of the sand,
> They danced by the light of the moon
> **—Edward Lear, "The Owl and the Pussycat"**

In Homer's epics the *Iliad* and *Odyssey*, a star-crossed affair between Paris and Helen (the people, not the seahorses) changed the course of history when incensed Greeks laid siege to Troy for ten years, snuck through the battlements in a horse of wood, and sacked the city, leaving behind nothing but ashes. In politics, the power of love can be awesome, indeed; on coral reefs, the urge to pair and reproduce is no less a potent force in evolution. As much as avoiding predation has driven the diversification of reef fishes, courtship and mating have played an equally central role.

Most reef fishes spawn in locations with prevailing currents that carry eggs away from the reef, into deep waters where predators and planktivores are less abundant and the chance for survival is higher. No morsel is more defenseless than a tiny fish egg or newly hatched larva bobbing in the sea. Only a miniscule fraction of planktonic larvae will survive this hazardous phase of their lives. In silvery-sided jack mackerels, for example, fewer than one offspring in 50,000 live beyond their second month.[50] Such prodigious losses have propelled fishes to find new strategies to reduce risks, and new forms of communication to coordinate care. As we have seen, toadfish and triggerfish males construct nests where eggs are guarded by one or both parents. Many damselfishes engage in this modest level of parental investment, often accompanied by striking color changes to improve the guard's camouflage. A few damsel species (in the genus *Altrichthys*) even defend their offspring after they hatch, patrolling tirelessly against predators. These attentive parents lack some child-recognition skills, however, as they will unwittingly serve as bodyguards for offspring of different parents, or even of different species. Genetic analyses in Philippine reefs showed two other (nonpatrolling) species have learned to drop their own toddlers into the defender daycare, just as cuckoos will lay eggs in another bird's nest to parasitize the parenting benefits.[51]

Once you have built a tiny nest on a sprawling coral reef, how do you manage to draw egg-laying females to it? Visual communication, practiced by parrotfishes and a great many brightly colored reef species, attracts females and allows them to judge the vigor of their male suitors. Some species invert the courtship, with males choosing females on the basis of size, as larger fishes lay more eggs than smaller contenders.[52] Vivid females can also be extra appealing to males, as intense colors signify a good diet and overall health. Two-spotted goby (*Pomatoschistus flavescens*) females with dazzling orange bellies, for example, attract four times the amount of male attention as goby gals who lack a glowing gut.[53] A number of fishes, particularly cryptic species, also rely on chemical attraction to seduce their partners. Those perfumes send important signals about the sexual receptiveness and location of a willing male (perhaps with a newly constructed nest site), or they serve as a calling card identifying the sender's gender. In Gulf pipefish (*Syngnathus scovelli*), who also have inverted courtship, females alone release a scent that alerts nearby males to their readiness for egg-laying.[54] Blind gobies (*Typhlogobius californiensis*), who cohabitate with sighted shrimps in a twist on the usual shrimp-goby tale, live in little caves off the California coast, waiting for a mate. When a visitor approaches the burrow's entrance, they sniff their caller's fragrance to distinguish between a same-sex rival and an opposite-sex suitor.[55]

Like gobies, blennies are small bottom-dwellers who scramble across sand and rocks, walking on their pectoral fins while searching for mouthfuls of small invertebrates or stray detritus. They are distinguished from their goby cousins by a dorsal fin with one or three, not two, lobes. Some 900 species can be found in oceans, and even some freshwater locales, where they adopt an impressive range of body sizes and colorations. Most species rely on camouflage for some protection, and many take up residence in small caves or holes for protection. Barnacle blennies (several species in the genus *Acanthemblemaria*) charmingly convert empty barnacle shells into their very own tiny houses, from which they peer out at the world with comically bulging eyes. That their eyes are further decorated with eyelash-like tentacles only adds to their cartoonish, come-hither appeal. Many blennies live out their days in extremely restricted territories, like seahorses, which can make it difficult to simply bump into a passing female, much less one who is reproductively receptive. But perfumes can help increase their chances.

In combtooth blennies (some 300 reef-dwelling species), males attract potential mates with colorful costumery, but they possess an added embellishment: twin scent glands near their anal fins. Male peacock blennies (*Salaria pavo*) engage in intricate courtship dances, with much raising and lowering of a rubbery, tangerine head crest that contrasts with their electric-blue striped body. Males defend territories around a nest installed in a small hole or crevice. To put females in the front row for the big performance, the males release powerful pheromones from their scent glands. Males with higher testosterone levels develop larger glands and are more committed to nest-guarding, suggesting the scent they release may communicate more to mates than simply the showtime.[56] Once eggs are laid in the nest, peacock and closely related redlip blennies (*Ophioblennius atlanticus*) have been observed smearing their eggs with a mucus from anal glands, a gel with antimicrobial powers.[57] Given that unguarded eggs can be destroyed by infection as readily as by a predator, this defensive behavior is another ingenious effort by parents to enhance their reproductive success.

It should come as no surprise that fishes can communicate by perfume. Marinating in water, they constantly shed mucus, urine and salts, amino acids and peptides, and so many other chemicals that one marine biologist described fishes as "leaky bags of body odor," an expression that really puts the ick in ichthyology.[58] Fishes at the peak of breeding season are extremely responsive to this mix of communication signals. Hormones circulating within males and females ramp up as the mating season approaches and amp up auditory, visual, and chemical enticements. Atlantic salmon (*Salmo salar*), among other species, even release some of those hormones into the water where they serve as enticing pheromone signals.[59] Female damselfishes and midshipmen (close cousins to toadfish) become more sensitive to male courtship songs when they are reproductively receptive, and both cycles are modulated by hormones.[60] According to Karen Maruska, from whom we heard in the preceding chapter about damselfish song, chemical signals have similar effects in fishes and human beings. "Steroid hormones, estradiol in particular, improve auditory function. That's actually true of people, too. If you look at women across the menstrual cycle . . . because of the levels of circulating hormones they have different auditory perception."[61] In effect, females have hormone-mediated hearing aids that crank up the volume when they are eager to hear serenading males. Male fishes (and boyfriends) may rejoice to know that when their partner turns a deaf ear on a lengthy monologue, it might be due to hormones rather than sheer boredom.

After courtship and egg-laying and fertilization, the lottery of life begins. Some eggs will be eaten, though fewer if they are guarded. Larvae hatch and are swept away from the reef to fend for themselves in open water, bobbing about near the surface where phytoplankton provide a verdant pasture during the early days of development. But eventually those larvae metamorphose into juveniles who must find their way back to a reef where they can settle into adulthood and start the reproductive cycle over. To do this, juvenile reef fishes listen and smell for clues that indicate a reef's location, and even its makeup. While still in the larval stage, reef fishes can orient to the sounds of a coral reef nearly a mile away[62] and swim vigorously to the source.[63] Cardinalfishes (small nocturnal reef dwellers named for their red colors) home in on scent plumes that drain from warm coral lagoons, sniffing the water with olfactory systems already keenly sensitive at an early age.[64] Damselfish juveniles may be even more selective, as they actively prefer waters laced with the scent of adult damsels, but crucially not the odor of close relatives, a cocktail of chemical cues pointing the way to healthy reefs that lack direct competitors.[65] Reef fishes can even detect smells that originate outside the ocean. In a study of clownfish settlement, researchers showed that larvae of at least one species (*Amphiprion percula*) of these adorable anemone-dwellers can detect not only a coral reef, but also a reef adjacent to a tropical island. They do this by discriminating the specific fragrance of leaves that have fallen into the water and use the scent cue when house hunting to choose only an anemone home that boasts an island view.[66] Location, as they say, is everything.

THE ART OF NEGOTIATION

> Only now are scientists thinking seriously about how parasites may be as important to ecosystems as lions and leopards . . . and perhaps the dominant force in the evolution of life.
> —Carl Zimmer, *Parasite Rex*

More weird behaviors and wonderful fishes are found in coral reefs than anywhere else in the ocean, save perhaps in the abyssal depths. Reefs are like rainforests, where so many species co-mingle and interact that the foremost pressures of evolution come from other living organisms, rather than the nonliving environment. Species are continuously coadapting, in feedback loops that spawn towering biodiversity and provoke fascinating behaviors. Reactions between prey and their predators, and courtship between males and females, are just two such evolutionary feedback loops. But around coral reefs, interactions can occur between almost any pair of species, and even the unlikeliest of companions can make for fine bedfellows.

Take sea anemones, for instance. They are a cantankerous lot, waving tentacles packed with stinging cells that fire poison-tipped harpoons into prey and interlopers with equal ferocity. But nestled amongst those tentacles, as cozy as a kitten on a couch, one can find a clownfish or two living quite contentedly. Thrust into stardom by the feature film *Finding Nemo*, about thirty species of these bright orange fishes with radiant blue bars enjoy a convivial mutualism with anemones throughout the tropics.

Also known as anemonefish, they benefit from the protection of their host's stinging tentacles and feast on some of the food ensnared by the landlord. In return, the fish's nitrogen-rich feces help fertilize the anemone. A clownfish's rental agreement includes pugilistically defending the anemone from tentacle-eaters like butterflyfishes. They also chase off three-spot humbugs (*Dascyllus trimaculatus*), small damselfish relatives who love nothing better than to couch-surf in their anemone pad without paying rent. Clownfishes win that tussle most of the time but eventually fall victim to their own success. As they fertilize their host anemone, it grows larger, eventually reaching a size where evicting a gate-crashing humbug is not worth the trouble.[67] Both fishes eventually reach a disgruntled détente, like grumpy neighbors in a large apartment building. Bah humbug, indeed.

Clownfishes begin their anemone residence early in life, shortly after settling on the reef from open water. As juveniles, their bodies are covered in a thick mucus, three to four times thicker than other fishes, which gives them an initial degree of protection against the stinging cells.[68] But they shortly begin rubbing themselves against the tentacles and pick up some of the secretions of the anemone itself. Soon, they smell like their anemone host, whose tentacles no longer find them to be foreign: no amount of contact will now fire the stinging cells. By the time they reach adulthood, the microbiome of their skin mucus is significantly different from that of clownfishes not hosted by anemones.[69] That difference fades if the clownfish (in this study, *Amphiprion clarkii*) is experimentally evicted, evidence that regular contact with tentacle mucus is necessary to recharge the fish's immunity. This mutualism with sea anemones drove a rapid diversification of clownfish, starting around 5 million years ago, once a few ancestral fishes adapted to life with their hosts: smaller bodies, shorter fins, and more skin mucus. Now, a unique cluster of species can be found in any coral-rich region, each one specializing in a select group of anemones, but genetically distinct from bands living on reefs a few hundred miles away.

Sometimes the benefits of a mutualism between animals can be hard to see, unless you know just how to look. Take a couple of other humbugs, known as whitetail and marginate dascyllus (*Dascyllus aruanus* and *D. marginatus*). These attractively marked fishes reach just 2 or 3 inches in length, and they associate with stony, compact corals sometimes called cat's paws. Resembling a cluster of stubby, branching fingers, these corals live in shallow and brightly lit waters where their symbiotic algae churn out sugars and oxygen. Quite a few humbugs will live between the fingers, depending on the size of the coral colony, and provide many of the same benefits as clownfishes to anemones: fertilization with feces, and defense against corallivores like the dreaded crown-of-thorns starfish. Once darkness descends, the humbugs settle down amid the fingers, but their sleep is fitful. Unlike most crevice-dwelling fishes, these humbugs vigorously flap their fins all night long: they are sleep-swimming.[70] Careful measurements reveal that the finning aerates the coral, just as their tissues are consuming oxygen in the water layer adjacent to the stony fingers. Without humbugs corals can use up 90 percent of the oxygen in that water, which is not replenished until morning light switches on the algae. But with a troupe of finning fishes providing ventilation, the water retains as much as 80 percent of its daytime oxygen. Soft corals, which are not frozen in place by a cement exoskeleton, pulse their tentacles for this very reason, something first noticed (but not understood) by French naturalist Lamarck some 200 years ago.[71] Hard corals with ventilator humbugs grow more quickly, host more algae with elevated photosynthesis rates,[72] and even produce more eggs than colonies without fishes-in-residence.[73] Although the gases wafted by sleep-finning fish are invisible, the benefits to the coral certainly are not.

Swirling above the canopy of the Amazon rainforest, a rainbow of birds collaborate to find fruits, watch for predators, and navigate through broccoli-shaped tree crowns. Mixed-species flocks are common there, so long as the benefits of partnership outweigh the detriments of competition. The same phenomenon plays out on the plains of the Serengeti, where a dozen species of mammals forage together, each browsing on slightly different plants, and all keeping eyes peeled for prowling lions. In coral reefs, too, mixed-species mutualisms abound when the advantages exceed the risks. A notable case is the diverse party that accompanies a day octopus (*Octopus cyanea*) on its rounds, one of the most complex examples of collaborative hunting and division of labor seen in the ocean. The hungry octopus scrambles over the reef, probing coral and rock crevices with slinky tentacles in search of hidden prey, from small fishes to sea snails, clams to crabs, marine worms to shrimp. Meanwhile, swimming in orbit, a compact solar system of partner fishes scour the surroundings, each engaging its own unique skill set.

Goatfish species like the yellow-saddle (*Parupeneus cyclostomus*) comb the bottom nearby, using sensitive barbels dangling from their chin to sniff for buried mollusks and crustaceans. Smooth cornetfish (*Fistularia commersonii*) hover a few inches above, sharp-eyed lookouts for any prey fleeing through the water column. Tailspot squirrelfish (*Sargocentron caudimaculatum*) hang about opportunistically, and a blacktip grouper (*Epinephelus fasciatus*) will join and use fin and body gestures to point out hidden targets. More than a dozen species have been seen in hunting packs with the octopus, and when foraging together the fishes were able to strike prey more frequently, and more successfully, than when hunting alone. But the pack's interactions are not always congenial. To show who is boss, the day octopus has been known to throw a punch. It balls up the tip of a tentacle into a fist, and with an explosive motion it can sock a fish right in the chops, or flank. The octopus uses punching as a partner control mechanism, warding off the squirrelfishes who contribute too little to the hunting party, for example, or warning a usually collaborative member against stealing prey.[74] The punching bag never retaliates, and fisticuffs do not break out among other members of the pack, who park their aggression when joining the hunting collaborative. This curiously selfless act of suspending hostility also characterizes one of the most spellbinding sites on any coral reef, where mixed-species flocks rival the rainforest in color and complexity: a fish cleaning station.

Fish are under constant assault, day and night, not just from hulking predators but also from tiny, vicious parasites. Worms, protozoans, leeches, and crustaceans may invade a fish's innards or gills, or colonize its skin and scales, to steal energy and nutrition like a tick on a dog. Fish go to great lengths to ward off these pests, including slathering themselves with bug repellent made from toxins secreted by their skin.[75] But on a reef there is no better way to get rid of parasites than to visit a cleaning station. There, like enthusiastic spa staff, a crowd of small fishes and an occasional shrimp or crab scour their larger clients, plucking parasites from the scales, fins, and even the interior of their mouths. Itchy customers go to great lengths to solicit the spa treatment. Lexa Grutter from the University of Queensland has spent years studying the interaction between clients and cleaners on Australia's Great Barrier Reef. She notes that customers often pose in "unusual postures where they open their mouths, they spread their fins out, they stay in the same spot and try to balance, and sometimes they end up upside down . . . and actually can look quite funny at times."[76] She speaks poetically of the relationship between the two fishes. "It is such an intimate interaction, when you think about it. Here's this fish that is just allowing this other animal to literally crawl over it, even climb into its mouth. I've seen it go into the mouth of a big giant potato cod, I've even seen it pop out through the gill opening. They'll clean their eyeballs and pick at their nostrils, it's just very intimate."

Giant grouper (*Epinephelus lanceolatus*) and bluestreak cleaner wrasse
(*Labroides dimidiatus*)

Among the most energetic cleaners are wrasses, cylindrical and highly decorated little fishes closely related to heftier parrotfish. Dr. Grutter has shown that their bright colors, often electric blues and yellows, signal their cleaner status to incoming fish and defuse the groomer's risk of being swallowed.[77] The small size of wrasses permits easy access to clients' mouths, where they scrutinize teeth, tongue, and palate with the zeal of a dental student. Cleaners also attend sharks, giant rays, and even sea turtles who accumulate an irritating crust of parasites during weeks on the high sea. After a few minutes of brisk attention, the customer is deparasitized and healthy, and the wrasses have full bellies. Spa treatment completed, the client is given a hot towel and shown the door, and the next fish in line glides into the salon. On reefs with readily available cleaning stations, groomed fishes are healthier, have lower stress levels, and grow larger, all benefits that outweigh the miniscule value of swallowing your masseuse.

Grutter's unique career was inspired in part by a childhood spent in Alaska with her mother and father, a deep-water fisherman. She vividly recalls gutting some of the day's catch and curiously examining their internal organs, only to find a few salmon and halibut repulsively loaded with parasites. Unnervingly, she notes how widespread parasites can be. "It's one of the most common life forms on the planet. We all have parasites, we're crawling with parasites, and we do all kinds of things to prevent it. As humans, we take showers and cut our hair and wash our clothes. Animals, they engage in grooming behaviors, they groom each other, cats clean themselves constantly. Fish, on the other hand, they can't really groom themselves, and so they have found other ways to deal with their external parasites . . . they've somehow worked out that these cleaner organisms are interested in these parasites as food." Cleaners can obtain quite a steady diet from their flea-ridden clients. During her tireless observations of the bluestreak cleaner wrasse (*Labroides dimidiatus*) Grutter showed that a single fish can pluck more than 1200 parasites in a day, an impressive feat for a fish only 4 inches long.[78] During that same day they may see as many as 2000 customers come through the spa door, with some individuals returning more than a hundred times.

Chief among the plucked parasites are gnathiid isopods (pronounced "nay-thid"), tiny crustaceans that behave like marine mosquitoes, stabbing a hole into a fish's skin and sucking blood. Unchecked, these vampires can drain an astonishing 85 percent of a fish's blood.[79] If the victims don't avail themselves of a cleaning service, they can perish from the isopods themselves, or indirectly from blood diseases they often transmit. Parasitized damselfishes swim more slowly and use more oxygen, for example, and juveniles with even a single gnathiid are more likely to disappear from the reef than establish a homestead. Grutter's long hours underwater have revealed just how important a cleaning station can be. "A rabbitfish, he went there every five minutes on average, for an entire day. The rabbitfish would be feeding on algae, and it would take a few minutes of bites and then off it went back to the cleaning station, got cleaned, came back and kept feeding."

While clients desperately need cleaning, and cleaners must eat, this intimate interaction is not without its dangers and deceptions. Cleaners, whether they are fishes or delicate shrimps, put themselves at risk when approaching a large fish. Their very lives depend on the restraint of the potato cod, who might fancy a wrasse as an appetizer. On the other hand, it turns out that parasites are not actually the cleaners' favorite food; they prefer to munch on the protein-rich mucus that envelops most fishes, and even the skin itself. If they begin nibbling at the mucus, though, or bite off a piece of skin, the client may get grumpy and depart in a huff, or even try to swallow the cleaner. Since either party can shatter the cleaning station's peace treaty, they have developed ways to reassure their willingness to play nice. Clients adopt funny, head-down poses to signal their readiness to be groomed. Cleaners often preface a session with a tickling dance, brushing the client repeatedly with their fins, to soothe the customer and discourage them from taking a murderous gulp. Grutter describes the value of this tactile interaction: "It's used to communicate with the client, and to manipulate the behavior of the client. They will use it to get clients to stay longer, and it also seems to be used to appease the client. For example, if it's a predator, they seem to do it more so, and they did it more when they had a hungry predator than when it was a well-fed one." A sort of safety zone develops around cleaning stations, where clients and cleaners treat each other with respect, and where even the clients desist from eating one another while waiting their turn at the spa.[80]

Still, cheaters can invade the station, slinking in for their own nefarious purposes. Wrasses dress in cobalt and sunflower to advertise their services, but those colors are readily mimicked. Among any cloud of Grutter's cleaner wrasses you can usually find a tiny bluestriped fangblenny or (*Plagiotremus rhinorhynchos*) or sabertooth blenny (*P. azaleus*), only just distinguished by their distinctive, eel-like swimming style. These little pests sometimes slip into a cleaning station unnoticed, and take a painful bite of scales, skin, and mucus. Before the chomped client can whirl in anger, they scurry away to safety. An uneasy balance exists, however, since too many biting blennies would permanently undermine the faith of customers in the safety of their neighborhood spa. As it is, infrequent visitors to cleaning stations where sabretooth blennies lurk typically will abandon the site, the very definition of the old saying "once bitten, twice shy."[81] Regular customers, though, will chase the offending blenny aggressively and return for more cleaning. Wrasses have learned these patterns and will leap to attend an infrequent visitor while shoving a regular customer off to the proverbial waiting room, confident that they are less likely to leave the spa in a sulk. In experiments where participants must adopt this so-called delayed reward strategy—servicing the antsy visitor before the faithful regular—cleaner wrasses master the optimal tactic faster than chimpanzees, capuchin monkeys, and even orangutans.[82]

Not everything is rosy on the coral reefs. Global climate change is warming oceans, and warm waters cause corals to suffer mass bleaching events. Corals expel their symbiotic algae when thermally stressed, losing the power of photosynthesis and turning white in the process. Grutter noticed that cleaner fish were also disappearing from those bleached reefs. "We're looking at the parasites, the client fish, the cleaners themselves, and the habitat. We can see the changes over time. Hot water tempera-tures don't just affect the corals, they affect the rest of the reef commu-nity in a cascade." Alarm bells first sounded in 1998, when unprecedented warming led to the demise of nearly one-tenth of the world's corals. Some recovery followed, but subsequent warming events killed around 14 per-cent of corals in the ocean and prompted a 20 percent increase in algae cover.[83] The one bright spot is in the famed Coral Triangle, ancient seas between Australia, Malaysia, and the Philippines, where more than 500 types of coral and thousands of fish species make their homes. In those highly diverse waters, coral cover actually increased slightly since 1983. Supporting nearly a third of all reefs on the planet, the shallow seas of the Coral Triangle may be protected by a long history of exposure to warm temperatures, and by the adaptability of their biodiversity. Here, for the time being at least, the myriad interactions between species may be saving the reef.

The reach of global warming is long, and it affects not only shallow water reefs, but even the cold waters of temperate and polar seas. There in those productive latitudes swim some of the world's most important food fishes. Immense schools of cod, haddock, pollock, and halibut feed millions upon millions of people around the world. They have been served with chips to British diners for centuries, been dried and salted by Nor-wegians, and made millionaires of many a New Englander. Their prodi-gious bounty has sustained civilizations, lured ancient mariners to cross oceans, and provoked wars on the high seas. And yet we are only just now discovering how these fishes live and learning to manage the fisheries that deliver their abundance to our plates.

7 SLOW FOOD

Cod, Haddock, Pollock, and Halibut

THE CONFLICT BEGAN on the bottom of a cold northern sea and swept two nations into three wars. In the early twentieth century, British fleets routinely steamed north of Scotland and crossed the North Atlantic to fish some of the richest waters in the world. Towns like Hull and Grimsby, sister ports lying athwart the River Humber on England's eastern coast, swelled into thriving cities awash in jobs and money. There was only one problem with this newfound source of wealth: it belonged to Iceland. When that young nation declared its national waters off-limits to foreign fishing vessels, open warfare broke out. But it was a most curious kind of war, in which few shots were fired, only one casualty was registered, and the worst offender was sentenced to a fine of just 5000 pounds. One Grimsby captain wittily recollected, "It was more mischievous . . . I don't think we realized the seriousness of it then. We would throw potatoes at the gunboats, you know, that is the way the fishermen viewed it. It was a bit of fun."[1]

The fish that inspired these so-called Cod Wars was the Atlantic cod (*Gadus morhua*). After independence in 1944, Iceland's new government recognized that a wintry land with little agricultural promise and few mineral resources had only one key source of national wealth: fish. Determined to protect their cod from overfishing by foreign fleets, Iceland declared in 1958 a ban on all international fishing within 12 nautical miles of the country's wild and rocky coastline. The news fell on the cities of Grimsby and Hull like a hammer blow. Here was the key to their prosperity, seized without warning through the impulsive whim of some upstart nation, whose own fishing fleet was dwarfed in size and technology by the modern trawlers of England. This was cause for hostilities, and the First Cod War was soon underway.

Trawler captains, notorious for snubbing authority, headed north to pursue their livelihoods, and Iceland's fish. Each boat dragged a single giant net, pulled from two ends like a hammock, and reeled in on steel cables by powerful winches. To this day, the end of any fishing net that holds fish is called the cod end. Initially Iceland's Coast Guard could do little to stop the trawlers. They were outnumbered by the fishing boats and outgunned by British naval vessels. High seas battles involved more shoving than shooting, however, with boats shouldering each other like wrestlers until one relented. Soon the Icelanders turned to more defensive, and creative, gambits. Adapting gear invented for clearing WWII naval mines, they dragged a cable cutter across the trawler's net lines, which snapped them like guitar strings and dumped the catch. Misinformation was another tactic borrowed from wartime. When the British navy began broadcasting the location of each netcutter, the Icelanders recorded these radio announcements and replayed them days later, bewildering trawler captains and rendering the naval transmissions useless.

A tentative truce was declared, granting English trawlers a reduced catch limit, but Iceland would soon enlarge its marine boundaries to 50 and then 120 nautical miles, provoking the Second and Third Cod Wars. Tensions boiled over, live shells were lobbed back and forth, a British seaman was injured, and an Icelandic engineer perished (repairing an electrical fault hours after a skirmish). Still, both sides recalled the encounters with a dollop of humor. Iceland's Coast Guard Commander Kristján Jónsson wryly recounted a maritime brawl in which his ship, the *Baldur*, rammed the bow of the naval frigate *Diomede*. "Our stern quarter made a pretty big hole, right into the Officer's Mess, knocking pictures of the Royal Family off the wall."[2] The British captain Robert McQueen, equally droll, listed the skirmish's casualties: "We managed to save the photograph of Her Majesty the Queen, but regrettably His Royal Highness Prince Phillip got lost in the action."[3]

By 1976 the warring nations grew weary of the conflict, and England found itself entangled in a double standard. "It was sheer hypocrisy on the one hand to be demanding the extension of 200 miles of continental shelf so Britain could get oil, but then deny the same facility to Iceland," said John Prescott, Deputy Prime Minister at the time.[4] An agreement signed in Oslo ended the Cod Wars. In the end, the British accepted fishing quotas lower than they had been offered just a few years ago, shattering the economies of ports like Hull and Grimsby. Boats were mothballed, crews laid off, and skippers like Richard Taylor—the trawler captain whom Iceland had fined—were forced to retire: "When our involvement with Iceland, fishing-wise, when it ceased, that was the end of an era in Hull."[5]

A few years later, after Iceland had stabilized its own fisheries, the president of their Trawler Owners Association visited the retired trawlers in England. "When I walked past the dozens of ships which had been permanently decommissioned, I felt that we must have done some irreparable harm," said Kristján Ragnarsson. "It confirmed my belief that we must realize, having been the force that drove them off our fishing grounds, we must take seriously our responsibility never to overfish . . . so that the young people of Iceland could be secure that the fish stocks would prosper in the future."[6] His British counterpart, Tom Neilson, gave a more stark assessment: "The sad thing is, you've put a hell of a lot of hard working and good fishermen out of work."[7]

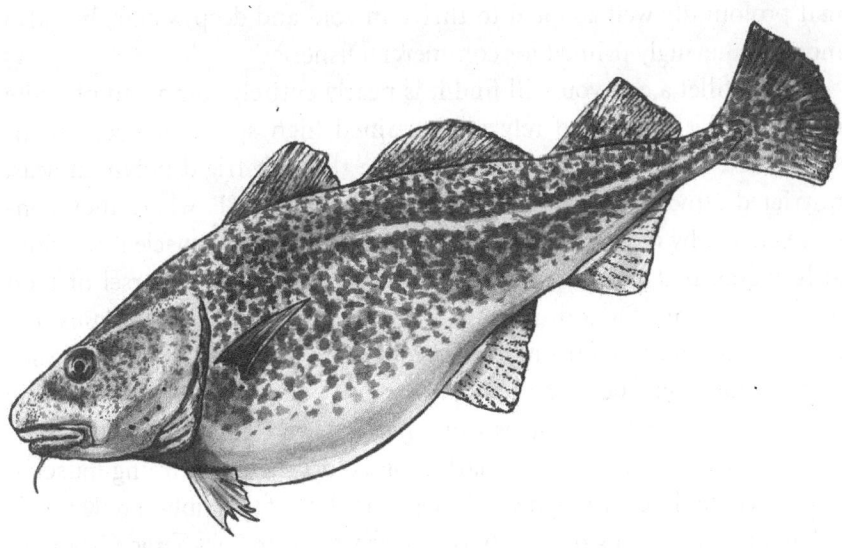

Atlantic cod (*Gadus morhua*)

THE MOST FAMOUS FOOD IN ENGLAND

> You can't get much meat for threepence, but you can get
> a lot of fish-and-chips.
> —George Orwell, *The Road to Wigan Pier*

Cod are the heroes of this saga, but they are less than heroic in appearance, sluggish and rotund with off-putting twin tentacles dangling from the lower jaw. Their flanks are gray-green, painted with a thin white stripe, and mottled with pox-like burgundy spots. On a plate, however, or wrapped in newspaper, they are unrivaled. Cod are flawless eating fish, yielding brilliant white fillets with thick flakes, a firm texture, and a delightfully mild flavor. Indeed, without cod there would be no such thing as fish and chips, arguably England's most famous food. When steam-powered trawlers first exploited massive schools in the frigid waters of the North Atlantic and the Barents Sea, a few scattered chippies—shops selling battered fish, fried potatoes, and mushy peas—swelled into a national industry. Today, more than 10,000 fish and chip shops in the United Kingdom dish out nearly 400 million meals annually and contribute a sizzling billion and a half dollars to the economy.[8] Cod habits and habitats, so different from those of coral reef fishes, have fashioned through evolution an animal profoundly well adapted to thrive in cold and deep waters, but also mouth-wateringly primed for commercial fisheries.

If you fillet a cod you will find it is nearly entirely composed of white meat. Unlike tuna, who rely on sustained high speeds to feed themselves, cod are a lethargic lot. They glide through frigid northern seas, restricted almost entirely to waters colder than 45 °F, where they conserve energy by moving slowly and deliberately.[9] White muscle is the sluggish engine that powers such languid cruising. When a morsel of food dances in front of a cod, perhaps a tasty capelin (*Mallotus villosus*, the polar equivalent of a sardine), it lunges with a burst of speed. Since they sustain that high speed for only an instant, cod have little need for the brown muscle that dominates tuna, herring, and other nonstop speedsters. Only a narrow ribbon of dark meat colors their swimming muscles, like the center line on a highway. The precise hue of this ribbon reflects the behavior of different stocks. Schools migrating from Spitzbergen Island to the Norwegian coast for spawning, a distance of some 500 miles, bear a stripe of espresso-brown muscle; shorter distance commuters to the Faroe Islands show only a line of cappuccino-tan.[10] Color even varies seasonally: in North Sea cod the stripe grows progressively darker as the spawning months of July and August approach, reflecting their intense hunting activity to fuel reproduction. Other than this narrow line, cod fillets are gleaming white, comprising 80 percent protein and almost no fat, and they are positively delicious.

Cod belong to a family of coldwater fishes called gadoids that includes pollock, haddock, whiting, and a few others. Their forms and behaviors are quite similar, though preferred ranges and depths vary. They swim in giant schools (though not as vast as those of herring and sardine), sometimes fittingly called herds, and rely on the impressive productivity of arctic seas to sustain them. Nearly all are important white-meat food fishes, caught by northern fisheries but distributed around the top of the globe. The most famous species, Atlantic cod, is divided into many populations: some around Newfoundland and the northeast coast of the United States; others around Iceland and Greenland, more in the North Sea waters around the British Isles and the Barents Sea above Norway and Russia. Across most of their range, cod are the apex predators within their favored, icy waters. Only minke whales and harp seals challenge them for this top spot. Despite their dominance, however, cod success and reproduction is famously sensitive. The slightest fluctuations in environmental variables can levy powerful effects that reverberate through the population, and all the way to fishery docks.

Fisheries science is an incredibly complex discipline, but at its heart lies a simple truth: you can't take more fish out of the sea than the population can replenish. Because of this maxim, scientists work tirelessly to understand every step of fish reproduction, from eggs to larvae, to juveniles, immatures, and finally spawning-ready adults. Each graduation from one age-class to another affects recruitment, the number of fish that join the adult population every year. And each step is extremely susceptible to minor changes in temperature, salinity, acidity, and the availability of prey populations, upon whose recruitment that of cod also rests.

Bjarte Bogstad is a fisheries scientist who has studied cod for years at the Institute of Marine Research of Norway. Though he grew up in an industrial town with no fishing vessels, he was drawn to the science of fisheries. Sporting owlish glasses and an affable grin (and a t-shirt proclaiming "In Cod We Trust"), he speaks rapidly and enthusiastically about his favorite fish. His Norwegian accent charmingly reveals itself on words with doubled consonants: "bigger" is distinctly pronounced with two syllables, as "big-ger." His institute's research is responsible for understanding a fish that contributes much to his nation's economy. For more than a thousand years, Norwegians have fished for cod, particularly in the spawning grounds of the Lofoten Islands, and dried them for storage and sale. "The spawning areas are so close to the coast that you could go out and have a good catch with a small boat. It was this fish that came close to the coast in big quantities, so it was fairly easy to catch with the old technology. In addition to the abundance, that was something that made it a good export."[11] Dr. Bogstad's institute is based in Bergen, a city practically built on cod. By the fourteenth century (and indeed until 1830), Bergen was the largest and richest city in Norway, thanks to an empire overseen by the Hanseatic League and founded on trading dried cod, a handy protein source that could be shipped all over Europe. Today, what cod eat and which environmental factors most affect their populations are central to Bogstad's charge.

As cod cruise toward shore for spawning, whether to the Lofotens or to the Newfoundland seaboard, they are gearing up for record-setting reproduction. "They are among the most fecund of all fishes," says Bogstad. "A female cod has typically a couple of million eggs, and they can spawn several times." Over her lifespan, a single female may release as many as 18 million eggs into the water; if every one were to survive, she would have given birth to more citizens than the combined populations of London and Chicago.[12] But they do not all survive; rather, losses at this life stage are catastrophic. Scientists from the Northeast Fisheries Science Center in the United States tracked egg numbers by dragging nets through the waters off Cape Cod. They found mortality reached a devastating 20 percent per day.[13] At this calamitous rate, more than 90 percent of her eggs would vanish in the week and a half they need to hatch.

Given such heavy losses, it is in a cod's best interest to release lots of eggs. A larger fish produces far more roe than a smaller fish, including one hefty female who shattered the records, according to Bogstad. "There was a fish of 54 or 55 kilos [121 lb.] recently landed off the coast in northern Norway," he relates proudly, "and this fish had 14 kilos of roe in it." That is 30 pounds of fish eggs, equivalent to the weight of a cocker spaniel. But there is a tradeoff: if the fish prioritizes growing to that stout size, then she may have to skip some breeding years in the process; alternatively, she can release fewer eggs earlier in life when she is more petite, although the effort will hamper her overall growth. Each cod population solves this equation differently (if one grants them the power of mathematical analysis). In the Irish Sea, for example, cod begin spawning early but release only a third of a million eggs in the first season. After four seasons, however, their total production can top 14 million roe. In contrast, Barents Sea and Icelandic cod have not even begun to spawn by the time their Irish cousins have reached their fourth mating season.[14] These differences are due in part to water temperatures, as the warmer southern waters boost growth rates of each Irish cod.

Once the eggs hatch, defenseless larvae emerge and begin gobbling plankton; the race is on to grow as quickly as possible into hardier juveniles. At this stage, temperature and timing are all-important. One of the favorite foods of a larval Norwegian cod is a copepod (*Calanus finmarchicus*, among others), tiny crustaceans whose numbers explode when springtime conditions are ideal, but unusual water temperatures can shift that peak by a few fatal weeks. In extremely warm years like 1960, copepods were most abundant around the first of April, and late-hatching cod starved.[15] But in the brutally cold year of 1981, the peak arrived in late May, and early hatching cod also starved. Cod's prodigious fecundity is probably a response to these unpredictable odds: when larvae and weather and copepods are all finally in sync, only then will enough of a female's 2 million eggs survive to guarantee the next generation.

As adults, northern Atlantic cod prefer capelin. In the Barents Sea alone, cod consume upward of 3 million tons of capelin every year.[16] These small schoolers feed on plankton blooms along the edges of polar ice sheets, but during warm years the ice retreats. Capelin must relocate to eat, a shift that can re-route spawning migrations, displace populations, and diminish survival. Many marine species rely on capelin, like adorable Atlantic puffins and doe-eyed harp seals, who target the silvery schools as they travel to breeding sites on the coast of northern Norway. When capelin populations shrink, cod and the other predators can go hungry, suffer increased mortality, and face seasonal breeding failures. As a balancing effect, however, the collapse of capelin lessens their predation on zooplankton, which begin to rebound.[17] Across the Atlantic in Newfoundland, capelin spawn on beaches where they hurl themselves onto pebbled shores with such vigor that they are often left stranded, much to the delight of voracious seabirds. Here, too, they are sensitive to the slightest of temperature changes: a mere 2 °F warming of surface waters can shift capelin distribution by hundreds of kilometers.[18]

Capelin make up some 60 percent of the annual energy budget of cod in some areas, so any change in the distribution of the former is bound to affect the latter.[19] When capelin populations are booming, cod eat well and grow quickly; but when they are scarce, cod suffer from hunger pangs and turn to other foods. Herring, sprats, shrimp, and all manner of shellfish are gobbled up by ravenous herds. But famished cod resort to a feeding strategy that is taboo in all but the most extraordinary human cultures (just ask Captain Cook's crew): cannibalism.

In a normal year, as many as two-thirds of all cod juveniles (up to two years of age) who die at sea are gobbled up by their very own brothers and sisters.[20] For a young adult cod, cannibalism may supply just 1 percent of their stomach contents, but in older fish that value swells to 5 percent or even 10 percent.[21] As described by Bogstad, "they become more cannibalistic as they get big-ger. They can eat fish, including small cod, up to half their own length." He pauses to recall the prize-winning cannibal: "We found once a cod of 107 centimeters who had a cod of 62 centimeters in the stomach." The total amount eaten fluctuates with temperature and food availability. When capelin populations plummet, or shift their distribution, their scarcity prompts even more offspring-eating.[22] During one particularly bad year for Barents Sea capelin, the cannibalism rate multiplied threefold.[23]

Cod, like capelin, respond strongly to small changes in water temperature. They prefer cold waters, typically between 36 and 50 °F but can survive at 30 °F (sea water freezes at 28 °F, because of the salt).[24] Atlantic cod off Newfoundland, however, painstakingly avoid the sub-zero Labrador Current that creeps down the coast from the North Pole. Remarkably, they swim in waters that are beneath the current, but slightly warmer: differences in salinity float the colder Labrador stream atop Atlantic Ocean water that measures between 2 and 4 °F. According to Canadian cod researcher George Rose, the cod "don't live in that cold water, it's too cold for them. It can get down to minus 2 [Celsius, or 28 °F], which really hurts."[25] By contrast, he calls the Atlantic's 36–39 °F "very favorable for cod." Over on the other side of the Atlantic, cod are avoiding water that is slightly too balmy: fishery data reveal the North Sea population has shifted to deeper, colder waters as ocean surface temperatures have risen.

Cod themselves are an important food source for two polar mammals—minke whales (*Balaenoptera acutorostrata*) and harp seals (*Phoca groenlandica*)—who are also affected by capelin abundance. Harp seals, also known as Greenland seals, tuck into some 100,000 tons of cod when capelin are abundant (by comparison, Brits fry just three-quarters of this amount into fish-and-chips).[26] But when capelin are scarce, cod are the next-best option, and the seals' cod harvest triples.[27] Shifting between food sources adds further pressure to cod populations themselves reeling from a lack of capelin prey. This arctic food web can be self-stabilizing, though, if given a chance. Take the years 1984–1986, for example. Capelin populations nosedived because of overconsumption of their larvae by young herring and cod, perhaps exacerbated by shifting sea ice or poor years for plankton growth.[28] Because of the scarcity of everyone's favorite food, whales and seals started eating more cod, and cod themselves began cannibalizing their own offspring. All this predation forced cod populations to decline and average adult weights to fall by half.[29] The smaller cod population in turn could no longer eat as many capelin as in pre-1984 years. And the capelin schools, freed from heavy cod predation, rebounded when their waters grew more productive, soon returning to the prodigious numbers that could again sustain an entire marine ecosystem.

If cod larvae survive starvation for lack of copepods, and cannibalism by hungry adults, they grow into juveniles and begin to change their ways. Eschewing the open waters where they spent their infancy, juvenile cod may migrate to inshore waters along the Newfoundland coast and the shores of Iceland and Norway. There they comb the shallows in a concerted effort to put on pounds and grow into adulthood, an eating contest that takes about three years. Until they reach adult size, cod juveniles are still at risk of being attacked by any number of hungry marine animals. Finding food while avoiding risk is a delicate balancing act, one that juveniles approach with a strategy referred to by marine researchers as "hunt warm, rest cool."[30] When the sun rises, and light suffuses into the cold sea, cod juveniles begin to dive.[31] Swimming toward the bottom, they slip into dark waters where they cannot readily be seen, biding their time in the relative safety of these cool depths. Their metabolism also slows, an added energetic benefit. Later, as the sun skids below the horizon and darkness cloaks the surface, juveniles ascend to shallower and more productive waters. All night long they feed, gobbling crabs and shrimp larvae and smaller fishes. Here, in the shallows, the warmer surface water heats up their small bodies and makes them faster, stronger, and more agile as they hunt their prey. Soon the sun has spun around the world (with apologies to Copernicus), the sea brightens, and juvenile cod descend to safe darkness and meditative cold, where they await the next night's feast.

One of the great benefits of cold oceans is that they are rich in oxygen. Polar waters, in comparison to tropical seas, hold nearly twice the dissolved oxygen, at least at the surface. Nevertheless, a gallon jug filled with air contains about forty times the oxygen in an average gallon of seawater, so fishes must use less oxygen, and they must absorb it more efficiently. To achieve the former, cod are steadfastly cold blooded and sluggish, with delicate white fillets. For the latter, they rely on counter-current blood flow across their gills, and on a familiar molecule: hemoglobin. As in human blood, fish hemoglobin greedily seizes oxygen from the outside world (albeit from water sliding through the gills) and dutifully transports it to heart, brain, and muscles.

Not all hemoglobin molecules are created equal, however, and various forms differ in the temperature at which they bind oxygen most effectively. If the veins of a fish in a 40 °F sea throb with hemoglobin suited to 50 °F water, it will be outcompeted by better-oxygenated rivals with properly tuned blood chemistry. Juvenile Atlantic cod have the ability to express two distinct hemoglobin morphs within the same fish.[32] When tested in a laboratory, the HbI1 type achieves peak oxygen stickiness at temperatures above 57 °F. A second morph, poetically named HbI2, hits its oxygen-binding stride in water below 50 °F. Such polymorphism can be found in fish species that travel great distances as they grow from larva to adult, to ensure their blood is well suited for the temperatures found in both environments.[33] But cod do not undertake such expansive journeys. Instead, they seem to retain the genetic code for several hemoglobin types so the right form can be produced no matter where they are born. In the icy waters around Greenland, Norway, Iceland, and Canada, up to 99 percent of juvenile cod carry the cold-adapted hemoglobin HbI2. Juveniles swimming along Ireland and southern Norway, meanwhile, bear the warm-water HbI1 morph. Each population is precisely adapted to thrive, feed, and grow in its own thermal neighborhood.

To reach adulthood, juvenile cod—powered by their site-specific blood chemistry—must succeed in feeding themselves and avoiding predators. They head into deeper waters over the continental shelf, where they devour mid-water capelin and other prey. Here too, the effects of water temperature play an outsized role in their lives. When food is abundant, an adult cod stays in warmer waters (45 °F) and snacks on capelin within 100 feet of the surface. The modest increase in heat shifts its swimming muscles up a couple of gears, permitting the cod to pack on the pounds with more vigorous hunting. But when food is scarce, its warmed metabolism burns quickly through the gains just made; in response, the hungry cod will descend to depths below 250 feet, where the cold (39 °F) dampens its metabolism and spares hard-won fat stores.[34] This strategy looks like a slower-speed version of the "hunt warm, rest cool" approach of juveniles: gain weight quickly when the buffet line is booming, and limit weight losses when meals are scarce. Cod, in effect, are picking the temperature that maximizes their overall growth.

When adults have bulked up sufficiently, to a satisfying length of 24 inches or more, they are finally ready to mate. This is no small undertaking, given the astronomical number of eggs involved. A female cod must abandon growth and direct all her energy to producing roe, not to mention migrating to the spawning grounds. Cod from Norway and Iceland quit their deeper feeding grounds and swim more than 500 miles to mate in nearshore waters. In part, they are driven to find water conditions that will accelerate egg hatching: in water of 47 °F, eggs will hatch in ten days, but they require four additional days at 43 °F, and more than twenty days when the watery cradle is a frosty 38 °F.[35] The less time an egg is bobbing helplessly in the water column, the greater its chances of survival, so migrating before spawning can be well worth the cost. Cod who live in less frigid seas, like those around Ireland, enjoy warm water without lengthy migrations, and the energy saved may explain why Irish cod are able to mature and spawn earlier in life.

Spawning itself begins with courtship, in which males display their readiness to egg-bearing females (disturbingly referred to as "ripe" in literature of a certain age). Males make low-frequency grunting sounds and flex their fins while crossing the female's path. Populations from different parts of the world even have different dialects: cod from the Americas make repetitive banging sounds, while their European brothers favor a lusty growl.[36] If aroused by these vocal enticements, a female cod follows her singing suitor as he inverts himself, and they swim belly-to-belly while simultaneously releasing sperm and eggs. Cameras on remotely operated submersibles show cod packed so tightly that water can scarcely be seen between them. Amazing patterns emerge from the twisting tangles of mating fish. Acoustic imaging of massive spawning grounds off Newfoundland revealed enormous columns of cod, broad and dense pillars stacked side by side like chips on a poker table, each formed by thousands of fish.[37] When seen from the surface, these breeding columns can carpet the sea for miles.

Such abundance even now is breathtaking to behold. Several hundred years ago, before fishing took its toll, the spectacle must have been mesmerizing. Even the most highly respected scientists could mislead themselves into believing that cod represented an inexhaustible resource. One was no less a luminary than Thomas Henry Huxley, a brilliant biologist known as "Darwin's Bulldog" for vigorously championing his friend's theory of natural selection in stuffy Victorian England. Huxley firmly embraced the idea that the natural world was inherently balanced, and imperturbable. With regard to Canadian cod fisheries, which already were sending back whispers of declining catches, he made a fatal pronouncement. "I believe, then, that the cod fishery, the herring fishery, the pilchard fishery, the mackerel fishery, and probably all the great sea fisheries, are inexhaustible; that is to say, that nothing we do seriously affects the number of the fish."[38] A report prepared by the Canadian Ministry of Agriculture, went further: "I say it is impossible, not merely to exhaust them, but even noticeably to lessen their number by the means now used for their capture . . . For the last three hundred years fishing has gone on in the Gulf of St. Lawrence and along the coast of our Maritime Provinces, and although enormous quantities of fish have been caught, there are no indications of exhaustion."[39] The two could not have been more wrong.

CRASH AND BURN: A TALE OF TWO FISHERIES

> Any attempt to regulate these fisheries seems
> consequently, from the nature of the case, to be useless.
> —Thomas Henry Huxley, *Inaugural Address to the
> London Fisheries Exhibition*

More than any other, cod is the fish that built the New World. Dried cod fed the Vikings on epic journeys to Greenland and Canada over a thousand years ago. Salted cod were a mainstay of the first-century Basque economy, although they had to dodge the ruthless Hanseatic League in northern European waters. At some point in the mid-1600s, Basque fishing boats established secret routes across the Atlantic and began exploiting fantastically rich cod fishing grounds on the doorstep of North America. A century later, when the French explorer Jacques Cartier became the first European to map the St. Lawrence seaway, he found more than a thousand Basque boats anchored there; unimpressed, he claimed the region for France. After the *Mayflower* pilgrims arrived, a hundred years after Cartier's declaration, they survived on the generosity of the continent's native peoples but also on the vast abundance of cod. Great fortunes were later amassed in what is now Newfoundland and Massachusetts, where treasured cod was commemorated on stamps and even currency. To this day in Boston, a lifelike cod sculpted of wood hangs in the State House of Representatives and casts an unblinking eye over the (only infrequently fishy) politics of that august chamber.

Early fishing boats in New World waters used lines and small nets suited to shallow seas, and they stayed near shore where young cod and columns of spawners could be fished. But things began to change with the industrial revolution. Coal and steam engines permitted boats to be larger, and they powered winches that could haul massive trawling nets. Trawls are either weighted to reach the bottom and dragged along on wheel-like bobbins, or pulled through the open ocean if the trawler's captain has located a midwater school. Motorized vessels could also travel much farther in search of their catch. Prior to the application of steam, cod in New England were harvested by picturesque schooners from Gloucester and elsewhere. Schooners were notoriously unstable, however, and accidents were legion: in one year more than 1600 Gloucester hands were lost at sea.[40] Once steam trawlers with their larger nets and indefatigable power began landing sixfold more fish than traditional boats, the iconic schooners were hastily retired.

As early as the 1890s, still in the opulent dawn of the steam-trawler era, reports began to surface warning of declining cod stocks. Huxley's optimistic pronouncements notwithstanding, the fact was that the fishing fleet could now catch cod faster than they could be replaced. In response to shrinking catches, the fleet scoured all known cod refuges, including the fabled Georges Bank. There, a shallow underwater platform extends out from Cape Cod, in places so near the surface that waves break upon its rocks at low tide. Georges Bank yielded some of the largest individual cod the world had ever seen, including a goliath measuring 6 feet in length and weighing more than 200 pounds. It beggars the imagination to think this giant was caught on a hand line. But as trawlers dragged nets repeatedly over the Bank and other productive cod grounds, they savaged the seafloor. Each net, as it rolled over the bottom, uprooted kelp, broke apart rock refuges, and inexorably, irreversibly destroyed the habitat that nurtured the great shoals of cod. Governments and ship owners refused to accept what was happening in front of their eyes: they were emptying the sea. From time to time, conditions would align, and huge catches would be landed. Everyone went on buying more and larger gear, and catching more fish, certain that the good times had returned. As Ralph Mayo, a US Fisheries Service biologist, remarked about their willful blindness: "You see some cod and assume this is the tip of the iceberg. But it could be the whole iceberg."[41]

Dr. George Rose, who commented earlier about cod's preference for cool waters, circles the decade of the 1970s as the end of cod in Canada. By that time fishing vessels from the collapsed Soviet Union and other postwar nations were "starved for protein" and desperate for new sources.[42] They settled on the rich cod grounds off Newfoundland, forming by night a "city of lights of trawlers and mother ships and all the rest, just sitting over the overwintering and spawning grounds." Previously, local fishing pressure had been constrained by what Rose wryly refers to as "berth control": the number of boats was limited by the number of harbor slips where they could tie up, so the fleet never got overly large. But the new ships were massive enough to stay at sea for days on end, fishing with gigantic nets and without respite, and never returning to a harbor. During the heyday of the 1960s, Newfoundland waters were yielding well over a million pounds of cod each year.[43] But in the face of this industrialized onslaught, the fish that built the New World did not stand a chance.

Echoing decisions taken by Iceland, Canada in 1976 declared national control of waters within 200 miles of its coast, effectively halting foreign fishing. This move might have reduced fishing pressure on cod stocks and sidestepped the looming disaster. Heavily subsidized domestic fleets, however, quickly filled the void and continued the prodigious harvest levels. Lamentably for the fish, and the communities that relied on them, those 1960s harvests represented the high-water mark of the fishery. Just a decade later, catch levels had fallen below half their peak levels, and they continued to sink like a stone at sea. By 1992, the northern cod had plummeted to just 1 percent of their historical levels.[44] By any measure, after hundreds of years of bountiful harvests, the cod fishery had utterly collapsed. In the summer of that year the Canadian government issued a complete moratorium, ending a 600-year-old cod fishery, and the careers of some 30,000 fishers.[45]

It did not have to be this way. As painful as catch reductions are for fishers, they should be less difficult to swallow than closure of an entire fishery. Overfishing and stock declines had happened elsewhere, but none were as catastrophic as in Newfoundland. In the 1980s cod populations nosedived in the Barents Sea, around the same time Bjarte Bogstad began studying the stock. "Cod is the most valuable fishery in Norway," he states succinctly, "but we were definitely overfishing it. There was a situation in the late 80s, things were going quite bad, a combination of, say, climate issues and very hard fishing." Faced with mounting evidence of a pending collapse, wise heads prevailed, and fishing communities swallowed the bitter pill. "Quite strong measures by the managers were taken [Norway and Russia manage the stock jointly], and the stock recovered quickly." Bogstad seems pleased with the way the fishery is now managed, and his role in the process. "The last decade or so we'd gotten all those things in place, and also had a bit of help from mother nature . . . and it seems like it is stabilizing at a good healthy level. And that also means that if we fish it more lightly, then there are more fish that get older or bigger, so the average size of the fish caught gets bigger ['big-ger']. So when you fish more lightly then it's more stable."

Today, Bogstad and his fellow Norwegian scientists work diligently to detect declines in cod populations and rein in fishing pressure before the industry gets walloped by mother nature. To predict the future, they rely on harvest data, acoustical soundings, remote vehicle surveys, and advanced computer models. But they also rely on a humble instrument found in every home in Norway: a thermometer. Oceanographic research spearheaded by Bogstad revealed that cod responses to shifting temperature can be delayed by years, a startling discovery that permits long-range modeling of future cod abundance. "The temperature of currents in, say, south Norway or the Shetlands would translate into warm or cold conditions in the Barents Sea a few years later, which would then affect how the cod was doing," he explains. "Then you get a different prediction horizon. If you can say that now it's cold here, [you know] it will be cold up there in three, four years and there will be less cod in six years." With disarming modesty, he notes that "expanding the prediction horizon is new." New indeed, and extraordinarily valuable to one of the most important industries in Norway. If water temperatures today can help predict cod abundance years in the future, then quota reductions that trim the fishing fleet's boats and crews can be planned well in advance and implemented carefully to minimize the hardship they might otherwise cause. Paying close attention to small shifts in temperature, it seems, can yield huge dividends for fish abundance and local livelihoods.

FILLETED, FROZEN, FLOWN, AND FRIED

In 1922, fifty years before Canada shuttered Newfoundland's cod fishery, an American inventor established a pioneering seafood company that would revolutionize the industry.[46] He was struck by the twinkle of an idea while working as a trapper in Labrador, where he marveled at sub-zero temperatures that flash-froze fish which later tasted as fresh as the day they were boated. "The first winter, I saw natives catching fish in fifty below zero weather, which froze stiff as soon as they were taken out of the water. Months later, when they were thawed out, some of these fish were still alive."[47] A native of New York, he soon realized this method could deliver fresh-tasting fish to the 6 million people living in his city, from as far away as northeastern Canada. His name was Clarence Birdseye, and Birdseye Seafoods would change the way the world ate fish. Soon the company, reconstituted as General Foods, would ship millions of pounds of frozen fish fillets—as well as berries, meat, and vegetables—across America, and around the planet. Now all they needed was a reliable source of fish, and they found it in haddock.

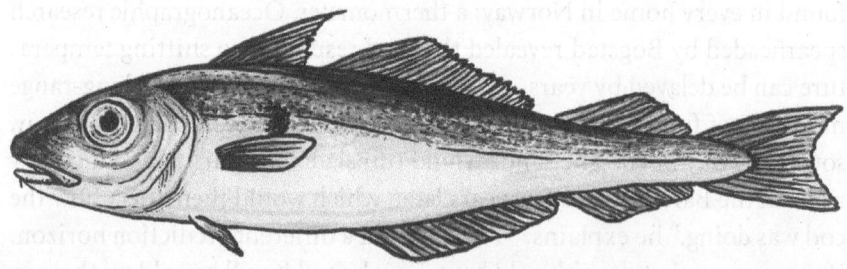

Haddock (*Melanogrammus aeglefinus*)

Haddock (*Melanogrammus aeglefinus*) is a close cousin of the Atlantic cod, and similar in many respects. Found from the Barents Sea to Iceland to Newfoundland, they overlap in depths and diets, though a few modest differences distinguish the two species. Haddock are much smaller, rarely reaching 3 feet in length, and are not as deep bodied: they look positively svelte next to a cod's barrel-belly shape. The lateral line, pearly white on a cod's flank, is replaced with a dark brown stripe in haddock. Between this line and the pectoral fin, they are marked with a distinctive dark blotch, like a thumbprint. Notably, haddock meat is more pinkish-tan than brilliant alabaster and has a slightly more fishy flavor, both traits making it decidedly secondary to cod as an eating fish. With the plunging decline in cod populations, however, fisheries turned to haddock. In England, the newcomer gradually replaced cod as the top seller in fish and chip shops. And after Clarence Birdseye's company perfected fish freezing, the entire world would become acquainted with a new and wholly modern seafood: the fish stick.

Food fishes must be able to reproduce at a rate high enough to sustain their fishery, and in this regard, haddock absolutely excel. A large female can release up to 3 million eggs.[48] She lays them in batches on the ocean bottom, a few hundred feet below the surface, where they are fertilized by males. The eggs become buoyant thanks to oil droplets studding their exterior and float to the surface, where they bob vulnerably for a week or two until hatching. As with cod, vast numbers of floating eggs can be lost to predators large and small. The developing eggs also are extremely sensitive to pollution. Industrial chemicals from mines and other sources can stick to the eggs and make them sink, forcing larvae to hatch earlier, in fewer numbers, and suffer higher mortality once out of the egg.[49] Studies of cod larvae have shown the ocean's rising acidity, a by-product of global climate change, also negatively impacts their early organ development and overall survival.[50] Oil spills, increasingly common in northern seas, further dampen the success of haddock reproduction. When floating eggs are exposed even briefly to raw oil, the larvae suffer skeletal deformations and damage to their heart that can be fatal.[51] Here the natural buoyancy of the eggs works against them, as the raw oil droplets adhere to the gummy egg surface, so even a brief exposure to petroleum can lead to sustained exposure and danger.

Despite the risks, natural and human-induced, massive numbers of haddock larvae hatch on the high seas and develop into juveniles. They gradually make their way toward the continental shelf, where they feed as they grow into adulthood. In this they are aided by a built-in compass. Ingenious experiments carried out bravely in the North Sea showed that haddock larvae all oriented themselves to the northwest: when a floating chamber experimentally shifted the magnetic poles, all the larvae in the chamber turned accordingly to the new northwest.[52] Their compass-following ability may help the larvae steer into favorable currents that transport them to their nursery grounds, northern seas where developing juveniles ravenously scour the rocky bottom for the clams and mussels, marine worms, sea stars, urchins, and brittle stars. When the myriad factors that affect hatching success, larval survival, and juvenile growth align favorably, haddock populations boom; when links in this reproductive chain are weakened, a crash can follow. Large fluctuations in haddock abundance are the result, which can span years or decades. Fishery catch data revealed colossal haddock schools in the 1920s plummeted just twenty years later, only to rebound dramatically by the 1950s.[53] For fishes like cod and haddock, conditions in the surface waters as well as on the bottom both affect the success of reproduction.

Haddock and their relatives are known to fishing boats as groundfish: they spend a substantial portion of their lives near the seafloor, where they forage and grow, court, and breed. Marine biologists call these fishes demersals, a term that encompasses rays and flounders, but also some of the world's most important food fishes. Trawl nets hauled across the seafloor are designed to catch groundfish precisely because of their inclination for the bottom. Haddock are not found at great depths, typically less than 700 feet. When groundfishes like cod and haddock were fished with hand lines, it was nearly impossible to make a significant dent in their populations. But once trawlers began dragging the seafloor, the sluggish behavior of these animals meant one boat alone could scoop up thousands in a single net. Remote cameras reveal that cod will swim vigorously to escape an onrushing trawl but tire within a matter of seconds; once their pace flags, they are swallowed by the mouth of the net.

Cod and haddock may be closely related, but they are not free from sibling rivalry. Their distributions largely overlap, although haddock prefer more northern waters. Where they occur together, they compete. Larvae of both species share the same feeding habits of gulping down copepods, particularly just before sunset.[54] Like siblings grabbing cookies from the same jar, the more voracious the larvae of one species, the less food there is for the other. Cod tend to take the lion's share of copepods, often forcing larval haddock to seek other food. In adulthood, that rivalry can be even more extreme. As Barents Sea cod grow larger, they graduate from capelin to demersal fishes, and their proclivities range from cannibalism to familial homicide: haddock can make up an astonishing quarter of the daily diet of large cod.[55] To some extent, a good year for cod is a poor year for haddock, and vice versa. The same applies to the effects of seawater temperature. When the surface waters of northern oceans warm, cod populations swell, but at the expense of haddock, which decline in numbers.[56] In the northwest Atlantic, however, neither population was able to resist the intense fishing pressure unleased in Newfoundland and New England: from a peak in the 1990s of more than 2 million tons landed every year, haddock and cod numbers nosedived to just 5 percent of those record harvests. In the case of cod, they have yet to recover. In the case of haddock, Clarence Birdseye's frozen fish empire had to look elsewhere, and they turned to pollock.

Several fish in the sea are known as pollock, though all are members of cod's clan. Haddock, whiting, hake, cod, and all pollock species are in a family (Gadidae) so important to global fisheries that its total landings are second only to the family of sardines, herring, and anchovies. The annual catch of Alaskan pollock (*Gadus chalcogramma*, also known as walleye pollock) from the Bering Sea alone routinely exceeds 3 million tons, bested only by 6 million tons of Peruvian anchovies.[57] True pollock is another fish so nice they named it twice, *Pollachius pollachius*, but its aliases include Atlantic pollock, European pollock, lythe, or even pollack. A close relative is *Pollachius virens*, itself known as a coley, coalfish, saithe, or even Boston blue. All three species (with their ten names!) resemble small haddock, with more or less speckling, and more or less pronounced lateral lines. They lack the haddock's distinctive dark thumbprint but have a similar underslung jaw, triple dorsal fins, and moderately deep belly. They grow quickly, reaching reproductive age in as little as three years, and lay huge clutches of eggs. On the high seas, pollock larvae pursue their own version of "hunt warm, rest cool" to accelerate their growth. When food is abundant, they stick to warmer surface waters and feed voraciously; when plankton abundance dips, however, they descend into colder depths where the icy chill slows their metabolism until plankton blooms return.[58] The result is a fast-growing fish that multiplies rapidly and plentifully.

Initially passed over by fisheries in favor of cod, pollock's stupendous powers of reproduction have made them a prime target of late. Their flaky, white meat is more strongly flavored than cod but still makes for excellent eating. Now one of the world's largest fisheries, total North Pacific pollock harvests can reach 7 million tons yearly, valued in the hundreds of millions of dollars.[59] The catch from the Bering Sea represents one-fifth of all fishery landings in the United States.[60] Today, most of the fish sticks, fish fingers, frozen fillets, and fast-food fish sandwiches sold are pollock, while imitation crab meat and minced fishmeal known as surimi are made from processed pollock. Their roe and rendered fish oil are widely sold. Medical uses for pollock have been investigated as well, including a promising adhesive spray manufactured from pollock gelatin that helps seal punctures in damaged lungs of people suffering from emphysema.[61]

Alaskan fisheries targeting wild-caught pollock are some of the best managed in the world. In the past, fishing boats could operate within a fixed season, a now-obsolete mode of stock management that inevitably provoked what is known as a "race to fish." In this mad dash, boats and crew work round the clock, regardless of risk or weather, to land as many fish as possible during the brief season. Pollock landings exceeded the target levels year after year, and the stock declined steeply in the face of repeated overfishing. Boat crews also suffered terrible rates of injury and even death as they labored in staggeringly dangerous conditions, hauling nets and gutting fish while icy walls of water smashed over slippery decks. Finally, a new policy of "catch shares" was implemented, in which a fixed amount of fish was awarded to each boat regardless of the amount of time taken to land it. Injuries and overfishing dropped almost immediately, and today the pollock stock—and the fishery it sustains—is well on its way to recovery.

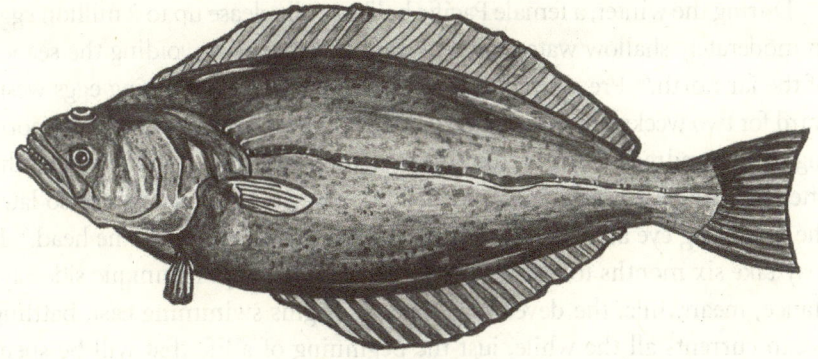

Pacific halibut (*Hippoglossus stenolepis*)

A HORSE'S TONGUE AS BIG AS A BARN DOOR

> He knew that at 50 below zero water from the mouth
> made a noise when it hit the snow. But this had done that
> in the air.
> —**Jack London, *To Build A Fire***

One of the finest examples of successful fishery management is the governance provided by the International Pacific Halibut Commission, which implemented a catch share system in 1995 for this gigantic fish. Pacific halibut is one of the largest fishes in the sea, with record-setters over 8 feet long sagging the scales at an incredible 500 pounds.[62] Such giants are dubbed "barn doors" for good reason, and some of these leviathans can exceed fifty years of age. Halibut are closely related to tropical flounders, those sideways-turned flatfishes with one eye that migrates across their head. Specifically, halibut are known as right-eye flounders, since the migrating eye ends up on the right side, which turns dark and faces away from the bottom. Occasionally this genetic code develops a bug, however, and anywhere from one in 20,000 Pacific halibut to an estimated 40 percent of California halibut (*Paralichthys californicus*) develop in reverse, with the migrating eye moving to the left side.[63] Naming of these flatfishes also develops a coding bug, with appellations like sole and flounder and dab being applied in restaurants with little regard for their scientific identity. Regardless of their name tag, most flatfishes spend their lives skimming the bottom of frigid seas, top predators on crustaceans, mollusks, and unsuspecting fishes.

During the winter, a female Pacific halibut will release up to 2 million eggs in moderately shallow waters over the continental shelf, avoiding the sea ice of the far north.[64] Prevailing currents, however, drag the floating eggs westward for two weeks until the larvae hatch, an aquatic conveyor belt transporting them hundreds of miles from the spawning grounds. About two months after hatching each larva has absorbed its yolk sac, and a month or so later the migrating eye arrives at its destination on the other side of the head.[65] It may take six months to a year for the fish to fully adopt its unique sideways stance; meanwhile, the developing juvenile begins swimming east, battling ocean currents all the while, just the beginning of a life that will be spent in perpetual motion. Despite its small size, the juvenile can log up to 2000 miles along the edge of the Aleutian Islands chain, and as far as northern Oregon.[66] An Atlantic halibut (*Hippoglossus hippoglossus*) can readily journey from Newfoundland all the way to the coast of Greenland. Along the way, the young halibut devours crabs and shrimp scooped from the bottom with its sideways mouth; by the time it reaches a couple feet in length, it will add juvenile pollock to its diet, as well as herring, capelin, cod, rockfish, and even octopus. Each winter, as sea ice advances and water temperatures plunge, young and old halibut alike move out into deeper, warmer water away from the continental shelf. Other flatfishes, like the winter flounder (*Pseudopleuronectes americanus*), are better protected in winter thanks to an antifreeze protein that courses through their blood.[67] But halibut, who lack such antifreeze, must flee the shallows when bottom temperatures approach freezing. In spring they stage a homecoming toward the shelf, re-dispersing over the flats that make for rich feeding grounds.

Halibut grow at a deliberate, unhurried pace. Juveniles spend three or four years mostly apart from adults, favoring shallow areas (less than 200 feet deep) with sandy bottoms that are well-defined nursery grounds. Around the age of five they can be considered young adults, though a few more years must pass before they even dream of mating. Males reach sexual maturity around eight years of age, while females do not begin reproducing until a full twelve years, practically the entire life span of an Alaskan pollock.[68] Given their leisurely development, halibut populations grow very slowly, a characteristic that mandates careful management of their fisheries. If too many are taken in a given year, the stock may require a decade or more to recover.

Commercial halibut fishing in the Pacific began in the 1880s, when the newly opened Northern Pacific Railway began shipping west coast fish to America's populous east coast. Commerce overwhelmed judgment, and landings virtually collapsed just thirty years later, prompting Canada and the United States to join forces in managing the stock. The world's first international effort to govern a fishery culminated in 1923, when the Pacific Halibut Commission was inaugurated, one year after Birdseye Seafoods was founded.[69] Careful consideration of stock size data, strict regulation of fishing effort, closure of nursery grounds, and the implementation of a catch shares program eventually stabilized the fishery and restored the population. Today, fishing for halibut is safer than ever, catch levels are sustainable, and prices are steady and high.

As in other groundfishes, the meat of a halibut is white and flaky, their muscles equally adapted for slow and steady swimming rather than high-speed dashes. Compared with a cod or pollock, however, the fillets are larger and more steak-like, reflecting the huge size of the fish and the added power needed to push a body of several hundred pounds through the water. Strong control of fishery landings combined with exceptional quality keeps prices high, and for the most part halibut is destined for affluent shoppers and expensive restaurants. Innovative products are being developed for the parts discarded by the fishery, however, to maximize the use and value of these majestic fishes. In Brazil, researchers at the University of Sao Paolo's Food Engineering Department invented a technique for converting surplus Atlantic halibut skin gelatin into an edible film. Their goal is to replace the flimsy plastic used for shopping bags with a bio-polymer that will degrade safely rather than polluting the environment.[70] Given that plastic bags washing into the sea pose a dreadful threat for marine life, from whales to basking sharks to sea turtles, a biodegradable alternative would be a triumph for materials science and the planet.

Fifty years ago another engineering triumph helped propel pollock and haddock into the kind of widespread popularity that slow-growing halibut cannot match. Because pollock were frozen at sea in great blocks, the unwieldy fish bergs could not be separated into fillets. Some ingenious engineer realized that the frozen blocks could be sawed into narrow, rectangular bars of flaky, white fish meat. Employees of Birds Eye Foods were asked to name the new product, and "fish fingers" were introduced to the world. In England, where fish consumption had been on the decline for years, towns that relied on fishing were revitalized by the new frozen food.[71] Fish fingers were launched in 1955, with the slogan "no bones, no waste, no smell, no fuss," and 600 tons were sold in their very first year.[72] Processing factories opened in the port towns of Grimsby and Hull, and jobs turning fish into fingers helped offset the collapse of employment caused by the cod wars. Today, as many as three-quarters of Brits get their first taste of seafood from a fish stick made from pollock, haddock, or even the venerated cod. Encouragingly, nearly all the fish that goes into these frozen delights is harvested sustainably, guaranteeing fish fingers will provide crunchy and flaky joy to hungry diners, and decent employment in traditional fishing towns, for years to come.[73]

8 INTO THE ABYSS

Barreleyes, Tripodfish, and More Deepwater Oddities

AUGUSTE PICCARD WAS A MAN DRIVEN TO EXTREMES. In May of 1931 the Swiss physicist became the first human to enter Earth's stratosphere, lofted in a pressurized gondola of his own invention.[1] The steel sphere, suspended from a hydrogen balloon, carried him to a record-setting altitude of 51,775 feet, nearly 10 miles above the planet's surface. From that height, he and his copilot peered through the gondola's porthole to observe something nobody had witnessed before: the actual curvature of the Earth. A few years later he realized that his design, resistant as it was to extremes of pressure, could be adapted to explore the sea, and he set his sights on a new invention. He dreamed that he could "float in the ocean depths just as I have traveled the stratosphere in a balloon."[2] Piccard drafted plans for a steel bathyscaphe that would carry explorers deep into the ocean using iron ballast to descend and tanks of gasoline (buoyant because of the liquid's low density) to float back to the surface. Shaped like a submarine, the *Trieste* packed the ballast and gasoline into a large cylindrical hull, while a small sphere mimicking his flying gondola was suspended beneath to house the crew. Off the coast of Naples in 1953, Piccard glided to a depth of 10,390 feet below sea level, another record.[3] But that was child's play compared with a new extreme his diving invention would soon attain, one that could never be exceeded.

Towed from Guam to a precise location in the Western Pacific on January 23, 1960, the *Trieste* slipped beneath the waves. Her pilot was none other than Auguste Jacques Piccard, son of the craft's inventor. The ocean was exceedingly rough, and many doubted they could successfully launch the mission. But once below the whitecaps, Jacques and his companion Don Walsh found still waters and began their descent. Their destination was the bottom of the sea, the deepest place on the planet, Challenger Deep canyon within the Mariana Trench. As they descended a guide rope to 500 feet, the sun's light began to fade, and they entered the ocean's "twilight zone," in their own words.[4] Piccard recorded his observations on what was called a Dictabelt. "With a big mercury searchlight we saw the water outside blue, clear, as usual . . . I could absolutely not see the beam of the light."[5] In earlier dives, they had witnessed delicate particles of organic matter that originated on the surface, which gave the impression of "numerous scattering particles streaming past the porthole like an upside down, very light snowfall."[6]

Disconcertingly, the aquanauts soon heard a popping sound, and a strong shudder rumbled through the ship: intense water pressure had cracked the glass on an exit window. Piccard commented coolly, "when we discovered that the window had been cracked I realized that it could eventually give some trouble, eventually even some terribly big trouble, to get out of the sphere."[7] Walsh was more succinct, describing the moment as "pretty hairy."[8] Undeterred, the two plunged on. Some five hours later, Jacques reached the bottom, his father's invention working flawlessly, and scrutinized the depth gauge: "At exactly 1:06 pm and the depth of 6300 fathoms read on our record gauge, we lightly, extremely lightly, touched the bottom of the trench."[9] Piccard and Walsh were resting on the seafloor at a depth of some 35,800 feet, nearly 7 miles beneath the surface.

Once nestled on the soft mud, Piccard reported that "by wonderful chance I could see a fish which I would call a sole . . . a kind of flat fish about one foot long, absolutely white with some part of which would be called silver."[10] It remains a mystery why the two explorers would invent such a sighting; it has long since been dispelled as a myth, since fishes cannot survive at such great depths.[11] In fact, the submersible's arrival stirred up the fine sediment so thoroughly that the two were engulfed in a dense, silty cloud that obscured their vision. Underwater pioneer William Beebe, who descended to 3000 feet in a cramped steel sphere around the time the elder Piccard was flying his balloon, had better luck with both vision and fishes. Beebe was the first in the deep sea to observe bioluminescence, the bluish light emitted by dark-water organisms, marveling that everywhere he looked, "some brilliant, animated comet or constellation would flash across the small arc of my submarine heaven."[12] During his dives in the

1930s, Beebe reckoned that two-thirds of the fish species were luminous, and nine out of every ten individuals emitted a glow.[13]

Back at the bottom of the ocean, the *Trieste* kept its crew safe from almost unimaginable pressure, the weight of water over their heads squeezing the spherical chamber with 16,000 pounds per square inch. Near-freezing water turned the sphere into an icebox, chilling the two men, but Piccard had learned to use the craft's CO_2 filters for warmth, briefly stuffing spent filter bottles "under our pullovers, and these gave a good heat which was extremely comfortable."[14] Apprehensive about the cracked windowpane, the two wisely cut short their stay on the bottom and departed after just eighteen minutes. On their way to the surface, the pair spied a few jellyfish out the fractured window, each glowing with a tiny point of bioluminescent light; otherwise, the sea was empty. After an ascent of just three and a half hours, Piccard and Walsh safely reached the support ship above, and were greeted as heroes. They were the first people ever to have reached the deepest place on Earth. Nobody would successfully return to the Challenger Deep for another sixty-two years.

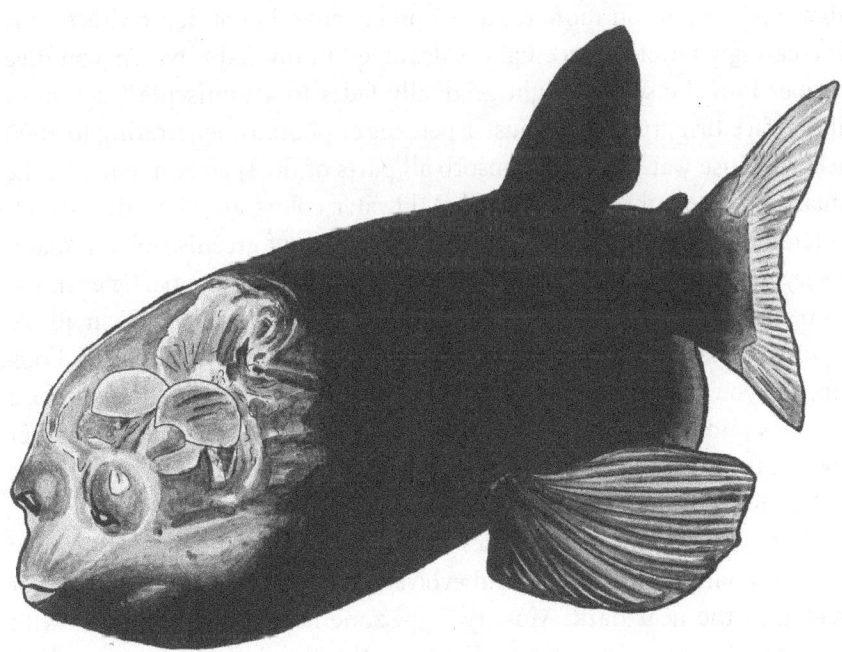

Barreleye (*Macropinna microstoma*)

THE FOUR-EYED MAN IS KING

> Darkness there and nothing more.
> Deep into that darkness peering, long I stood there, wondering, fearing
> Doubting, dreaming dreams no mortal ever dared to dream before.
> —Edgar Allen Poe, "The Raven"

Only a thin layer of the ocean enjoys the kiss of sunlight, as Jacques Piccard saw firsthand. Depending on water clarity, daylight penetrates to only 600 feet or so; faint twilight reigns to about 3000 feet, beneath which the ocean is enfolded in total darkness. Between those two depths the temperature plummets too, the sun's warmth fading rapidly. From there to the bottom, a thermometer will register just a few degrees above freezing, around 36 °F. In one of the sea's great contrasts, the sunlit layer is home to approximately nine-tenths of all marine life, and yet comprises a scant 5 percent of the ocean's volume.[15] Hunters in this thin layer of abundance rely on vision more than any other sense, but at depth this tool is increasingly blunted since light is degraded in myriad ways. As you dive deeper into the sea, sunlight gradually fades to a miniscule fraction of its surface brightness, with just 1 percent of photons penetrating to 1000 feet. Because water does not absorb all parts of the spectrum equally, the quality of the light shifts as well. Midwater colors are skewed: reds are filtered first, then yellow, and soon only shades of greenish-blue remain. Everyone at depth has a case of the blues. And as minute particles in the upper ocean absorb any photons scattering from side to side, soon all the light that reaches deep waters arrives only from directly overhead. Look up, and you can just see a glimmer of the sky, and perhaps a silhouette of a fish or squid; look sideways or down, and all is inky black. The only other light at these depths is bioluminescence emitted by fishes and shrimps and jellies that glow like pale blue fireflies.

In response to the peculiarities of ocean twilight, fishes have developed phenomenally sensitive eyes and evolved downright bizarre solutions for seeing in the near-dark. Most twilight zone fishes have large eyes with enormous pupils to admit more light, like the descriptively named owlfish (*Pseudobathylagus milleri*), just 6 inches long with an eye nearly an inch across. Fish peering into near-darkness have extraordinarily sensitive retinas, with large light-sensing rods for capturing scarce photons. Because larger rods reduce acuity—the ability to see fine detail—in some fishes the retinas feature discrete areas packed with smaller, tightly grouped rods, like a built-in magnifying glass.[16] All those detectors are tuned to the blue-grey color of downwelling light, though some midwater denizens like dragonfishes add an extra rod type capable of perceiving shortwave radiation that also penetrates the gloom.[17] Silver spinyfin eyes (*Diretmus argenteus*) possess as many as thirty-eight rod pigment types, compared with just one in terrestrial animals, to better capture the deepwater spectrum.[18] In many midwater fishes, the rods are arranged in layered banks to increase their sensitivity to dim light. In the eyes of escolar (*Lepidocybium flavobrunneum*), up to six banks are packed onto the retina, yielding an astounding density of 1.5 billion photosensors per square inch (25 times more than today's most sophisticated digital cameras).[19] Those rods deliver signals to a much smaller number of nerve endings, a phenomenon known as spatial summation that drastically increases the retina's sensitivity. And if a photon should manage to pass all those rods without being detected, twilight fishes like the escolar and many others possess a *tapetum lucidum* mirror behind the retina to bounce the strays back for a second viewing.

All these adaptations allow fishes of the twilight to hunt in the dimmest of light. They improve their chances by staring up into downwelling surface rays to catch a glimpse of a prey's silhouette. The eyes of many midwater fishes are located atop their heads; like perpetual optimists they are forever looking up. One of the most bizarre of these sky-watchers is the barreleye (*Macropinna microstoma*), a bulbous and slow-moving fish who appears to have a large, smooth forehead and tiny, puckered eyes. On closer inspection, the "eyes" turn out to be nostrils, connected to a finely tuned olfactory system. The true eyes, barrel shaped marvels of engineering, are lodged within the fish's forehead. No lids betray their presence, and no pupils are externally visible. Instead, these tubular eyes peer through a forehead composed of utterly transparent tissue, like twin telescopes aimed through a glass skylight. Broad light-capturing lenses are permitted by the tubular design, much larger than if the eyes were spherical, and the double-barrel design provides excellent depth perception. They can even shift their eyes, twisting the tubes inside their foreheads until they can see straight ahead.[20] This orientation helps them detect predators and also attack their favorite prey, ribbon-like jellyfish relatives known as siphono-phores. Looking up, the barreleye can detect a siphonophore's silhouette, but when it closes in to feed on the tentacles, the forward view helps them pick delicately at their meal to avoid being stung.

Other fishes have tubular and upward-pointing eyes, like the appropri-ately named telescopefish (*Dolichopteryx longipes*), but these creatures have solved the problem of forward vision in an entirely different way. Their eyes protrude slightly from the head and are fixed in position, unable to rotate forward. Instead, to detect the faint blue bioluminescence of prey and predators swimming in front of them, telescopefish possess an extra lens on the side of each squat, tubular eye. Light from ahead of the fish passes through this lens, then bounces off an internal mirror to strike a secondary retina draped on the opposite wall: the eye can see above and in front of the fish simultaneously.[21] Perhaps these marine marvels should instead be called periscopefish. The mirror is unique, composed of stacks of silvery guanine crystals arranged almost like a lens, so the reflection produces a well-focused image. Such eyes were the first ever discovered in vertebrates, an entirely novel visual system that can form images with a mirror.

Fishes that can be seen by predators lurking below them are under powerful pressure to hide their silhouettes. Numerous twilight zone fishes have evolved light-emitting photophores on their belly that glow with a color and intensity matched precisely to the light filtering from overhead, obscuring their outline when viewed from below. Photophore brightness must be dynamic as well. When radiant lanternfishes dive into darker waters they dim their underbelly glow; as they ascend, nerves connecting eyes to photophores increase the brightness, rendering them practically invisible at all depths.[22] So widespread is this adaptation that it can be found in more than forty families and hundreds of species of fishes, not to mention crustaceans, squid, jellyfish, and myriad other denizens of dimly lit depths.[23] Remotely operated submersibles conducting deep-sea sur-veys with highly sensitive cameras found an astounding three-quarters of 350,000 animals displayed bioluminescence (not far from Beebe's infor-mal estimate).[24] In some species, photophores produce light through a chemical reaction nearly identical to fireflies: a protein and an enzyme, devilishly called luciferin and luciferase, combine like the two liquids in a glow stick, and energy is emitted in the form of light. Others rely on sym-biotic bacteria to produce the light, and their photophores are adapted as chambers to keep these helpful bacteria alive and aglow.

A special case of bioluminescence gives lantern sharks their radiant nickname. These diminutive deepwater sharks, distant relatives of great whites and hammerheads, are attractively dappled with blue photophores, the proverbial wicks making their namesake lanterns glow. In one family, prosaically dubbed the "dwarf mesopelagic sharks," photophores are restricted to the belly and their brightness fades around the sides, a phenomenon known as counter-shading or better still, counter-illumination.[25] When seen from below against overhead light, the adaptation produces a perfect camouflage for this pelagic, midwater family. A second family of lantern sharks contains mostly demersal, or bottom-swimming species, for whom counter-illumination is unnecessary. Instead, an indigo glow is employed by amorous suiters to signal in the darkness for prospective mates. Photophores are arrayed in complex and species-specific patterns, in patches on the flanks, eyelids, and tail. Receptive partners search the blue light district for a familiar pattern of lights and slide in quietly to make an acquaintance. A couple of these shark species even use lamp light to advertise poisonous spines in their dorsal fins, highlighting them with a few strategically placed glowing spots.[26]

Unfortunately, illuminated lotharios also attract the attention of hungry predators, lurking wide-eyed in the gloom. In response, many bioluminescent fishes gamble only on intermittent flashes of light, timed to minimize their risk of being eaten but maximize the chance of finding a mate. Researchers assessing fish swimming speeds and mate densities estimate that a cautiously flashing female can find a receptive male in as little as two to four hours, but still these are risky hours, and a cat-and-mouse game of evolution and counter-evolution is underway.[27] Predators have harnessed photophores themselves, to shine like a spotlight and illuminate prey, turning their own tricks against them. Not to be outdone, prey developed darker and darker skin: deep-sea fishes are among the blackest animals on the planet, achieved through tiny melanin crystals arranged on their skin so as to side-scatter any incoming light.[28] At the very bottom of the ocean, many fishes have red-tinted skin that appears black when struck by blue light, because the skin has nary a particle of blue to reflect the beam. But in a final twist, dragonfishes (*Malacosteus*) evolved a red "sniper scope" to outwit these rosy-skinned fishes. Beamed from special photophores beneath their eyes, the scope's red rays are tuned to reflect off the crimson-tinted fishes and light them up as brightly as a drive-in movie screen.[29]

When light fails to safely attract mates, fishes turn to perfume. Many species, such as the twilight zone–dwelling hatchetfish (*Argyropelecus hemigymnus*), rely on their sense of smell and release of pheromones to detect and summon partners for reproduction.[30] Underwater perfume likely provides the first clue a potential mate is in the neighborhood, as it travels much farther than the dim blue light of photophores.[31] Hatchetfishes pair exceptionally sensitive noses with their own form of double vision: oversized eyes that bookend severely flattened heads boast complex retinas with twin areas of high sensitivity, one at the bottom for receiving overhead light and one near the back that can view potential mates ahead.[32] But when swimming in total darkness, olfaction provides a handy system of distant detection. Abyssal grenadiers, known ingloriously as rat-tails (a reference to tails tapering to a point rather than a terminal fan), find food by scent alone, in utter darkness. Cameras monitoring food baits in the North Pacific, suspended at depths below 19,000 feet, revealed that grenadiers arrived within minutes and began feeding ravenously. Baits were discovered sooner when there was strong current to scatter its odor, with three-quarters of the diners arriving from the down-current direction, clear indication that the hungry fishes were following the tantalizing aroma of dinner.[33]

Sometimes it is not dinner that lures diners with a wafting scent. In deep-sea anglerfishes it is the cook herself who entices patrons with a different sort of hunger, reeling them in with an irresistible perfume. Anglerfishes are a taxonomically sprawling group (more than 320 species; bioluminescent courtship yields surprisingly high rates of speciation) composed of squat, lumpy, gargoyle-like fishes equipped with enormous mouths and distensible stomachs, and bristling with a variety of odd appendages.[34] Most notable is a unique fishing apparatus composed of a bent rod-shaped stalk (itself derived from the foremost dorsal fin spine) tipped with a lure called an esca. Often bioluminescent, the esca dangles in front of a toothy mouth where it attracts both prey and partner. Anglerfishes live in low density, and their sluggish swimming abilities can make finding a mate difficult. Instead of searching the sea, females release a powerfully appealing pheromone to draw mates to them. Males home in on the scent using their oversized nostrils (reputed to be the largest of any vertebrate), then follow the light of the dangling esca until they latch on to a partner.[35] Literally. Males are a mere fraction of the size of an adult female, and when first discovered they were thought to be tiny, external parasites. Which, in truth, they are.

Many anglerfish males have nonfunctional mouthparts and will die if they do not find a mate. Once joined in watery matrimony, the male attaches to the female's body, dissolves her skin with a salivary enzyme, and connects his circulatory system to hers. Like a leech, he draws nutrients directly from her and will never live a solo existence again. Anglerfish with parasitic males (about half the known species) have immune systems peculiarly adapted to this curious behavior. Unlike people, in whom organ grafts are often rejected with disastrous consequences, these anglers lack the histocompatibility pathways that flag foreign tissues for destruction by T-cells: they literally cannot distinguish between their own flesh and that of their mates.[36] Meanwhile, the male's gonads are stimulated during fusion with the female, and soon he is able to release sperm into the water to fertilize her eggs. In a few species, the female decorates herself with as many as eight males, each contributing to reproduction, albeit to no other chores of domesticity like swimming or foraging.[37] One wonders if not a

Devil anglerfish (*Lophius vomerinus*) female and parasitic male,
who is attached under her jaw

few human females might assess the role of their males as being similarly limited in supportive scope.

MORSELS, TIDBITS, AND CORNUCOPIAS

When Piccard peered out of his bathyscaphe as it descended into the darkness, the craft's exterior lighting illuminated motes of organic matter sinking ever so slowly from the surface. He called those morsels "snow," a term coined by earlier explorers. Fishes and other animals of the deep sea rely on these particles, as a mouse relies on crumbs brushed from a café table. More trash than snow, the steady drizzle of particles includes dead algae, cast off exoskeletons, deceased copepods, fish scales, and feces from schools foraging in the sunlit shallows. That snow is relentlessly assaulted as it descends, by all manner of organisms from bacteria to zooplankton, shrimps to squid, and countless fishes. Paucity of food means fishes survive only at low densities, far lower than their shallow-water relatives; that deepwater scarcity presents challenges for prospective mates and demands for hungry predators. If you are going to forage in deep, black waters, you are going to need some skills and adaptations, the likes of which almost never arise in fishes of the shallows.

Attracting prey with a simple light, a glimmer they may mistake for a tasty shrimp or other tidbit, is a popular approach. Our lovelorn friends the anglerfishes hang from their foreheads a glowing esca as bait, with forms unique to each species, ranging from a simple blob to fanciful shapes that resemble brushes or the gnarled fingers of a witch's hand. The lure's lamplighters are symbiotic bacteria that reside within the esca, their glow echoing off mirror-like guanine crystals that line the esca's walls.[38] Bacterial mutualisms also are harnessed by ponyfishes (*Leiognathus*), and the fittingly named flashlight fishes (*Photoblepharon*) who sport bright, forward-facing headlights under each eye.[39] Once the bewitched prey is drawn within reach by an anglerfish's lure, the innocuous-looking owner throws open its enormous mouth like an attic trap door, sucking water and the hapless victim into its gaping maw. The largest mouths in the marine realm belong to deep-sea fishes, sometimes comically dwarfing the diameter of their own bodies, a toothy opening so cavernous they can even swallow a fish larger than themselves.

Among bigmouth fishes the pelican eel (*Eurypharynx pelecanoides*) is the most surreal, straight from the pen of Salvador Dali. Its ribbon-like body connects to a rounded head that nearly splits in half, like a walnut, when the jaws are unfolded. So unusual is this eel's architecture that it has inspired new designs in 4D printing, solid structures fabricated in three dimensions that fold into new conformations, like origami, in response to heat or light.[40] In case the snack is not entirely flushed down the throat by a tidal wave of water, huge dagger-like teeth foil quick getaways, a terrifying adaptation brandished by bristlemouths (an excellent moniker in its own right) as well as anglerfishes. The stomach is distensible to accommodate the recently gulped dinner, a common adaptation among dark-water predators, and it can swell to preposterous proportions, turning the fish into a veritable football with teeth and tail. And because many of the captured prey are bioluminescent themselves, stomachs of most abyssal predators are lined with jet-black pigments. Once a shrimp or fish has been seduced by dreamy light, stabbed or gulped whole, and swilled into the ebony stomach, no light escapes to mark its end nor illuminate the clandestine killer. A few predators go one step further, like the golden sweeper (*Parapriacanthus ransonneti*), who gobbles up glowing copepods by the hundreds then mobilizes their enzymes to power its own underbelly glow, a light source elaborately titled klepto-protein bioluminescence.[41]

If a meal is hard to come by, and starvation is to be avoided, then a fish must expend as few calories as possible. Efficiency is rewarded at great depths: anglerfishes, for example, are models of patience as they float motionless, waiting for prey. Tripodfishes stand on the bottom, perched on three slender stilts formed from pelvic and tail fin spines that lift them above the silty floor to just the height preferred by their favored lunch of shrimp. They are hermaphrodites, capable of mating with whomever happens to stilt-walk by, and thereby conserve energy otherwise spent on chasing down a suitable mate.[42] Many deep-sea fishes must travel to find food, however, and here evolution has honed locomotion to achieve maximum forward progress with minimum effort. Because dark-water fishes do not rely on visual cues when hunting, there is no chasing of quarry that requires bursts of speed, or quick cornering. Predators instead are languid, biding their time, keeping their metabolism and oxygen consumption low while relying on unhurried swimming until they bump into a prospective prey item (sometimes quite literally).[43]

Grenadiers, those rat-tailed bloodhounds of the deep, earned their unpleasant epithet for their eel-like tails (decidedly not rat-like) and an undulating swimming style that is the pinnacle of efficiency. Hydrodynamic studies have shown that eel-like swimming can propel its pilot four to six times farther than a typical fish while expending the same amount of energy.[44] Tails undulating like ribbons are not exclusive to eels and rat-tails; many deep-sea fish have adopted them, including the ancient frilled sharks (though no sharks inhabit the true abyss).[45] These living fossils gradually lost the lower lobe of their ancestors' bi-lobed tails until all that remained was a long thin ribbon. Sharks of the deep sea, whether ribbon-tailed or original flavor, also swim more slowly than close relatives who hunt in shallower waters, further conserving meager energy supplies.[46]

All these adaptations—bioluminescent lures, bi-directional eyes, low metabolism, and eel tails—are driven by the extreme scarcity of nutrition in the dark depths of the ocean. But there are a few places at the bottom of the sea where scarcity is but a memory, and a cornucopia of food is freely available to those able to dine at some very peculiar tables: whale carcasses, seamounts, and boiling water volcanoes.

Every few days a whale dies of natural causes, taking its last breath at the surface of the open ocean. Drawn by the inexorable hands of gravity and density, its heavy body soon dips beneath the waves and sinks slowly into the blue-black gloom. Eventually, it comes to rest on the bottom. In days of yore, before the whaling industry brought slaughter to the seas, large whales may have numbered more than 2 million.[47] Even a conservative estimate of their mortality (around 4 percent annually) suggests that as many as 200 whales may have expired every day and sunk to the bottom.[48] In a food desert like the deep seabed, the arrival of a single whale's carcass weighing 40 tons is the equivalent of having a cargo plane full of fried chicken land at your fasting retreat. For animals eking out a famished existence in the deep sea, this cornucopia from above represents the equivalent of two centuries of organic snow, a concentrated banquet for those who can find it.[49] Across all oceans, whale falls may contribute as much as 9 pounds of food to every acre of the seafloor, every single year.[50]

When a whale hits the bottom, its arrival triggers a scramble of diners. Among the first to gather are sleeper sharks (*Somniosus pacificus* and relatives), snub-nosed and slow-moving predators that feast on whale falls over the continental shelf.[51] Ravenously they latch on to the giant's flanks, then furiously spin their bodies to rip loose ragged chunks of flesh. Smaller rat-tails slip between writhing sharks to slice a piece from the cadaver, or scrounge stray morsels scattered by others. Snubnosed eels (*Simenchelys parasitica*), who often parasitize living fishes by lodging themselves in shark hearts, may be among the most numerous feeders.[52] Eel-like hagfishes—slimy, jawless relatives of lamprey—burrow into the whale like drills, then eat from the inside out. These early scavengers take 100 pounds of flesh per day from the carcass, for weeks on end. Smaller and smaller fishes, and many invertebrates, nibble their own tidbits until eventually only bones remain. A single great whale may sustain such hordes of scavengers for a year or more before its flesh is stripped clean. Even then, a skeleton chock-full of nutrition remains.

Exposed bones are swarmed by crabs, shrimp-like amphipods, and marine worms, which themselves are snatched up by sharks and other deepwater predators. Years later, when only the hardest and most inedible parts are left, still digestion by the deepwater community continues. A rich assemblage of bacteria relying on sulfur for energy (more about this in a moment) invades the skeletal shards. These microbial colonists slowly break down marrow lipids that account for half the mass of whale bones, a process that can stretch to fifty years until the last vestiges of a great whale finally turn to dust. More than 400 species of fishes and invertebrates rely on whale falls, more than are found on any other patch of ocean floor, and many of which occur nowhere else.[53] Some ichthyologists have speculated that shallow-water organisms may have used whale carcasses as stepping stones, guiding them ever deeper into the ocean and permitting the first colonization of the abyssal depths by species formerly restricted to coastal, nutrient-rich waters.[54]

Wherever nutrients accumulate in the ocean, one will find eruptions of fish diversity and abundance. Nowhere is this more true than around seamounts. Rising like desert mesas from the seafloor, these undersea mountains range from narrow ridgelines to massive tables of bedrock. Though they never crest the surface, the local gravity of giant seamounts draws water toward them, pushing up a bump on the sea that can be spotted from space. Satellite altimetry surveys estimate Earth's oceans may be studded with as many as 100,000 giant seamounts taller than 1000 feet; smaller undersea hillocks and knolls likely number in the tens of millions.[55] No matter how numerous, each mountain is a nutritional hotspot that can sustain massive fish schools. Seamounts divert prevailing currents to create local upwellings that draw nutrients and much-needed oxygen toward the mountaintops, and block sinking organic snow from falling further into the abyss. They also impede the daily vertical migrations of zooplankton, trapping millions of these tiny animals at the summit during dawn dives.[56] Peaks and flanks of seamounts are colonized by thickets of cold-water corals and diverse communities of sponges, anemones, crabs and shrimp, sea stars and urchins, squids, sea cucumbers, and marine worms. Midwater grazers and planktivores find ample food supplies on and above these marine monuments, browsing on the rich buffet line of benthic invertebrates or sieving stymied plankton from the water.[57] Nearly a thousand types of fishes are sustained by this buffet, many completely new to science, and a single site can host hundreds of species and millions of individuals.[58] Massive shoals of small fishes eddy above seamounts, much more populous than in the open ocean, and their abundance attracts the attention of huge schools of predators who hover nearby.

Perhaps the most famous of these looming predators, celebrated worldwide as a delectable menu item in seafood restaurants, is the orange roughy (*Hoplostethus atlanticus*). Formerly known to marine biologists as a slimehead, this formerly hyper-abundant fish was subjected to an all-too-effective rebranding campaign when colossal schools were discovered around seamounts off the coast of New Zealand in the late 1970s. There, schools of orange roughy congregate to feast on the prawns and squids and smaller fishes drawn to the bounty of nutrients amassed around these underwater mountains. Like other denizens of deep and cold waters, orange roughy are sluggish, with a slow metabolism and lethargic swimming style. They are some of the oldest fishes in the sea, routinely living beyond 150 years and perhaps even reaching two and a half centuries.[59] In other words, the fish on your plate could be older than the United States of America. Concomitantly late to reproduce, an orange roughy will not mate until it attains twenty or even forty years of age. All this means that enormous schools of adult orange roughy took an enormous amount of time to reach that abundance. Once fishing began, with trawl nets deployed relentlessly on and around their favored seamounts, populations plummeted and never recovered. Those nets irreparably damage the rich but delicate ecosystems that drape the flanks of ocean mountains, cutting off the legs of the buffet table that supports all that fish life. While recovery is possible, if fishing pressure is firmly controlled, seamounts are too isolated to enjoy rapid recolonization by fishes, and evidence suggests even partial recuperation will take decades or centuries.[60]

It is hard to blame fishing communities, however, for delving into the proverbial pot of orange gold at the end of a watery rainbow. Around a single seamount off Tasmania known as Saint Helen's Hill, orange roughy catches netted trawlers US$17 million in just three weeks.[61] One net run could haul in 50 tons of fish, thanks to the slimehead's habit of clustering together as tightly as do cod. For New Zealand, here was a rich new source of revenue, one that in peak years landed 60,000 tons per year and annual revenue in excess of $65 million.[62] But fish can only be caught sustainably if their reproduction is able to replenish the stock, otherwise you are sacrificing the future by eating up the product of history. And the outlook for orange roughy swiftly turned dim, sacrificed on the barrelhead of local and national financial gains. Just as readily as fish filled nets and dollars filled bank accounts, so too came the rapid decline. In New Zealand catches nosedived after boom years in the mid-1980s, and by 1997—not twenty years after the fishery had opened—orange roughy populations had plummeted by 80 percent.[63] On the other side of the world, an orange roughy fishery in Ireland opened in 2001, catches peaked just one year later, and by 2005 the fishery had utterly collapsed.

In contrast with the North Sea's extremely fecund cod, who may release more than a million eggs in a breeding season, an orange roughy adult normally produces only a twentieth of that lavish output.[64] Despite the apparent richness of seamounts, the limited reproductive potential of fishes who thrive at these sites, combined with their glacially slow growth to maturity, makes fisheries collapses as seen in Ireland and New Zealand inevitable. Despite the richness of seamounts, those fish simply cannot reproduce quickly enough to offset the arrival of high-tech trawlers, and they have no means to accelerate their procreation. But deeper still, hidden from the eyes and fishing nets of the world, lies another constellation of nutrient hotspots where animal life is unexpectedly abundant. And while seamount fishes have not yet evolved a way to accelerate their reproduction, such advances have been invented by other deepwater species, some of whom discovered an extraordinary use for an abyssal environment so strange that it is quite literally unlike anywhere else on planet Earth.

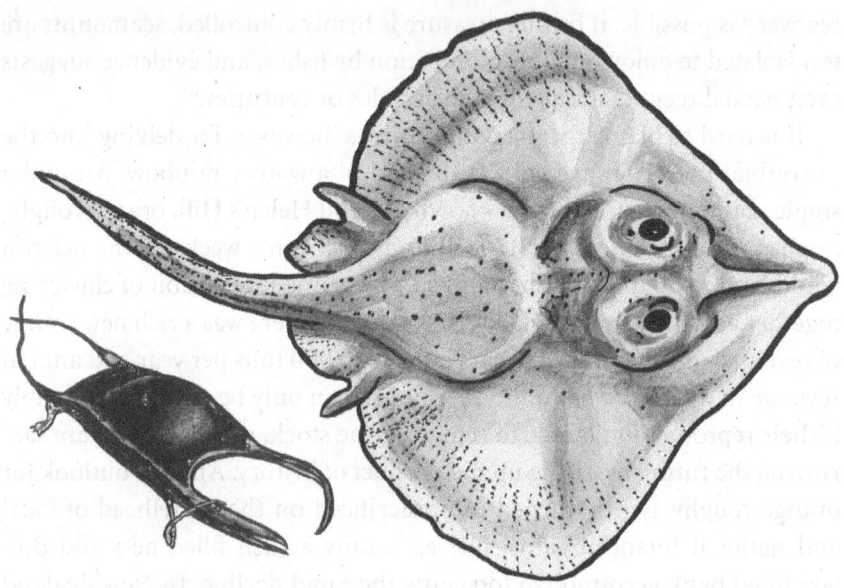

Spiny skate (*Bathyraja spinosissima*) and egg case

FREEZING COLD, AND BOILING UNDER PRESSURE

> Double, double toil and trouble;
> Fire burn, and caldron bubble.
> —**William Shakespeare,** *Macbeth*

On land, the record-holder for longest pregnancy is a female African elephant, who carries her calf for twenty-two months; Asian elephants are close runners-up, with gestations lasting eighteen months or more.[65] But both trail the pack when marine animals are added: the eggs of some deep-sea fishes require an astonishing four years to develop, more than twice as long as an elephant. One such species is the spiny skate (*Bathyraja spinosissima*), whose four-horned egg cases—the curious mermaid's purse that distinguishes skates from rays—can repose on the seafloor for 1300 days or more before a tiny skate emerges. Development is slowed in the coldest waters to as long as 1500 days, a perilous delay that gives egg-eaters several extra months to snatch and devour a forsaken purse. Temperature plays a strong role in the development of many animal eggs, from turtles and alligators (in which gender is determined by egg temperature) to all manner of fishes. But in slow-developing species like the spiny skate, there is severe evolutionary pressure to find warmer locations that can accelerate egg development. The sooner your young emerge, the more likely they, and your species, are to survive. What the skates hit upon several million years ago were not discovered by humans until 1976: underwater volcanoes spewing boiling water.

KC and the Sunshine Band's "Shake Your Booty" was turning up the heat in discos around the world when a doctoral student named Kathleen Crane went to the Galápagos Islands to investigate deepwater temperature anomalies. Her dissertation supervisor, not optimistic about her chances, advised her to "find another thesis topic . . . you'll never find hot springs on the seafloor."[66] But on June 2, after more than a month at sea, her determination was rewarded as their sampling device struck thermal pay dirt. "The Deep-Tow swept over warm water emanating from the fissures in the rift valley below. We recorded a temperature increase of 0.1 °C (which was a phenomenal increase given the Deep-Tow's height above the bottom), and we trapped bottom water in Ray's bottles. For three days I did not sleep more than two hours . . . I knew we were on the verge of something momentous."[67] Crane and the team had found the world's first hydrothermal vent.

At the Galápagos site and many discovered since, deep cracks in the Earth's crust reach down toward molten magma several hundred miles below. Seawater percolating through these fissures is heated, as in the boiler of an apartment building, and steams its way back up to the ocean floor. Along the journey, minerals from the crust and magma dissolve into the superheated water, metals like iron and manganese that bind with sulfur to form compounds known as sulfides. When water jets from the crack it may be as hot as 400 °F or more. Only the extreme pressure weighing on the bottom of the sea prevents that water from boiling. As the scalding plume disperses, it cools rapidly: shift just a few yards from the hydrothermal vent, and temperatures approach normal. It is in this delicate transition zone, between superheated and super frigid waters, that the spiny skate stations its egg cases, where they are warmed by a natural incubator that hastens their development. Baby skates, snug in their hydrothermal cribs, hatch a little earlier than their cold-water cousins, and the tightrope act of raising your children near a roaring volcano is rewarded.

While a few other creatures use the plumes for their heat alone (some flatfishes lie in wait for an unsuspecting squid or fish to swim too close and poach themselves to death), it is the mineral soup spewed by these hydrothermal vents that supports, as if by magic, a thriving community. Until 1976 it was thought that virtually all life on Earth, or more precisely the energy of all life, must derive from the sun. Plants grow through photosynthesis, as do phytoplankton; grazing animals survive by eating those green solar factories; predators can succeed only because prey have plants or algae to eat. Even decomposers like fungi and hagfish are gnawing on the decayed remnants of sunlight. But the discovery of hydrothermal vents changed all that. Miles below the surface and thousands of feet beneath the twilight shallows, where the only food is an ephemeral drizzle of degraded organic material, astonishing hamlets of life were unveiled. Marine worms poking red heads from giant cream-colored tubes cluster by the thousands around the vent's mouth. White crabs, hairy as yaks, crawl one upon the other like so many ants in a colony. Farther from the fissure are sea cucumbers, long-legged crabs, and round-headed miniature octopuses. To watch video from remote vehicles navigating around hydrothermal vents is to be staggered by an exuberance of life that can scarcely be comprehended.

What puzzled scientists was how that abundance could be supported, so far from the sun's rays. The answer to the conundrum lies in the tiniest of organisms: bacteria. While photosynthesis in a plant fabricates sugars from atmospheric carbon with the power of sunlight, a few special deep-sea bacteria can power the manufacture of sugars with energy bound in chemicals pouring from the hydrothermal vent. Earth's minerals spew forth as high-energy sulfides, typically metal atoms bound to sulfur. Huge chimneys, structures that can tower more than 100 feet tall, are formed from the clouds of minerals that pour from a vent like smoke from a chimney. Most of these smokestacks vent billows of black water, rich in sulfides of zinc, lead, and iron, giving them the nickname of black smokers.[68] After gushing from the chimney those heavy metal particles sink and settle around the vent mouth, concocting fantastical spires and twisting pinnacles, geochemical sculptures reminiscent of the visionary Spanish architect Antonin Gaudí.

Bathed by the black clouds, enterprising bacteria split metals from their sulfur partners, releasing a jolt of electricity they can harness to make sugars and starches from carbon in the surrounding seawater. These unique bacteria are known as chemoautotrophs, a tongue-twister signifying they make food ("troph") by themselves ("auto") solely with the energy in chemical bonds ("chemo"). But bacteria need a place to live, an apartment to settle in while they play with their chemistry sets, and they have willing landlords in deep-sea worms that evolved over millions of years into the perfect hosts.

Clustered around any hydrothermal vent, colonies of colonial tube worms bristle like white skyscrapers. The giant tube worm (*Riftia pachyptila*), first discovered when the bathyscaphe *Alvin* (modern cousin of Piccard's *Trieste*) explored the newfound Galápagos hot water plumes, is one of a handful of tube worm species now known to thrive near hydrothermal vents. Distant relatives of earthworms, these 6-foot invertebrates construct a thin-walled, cylindrical case for themselves made of proteins and chitin.[69] The tube protects the resident's delicate tissues from attack, but more importantly it allows the worm's feeding apparatus to stretch away from the seafloor and up into the water column. Creatures like *Riftia* are descended from filter-feeding worms that sift tiny organic particles from the water with netlike tentacles mounted around their mouths. In hydrothermal worms, the tentacles have evolved into a carmine-colored branchial plume, an external gill folded like a taco that they use to capture a novel kind of deep-sea food: metal sulfides. Occupying most of the rest of the worm's body is a blood-rich organ called the trophosome, a veritable apartment where chemoautotrophic bacteria comfortably reside.

In the worm's blood, highly specialized hemoglobin molecules travel to the branchial plume, pick up a load of sulfides, and transport these invaluable packets of energy to the overcrowded apartment (which harbors nearly a trillion bacteria per gram).[70] Inside the trophosome, the chemistry begins: bacteria grab carbon dioxide, also transported by the blood, and manufacture a steady stream of small sugars, as well as key organic acids and proteins synthesized from nitrogen.[71] The worm never eats a mouthful but flourishes nonetheless, thanks to a life-sustaining soup of nutrients fabricated by its symbiotic residents and absorbed without lifting a fork.

Factories of chemoautotrophic bacteria take up residence not only in worms; colonies have been found cohabitating in shrimps, mussels, and even snails.[72] One in particular, the scaly-foot snail (*Chrysomallon squamiferum*, aka sea pangolin) hosts bacteria in nano-tubes jutting from scales on its foot, and as sulfurous waste puffs from each tube, it reacts with dissolved iron in the water that precipitates onto the snail's foot as iron pyrite (otherwise known as fool's gold).[73] The result is a snail fed by bacteria and encased in iron-plated armor, a double benefit. Still another beneficiary of bacterial exuberance are yeti crabs (*Kiwa hirsuta*) who host them on a distinctively thick, hairy coat. Comically nicknamed "Hoff crabs" for their indirect resemblance to the densely hirsute chest of bygone actor David Hasselhoff, these crabs farm luxuriant colonies of bacteria on their fur, then harvest them with a claw modified as a comb. So abundant are colonies of Hoff crabs they literally carpet the seafloor, draped over every promontory, boulder, and swale.

Surprisingly, fishes are relatively rare at such cornucopias of life, eschewing it seems the dangers of swimming near a boiling volcano of water spiked with hydrogen sulfide.[74] Still, one known as a pink eelpout (*Thermarces cerberus*) for its elongated shape, flat tail, and sulkily down-turned mouth, wriggles between clustered vent worms, gnawing on limpets and other mollusks encrusting their tubes.[75] How the eelpout can live where earthly fires threaten to poach it at every turn—its scientific name even references the multiheaded dog guarding the Greek underworld—is anyone's guess. Hydrothermal vents continue to yield surprises, and strange new animals are routinely discovered at these bizarre locations; after thirty years of study, two new vent species are still being described every single month.[76] Although isolated by mile after mile of cold, nutrient-poor water, those animals appear capable of colonizing distant hydrothermal vent fields. Genetic evidence suggests vent specialists may disperse by leapfrogging, probably in their larval stage, using favorable currents to reach vent sites like stepping stones across inhospitable seas.[77]

All these extraordinary organisms, including tube worms, navigate a delicate and dangerous balance between the cornucopia of life afforded by superheated vent water and the freezing, hyper-pressurized sea all around them. Auguste Piccard and his bathyscaphe withstood the latter two extremes, but fishes lack the protection of a pressurized capsule. For them, the immense water pressure of the deep sea imposes harsh consequences at the cellular and subcellular level. In fish cells, fatty acids that normally keep cell membranes pliable become rubbery and inflexible under high pressure, treacherously impairing their function. Imagine if pancreatic cells could no longer pump digestive enzymes across a nearly impermeable membrane—the fish would starve. As depth increases, fish (and bacteria) dial up the proportion of polyunsaturated fats in those lipids.[78] Unsaturated fats are more flexible and fluid under pressure, thanks to kinks in their carbon backbone; for a home kitchen example, compare solid butter (rich in saturated fats) at room temperature with olive oil (mostly unsaturated) that remains liquid. Deep-sea sharks, who like their shallow water cousins rely on a fat-rich liver for buoyancy, also show increased levels of unsaturated fatty acids to keep organs and cells supple.

Housed within those liver cells, and all the body's cells, proteins are hard at work. But the crushing pressure of the abyss also squeezes proteins during their formation, bending them dangerously out of shape and hobbling their performance. Faulty proteins are another swift highway to death, so the evolutionary pressure to solve this malfunction has been acute. Fishes of the deep sea rely on a chaperone molecule to assist protein folding and performance, as a yoga instructor might help an inflexible newcomer to stretch without injury. Called trimethylamine oxide, or TMAO for short, this mouthful of a molecule prevents extreme pressure from pinching shut a protein's active site, the place where it binds to target molecules as precisely as a lock accepts a key.[79] While modest levels of this compound are common in shallow-water fishes (where it is responsible for the characteristic "fishy" smell), its concentration soars with depth, protecting the proteins of deepwater fishes and ensuring they keep swimming, seeing, breathing, eating, and surviving.[80] Too much of a good thing, though, can be problematic. In the case of abyssal fish, this helpful chemical may actually set a hard limit on the maximum depth at which they can survive.

Among the deepest fishes in the world are the Mariana snailfish (*Pseudoliparis swirei*), and an as-yet unnamed relative, both having been discovered at depths of 27,000 feet.[81] A snailfish's body is so loaded with TMAO, however, that it risks death by drowning, or at least by osmotic imbalance. Fish are normally less salty than seawater, causing them to lose water to the ocean by osmosis, the diffusion of a fluid. They replace that lost water by drinking saltwater, often while gulping prey, then eliminate excess salt across the gills. If a fish finds itself saltier than the surrounding water (as happens when oceanic salmon run up freshwater rivers), osmosis brings water into its body, which must be pumped out by its kidneys. If the pumps cannot keep up, the fish can absorb so much water that it would drown in its own juices. Osmotic calculations suggest that the amount of TMAO required to protect proteins at depths much beyond 27,000 feet is so great that no fish on Earth could survive the inrushing water.[82] Overwhelmed kidneys would fail, and the fish would perish. The Mariana snailfish, it seems, will meet few rivals for the title of world's deepest fish.

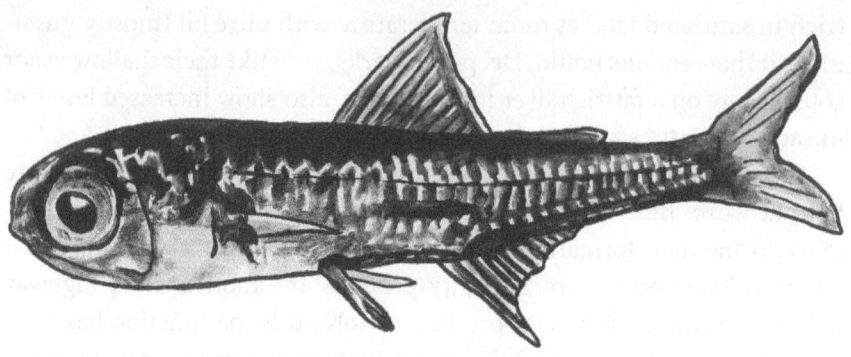

Spotted lanternfish (*Myctophum punctatum*)

THE BEST OF BOTH WORLDS

While only a fraction of the planet's fishes can withstand the harsh extremes of the deep, and numerous species can enjoy the shallows, there are some who have chosen an intermediate path. Vertical commuters, these fishes seek the benefit of both worlds: the relative safety afforded by hiding in the dark depths, and also the abundant food on offer in the sunlit shallows. To enjoy both realms, they make impressive daily journeys, swimming up to the surface at dusk and descending well below the twilight zone at dawn, when visual hunters emerge to terrorize shallow waters. One such daily migrant is the innocuously named cookie-cutter shark (*Isistius brasiliensis*). Diminutive but dangerous, these slender little sharks rarely exceed 2 feet in length. Their mouths are oddly shaped, more rounded than in their relatives, and bear slightly fleshy lips that give the appearance of having recently enjoyed a Botox treatment. Kissing, however, is not their specialty. Instead, these lips latch onto the flanks of larger fishes, whereupon a muscular pharynx applies strong suction, saw-like teeth slice into flesh, and the shark's body swivels violently until a circular plug is removed and swallowed.[83] Fish, whales, and even scuba divers who have suffered such an attack bear distinctive, hockey puck-shaped divots where a painful mouthful was excised.[84] Cookie-cutters camouflage themselves against predators using bioluminescence to break up their silhouette, but they add a deceptive twist. Interrupting the broad apron of gleaming blue is a dark collar, crossing the neck. When seen from below, this stripe mimics the outline of a slender, much smaller fish or eel, luring large predators in for a closer look, and a date with the cookie monster.[85] After eating, hesitant to get caught with its proverbial hand in the proverbial jar, these sharks dive to great depths when daylight arrives. Some specimens have been recorded as deep as 12,000 feet, where they can escape the wrathful vengeance of their latest victim.[86]

Lanternfishes, in whose bioluminescent glow we recently bathed, also have enjoyed tremendous success adopting the vertical commuter lifestyle. This constellation of approximately 250 species ranks among the most abundant vertebrates in the ocean: they account for some 65 percent of the deep-sea fish biomass that may tip the scales at nearly 5 gigatons (more than 11 trillion pounds).[87] Massive schools use the vertical range of ocean environments to their best advantage. By night, they ascend to the surface where they dine voraciously on copepods and other plankton, relying on huge eyes to find prey by moonlight. That rich and abundant food helps them grow quickly, and after one or two years they are ready to reproduce. As dawn breaks, they creep down into the twilight depths and flick on their belly photophores to obscure them from deepwater predators.[88] In comparison with orange roughy and other deep-sea fishery targets, lanternfishes may represent a sustainable source of fish protein, as long as their fishery is properly managed. Though the flesh is too oily to be favored by gourmands, when ground into fish meal it is delectable to salmon, tuna, and other predatory fishes reared in aquaculture ponds (about which we soon will learn more).[89] If we are to save our seas and feed a planet at the same time, deepwater lanternfishes may light the way.

Life in the deep sea is confronted by pressures, literal and figurative, that are found nowhere else on the globe, at least not today. The primordial environment around hydrothermal vents has not been seen on the planet for several billion years, when volcanoes filled a nascent atmosphere with sulfides and oceans were scalded by an unshielded sun. In 1929, John Haldane of the University of Cambridge shook the scientific world when he proposed a new theory for the origin of life on Earth: organic molecules spontaneously assembled within a "hot dilute soup" of sugars and proteins swirling in the "primitive oceans."[90] He posited that "the first living or half living things were probably large molecules," and then envisioned the appearance of the first primitive cell on the planet. "The cell consists of numerous half-living chemical molecules suspended in water and enclosed in an oily film." Fifty years later, the discovery of deep-sea vents prompted NASA chemist Michael Russell to suggest abyssal sites as candidates for this hypothesized origin.[91] Researchers at University College of London in 2019 added fatty acids and fatty alcohols to a simulated hydrothermal vent, and lo and behold cell membranes emerged from the stew, Haldane's selfsame "oily film."[92] Fossils in oceanic crust rocks dated to 4.2 billion years ago have revealed what may be the earliest evidence of a living organism: "microscopic tubes and filaments" that resemble microbes living on today's hydrothermal vents.[93] Life on planet Earth may very well have originated as chemoautotrophic bacteria that first survived on minerals gushing from hydrothermal vents, then evolved into more complex organisms which eventually colonized the seas by vaulting from one vent field to another.

Skipping from one oasis to the next, across inhospitable seas, is a common strategy in the world's oceans. Seamount fishes leapfrog between far-flung mountaintops, whale fall specialists do the same, and dazzling coral reef fishes release eggs into currents that deliver them to distant yet comfortingly familiar reefs. It is a strategy that can lead an animal across the seas, hopscotching its way for generations from one favorable locale to the next, traveling thousands of miles from home. And not only fishes have taken advantage of this tactic to colonize new seas. Human beings in the dawn of our history followed stepping stones of their own, crossed an ocean, and ended up settling a vast new world.

Part III
WHERE MOUNTAINS MEET WAVES

AFRICA COLONIZED THE WORLD 150,000 YEARS AGO, when hominids traveled by land to Europe, the Middle East, and Asia. But how did humans settle North and South America, two continents detached for millions of years from the cradle of civilization? Once it had been supposed that early people trekked from northeast Asia during the last ice age, when sea level was hundreds of feet lower, crossing a formerly submerged land connection called Beringia to populate a vast emptiness by foot. Archaeological evidence shows the plains of North America were occupied soon after the glaciers retreated, opening the proverbial gates to the New World some 13,000 years ago.[1] But a discovery in southern Chile flattened that theory like a glacier bulldozes a spruce forest: there, in a site known as Monte Verde, people had been thriving for several thousand years *before* the ice sheets melted.[2] Maybe early peoples invented cross-country skis to cross miles of glaciers, but a more plausible theory soon emerged. Hominids did not walk to the New World, but rather paddled the salty seas and hopscotched down the Pacific coast all the way to Tierra del Fuego, following a ribbon of marine abundance fancifully dubbed the Kelp Highway.[3]

Seafaring has a long history: ancient mariners from Southeast Asia reached Australia some 50,000 years ago, and modern humans boated from what is now Japan to the Ryukyu Islands 10,000 or 15,000 years later.[4] Hominids may have walked out of Africa, but when they reached the Pacific Ocean, they fashioned boats and piloted them across the waves. Voyaging by water overcomes many of the obstacles of journeying by land. Mountains need not be climbed, nor raging rivers forded; snakes and sabre-tooth tigers do not lurk behind every wave. And critically, familiar foods are abundant and readily gathered. Where there are dense groves of kelp, rooted in the seafloor and tickling the surface with green fronds, there is sustenance. Abalones, oysters, mussels, and snails all can be foraged with ease, using only a pair of hands. Spears, nets, and hooks can be wielded to catch fish such as sheephead and sea bass, who swim in thick schools among the kelp. Nearby seabird rookeries provide copious quantities of eggs, while seal and otter colonies offer abundant meat.

Bob Steneck, an accomplished marine biologist, has studied the kelp highway for years.[5] "The very first people who got to Daisy Cave [on California's Channel Islands] were eating bivalves—you know, clams and snails—and then over time they developed fish hooks and other things that allowed them to stay on the coast. You could actually be a migrant to the New World, living well on easy resources that are along the coast."[6] Tracing the migration route farther south, Steneck reveals what botanical archaeologists discovered: "In the Monte Verde site in Chile . . . [they] were able to show *Durvillaea*, a type of kelp, in a cave that was consumed by the earliest human colonists in Chile 16,000 years ago. So direct evidence that they're actually literally using the kelp." Paddling away from Asia, primordial humans could have easily followed the crenellated coast of Beringia, glided past the fjords of Alaska, and bobbed down to southern California, reliably feeding themselves the entire way on the bounty of kelp forests, one of the most productive ecosystems on the planet.[7]

Even where kelp gives way to rock and sand, oceanic shorelines boast abundant resources, for people and for fishes. Compared with open ocean waters, where nutrients come only from the steady churn of photosynthesis, productivity near the coast is supercharged by nutrients flowing from streams, marshes, and the shoreline itself. Where rivers disgorge their upland waters into the ocean, the estuary teems with life. Nearshore fishes like tarpon feast on the crabs and snails and plankton that flourish where salty water meets fresh. Long-distance migrants like salmon gather in these brackish waters, preparing for the journey upstream to breed. Away from the great river mouths, rocky shores host tide pools brimming with shellfish, urchins, seaweeds, and fishes who have adapted to the coast's crashing waves and wild swings in temperature and salinity. Offshore breezes here also blow surface water out to sea, which is rapidly replaced by local upwellings that further fertilize these waters. Even bird and seal colonies make a donation: their fishing triumphs are soon released as abundant, nitrogen-rich guano that washes into the ocean with every rainfall.

In tropical latitudes, where mangrove forests replace kelp, early mariners would have encountered equally concentrated sources of food: crabs, mussels, fishes, and seabirds all congregate in mangroves. Their tangled and interwoven roots stabilize soft shores, trap nutritious leaves and fruits falling from above, and serve as a sheltered fish nursery for snappers, barracudas, and dozens of other species. Beyond and between the mangroves, where these tropical waters wash over shallow bottoms, saltwater flats can be found: mazy patchworks of marshlands, inlets, tidal streams, and seagrass meadows. Seagrasses, the only true plants in the ocean, carpet sunlit shallows and act as fishy stepping stones between mangroves and coral reefs. This lush underwater lawn—another of the planet's most productive ecosystems—supports an abundance of crabs, snails, sea turtles, manatees, and juvenile fishes, including many commercially important species.[8] Full-grown reef fishes often commute from nearby patches of coral, transporting valuable nearshore nutrients along with them. Meanwhile, over the salt flats, all manner of predators and prey slip through a thin skin of brightly lit water, like the bonefish and tarpon so widely sought by anglers.

Exuberant abundance along shorelines and in nearshore habitats may have nourished seafaring humans and undoubtedly supports an extraordinary diversity of fishes. Opportunities and challenges abound in these diverse, productive, and entrancing environments, prompting a suite of adaptations that include mouth breathing, water archery, and mysterious migrations over thousands of salty miles. Shorelines are where the ocean ends, but it is where life on Earth began. Let us pick up a paddle and follow the Kelp Highway, exploring the world's coastlines and the fishes who call the edge of the sea home salty home.

9 FLOWING RIVER, POUNDING SURF

Tarpon and Other Coastal Cruisers

IT'S HARD TO KEEP YOUR FOOTING ON A BOAT IN THE DARK. Groping your way to the stern, you kick a heavy cooler, mutter a low curse. Maybe flip-flops weren't the best choice for a day of fishing. But that was all you could rustle at three o'clock in the morning, when you gathered up your gear and scurried to the docks. A salted breeze tugs the brim of your hat, a Cubs cap from the World Series, faded after a few years of protecting you from the Florida sun. Your boat, a 30-foot open skiff, skims over opal waters, the outboard motor humming smoothly. Ahead, the faintest blush of orange paints the undersides of the clouds. Now you can just make out the horizon, your destination lies ahead. Wilf is driving, a buddy who grew up in central Florida, and is one of the best: in twenty minutes he'll cut the motor and pick up a pole. Stealth is of the essence when the quarry is elusive and easily spooked. It's time to ready your rod, fly line, leader, tippet, and flies. You are entering the salt flats, where one of the world's feistiest fishes awaits. Today you will try to outwit and outmuscle a tarpon.

Leaving the rivers and sawgrass behind, you slide onto the salt flats. Freshwater streams deliver a steady supply of nutrients that spark an abundance of marine life and draw fish to rich hunting grounds. Wilf hoists the pole, slips it smoothly into the still water, gives a skillful push. He learned his trade watching James Holland, a legendary angler who landed the world's largest tarpon caught on fly: 202 pounds. For the next few hours you'll crisscross the flats with only Wilf's push pole, slowly and silently. The shallow water is ideal for poling, a mix of coral-white sand, some streaks of river silt, and irregular patches of lime-green turtle grass waving just beneath the surface. In the distance, fringing the flats, are ranks of red mangroves, their roots jutting into the water from dense clusters of shiny green leaves. A trio of pelicans flap by, skimming noiselessly over the calm surface, each wingbeat making pairs of concentric ripples like footsteps. You pull a couple of flies from your tackle box, and your hefty 12-weight rod; a clinch knot you could tie in your sleep, and the fly is secured to the tippet. Despite a rough start, all is ready, and now you sweep your eyes across the flats, looking for a mouth, tail, or belly. As your grandmother always said, you have to know a fish to catch it. These days, the fish you know most about is the tarpon, the one they call the silver king around these parts.

You know they are fond of shallow water and swim so close to the surface that dorsal fins or tails can be spied slicing the surface. That's one way to find them, and you scan the flat water now for the telltale wakes. Males can weigh 100 pounds, but it's the females who are the real champions, reaching twice that size or more. You know, too, that silver kings can be mouth-breathers: they regularly thrust their heads from the water to gulp a mouthful of air, especially in the early morning, then pivot their bodies beneath the surface in an unmistakable rolling motion. And suddenly, you see one, eleven o'clock off the port bow, eighty yards out. One roller, then another, and a third. You signal wordlessly to Wilf, who poles in silence toward the spot. Lifting your rod, you pay out a hundred feet of line; now you can make out a parade of dark shapes, two or three dozen, following each other over sandy shallows. You pick out the third fish in the daisy chain, a gargantuan female, and aim for the tail of the tarpon just in front of her. A couple of false casts and you've measured the distance. Then a few expert flicks send your line into the air, graceful arcs of yellow, and you drop the fly just in front of her. For a half-second, the world comes to a halt.

Wham, you feel the hit! Good grief, it all but yanks the rod from your hands. Instinctively you pull the butt toward you, setting the hook. She's powerful, you've latched onto a leviathan. You grip the rod as she takes off on a stormy run, and then you're treated to a sight from the dreams of every angler. A massive tarpon explodes from the water, huge silvery scales shimmering in the tangerine light of dawn, head whipping from side to side, her tail thrashing a twinkling waterspout into the air. Wilf whistles softly. She splits the water, and you feel her pause. You go on the attack, pumping the rod to the side and reeling, keeping the fish off

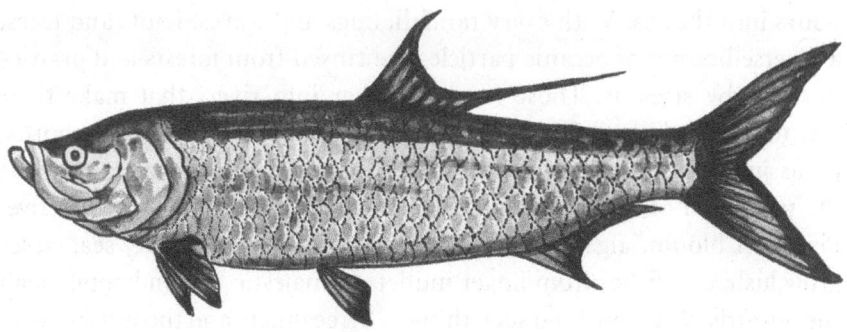

Atlantic tarpon (*Megalops atlanticus*)

balance while drawing her ever close to the boat. When she slows, you pull; when she runs you let the reel spin, and wait. Every subtle shift in her tactics is a clue, a tell, showing you the right counter-move if you are paying attention. You've been learning the art of this duel for years now, and you know enough not to let her get her second or third wind, or the fight could stretch to hours. Her sandpapery mouth can even fray your leader, leaving you with neither fish nor fly. Three more times she bursts acrobatically from the ripples and jackknifes in midair, then she runs at top speed, taking your line with her. But gradually you coax her back, regain your line, and bring her to the boat. Forty minutes of fierce concentration have skidded by without your noticing, a dawn of pure adrenaline and euphoria. Looking over the starboard gunwale, you see royalty clothed in scales: a glittering torpedo that must weigh over 150 pounds. Rather than haul her out of the water, you check that this silver queen is uninjured and skillfully release the hook. The game has been played out, and you handle your adversary with respect. She rolls, glares balefully at you with a giant eye, then twirls and swims nonchalantly away. You exhale. Wilf smiles. It's been a mighty good morning on the flats.[1]

A RIVER RUNS TO IT

> Every yard of river seemed to hold a rolling fish.
> —**Zane Grey**, *Byme-by-Tarpon*

While fishes of the abyssal depths must wait, peering skyward in hopes of spying a falling morsel, coastal fishes like tarpon live where food literally pours into the sea. With every rainfall, bugs and leaves, fruits and feces, a diverse bounty of organic particles are rinsed from forests and prairies into nearby streams. Those rivulets gather into rivers that make their way to the world's great oceans. There, the flow of nitrogen, phosphorus, potassium, and other essential nutrients mixes with salt water and fuels the profusion of life that makes saltwater flats and inlets so productive. Plankton bloom, algae abound, clams and mussels prosper, seagrasses grow lush, and fishes from finger mullets to majestic tarpon happily reap the rewards. But there is no such thing as a free lunch, and those fishes who school in the coastal shallows face risks unknown to open-water species.

While Atlantic tarpon (*Megalops atlanticus*) hungrily patrol the thin layer of water blanketing saltwater flats—along eastern shores of North and South America, and the west coast of Africa—they are themselves hunted by blacktip sharks and bull sharks (*Carcharhinus limbatus* and *C. leucas*). These speedy and rapacious relatives of great whites also pursue Indo-Pacific tarpon (*Megalops cyprinoides*), the only other species in the tarpon family, who swims the waters of eastern Africa, Central America, and Southeast Asia. When predators attack in the salt flats, there is nowhere for prey to hide, nowhere to escape: in a watery film sandwiched between sky and sand, tarpon have few options. But evolution has helped them develop some unique advantages. A hunted tarpon relies on armor, eyesight, the ability to tolerate conditions few others can even survive, and if all else fails, a turbocharged burst of pace that rockets the sleek fish to safety.

Tarpon evolved more than 140 million years ago and have changed little since: genetic evidence suggests they may be among the most ancestral of all bony fishes, their successful body plan remaining stable for many millennia.[2] Their bodies are elongated and flattened cylinders with deeply forked tails, adapted for speedy travel. Stretching to 8 feet in length, their flanks are packed with powerful swimming muscles, explaining how they can battle an angler for hours on end. Draped over elegant frames like intricately jointed plates of armor are huge, chrome-plated scales. So tough are those scales that only the sharpest teeth can pierce them, and even strong-jawed predators like sharks may bite at an awkward angle only to find their lunch skid elusively away. Tarpon earned the moniker of silver kings from these gleaming scales, silvery enough that they are used attractively in jewelry and ornamental artwork throughout Central and South America.[3]

Foremost among a tarpon's defenses is its ability to thrive in waters that few other fishes can abide. Nearshore environments like salt flats and mangrove inlets are highly variable in salinity, thanks to the uneven drainage of freshwater rivers. Most marine fishes are adapted to a stable oceanic salinity of around 35 parts per million of salt, but on the flats the water can veer from 10 to 50 ppm, thanks to freshwater floods during rainy months or evaporation during dry periods. Tarpon, as well as bonefish (*Albula vulpes*), bluefish (*Pomatomus saltatrix*), and ladyfishes (a handful of species, kith and kin to tarpon) are among the rare fishes who can handle such violent swings in saltiness. Larvae of all these species—glass-like ribbons that resemble eels—are extremely sensitive to salinity and must spend their first few months of life in the open ocean, where saltiness is unwavering. But as juveniles they migrate inshore and settle in the salt flats, mangroves, and other shallow coastal waters where their newly developed kidneys allow them to tolerate salinity fluctuations. When coastal water is extremely salty, fish kidneys work overtime to extract salt from the blood and release highly salty urine; when freshwater floods in and salinity plunges, the gills actively pull salt from water, keeping the blood in balance.[4] Salmon and eels and their migratory ilk are the most remarkable of all, as we shall soon see, capable of transitioning from the salty open ocean to pure freshwater. Tarpon can occasionally venture upstream into fresh rivers, but only briefly, and most stick to coastal areas where they are within their salinity comfort zone. But the majority of oceanic fishes cannot safely enter such waters, and when confronted with salt levels too high or too low, like Goldilocks and the Three Bears' porridge, they give up and turn back to an ocean that is "just right." Not unlike science fiction aliens in futuristic space crafts, tarpon swim behind an invisible force field of salinity that blocks many predators from breaching their sanctuary.

More than one force field guards inshore fishes from danger. While salt imbalances affect fishes slowly, and rarely fatally, nearshore waters are protected by a second invisible shield, one that can kill in a matter of minutes. As the sun shines on chartreuse meadows, those seagrasses pump out oxygen, suffusing the shallow waters with the life-giving gas. But come nightfall, photosynthesis ceases, while all manner of living organisms—from zooplankton to fishes to bacteria—continue breathing, slowly using up the accumulated oxygen. By dawn, the water can be nearly anoxic. Any oceanic fish that ventures into such waters is entering a death trap: a few minutes in anoxic waters and gills cease delivering oxygen to muscles and nerves, lethargy and paralysis ensue, and if well-oxygenated pockets cannot be found the fish will die of asphyxiation.

On planet Earth there are some 400,000 miles of shoreline (although if you enjoy the mathematical allure of fractals, then coastlines are infinitely long: the closer you look, the more crenellated they appear). Tarpon colonized these commodious inshore environments early in their evolution and discovered ways of surviving within the force fields of salinity and anoxia. Swimming toward shore, a tarpon is like a mountain climber ascending the slopes of Everest. With each step, there is less oxygen. Alpinists make their movements as efficient as possible; no wasted effort can be allowed. Tarpon are the same, swimming slowly and efficiently, staying within the range of exertion permitted by surrounding oxygen levels. As oxygen declines, they flap their opercula to flush ever more water across the gills. At this stage, mountaineers rely on "pressure breathing," deep exhalations that flush carbon dioxide from the lungs and allow more oxygen uptake. But eventually, such techniques fail to offset declining oxygen, and the climber must pull another trick out of their bag. Literally. At very high altitudes, all but the most extreme mountaineers will grab a cylinder of compressed oxygen from their backpack, don a face mask, and draw strength one breath at a time from the supplemental bottle. For tarpon, that emergency reservoir of oxygen is constantly at the ready: just a few inches over their dorsal fin, above the surface of the anoxic water, lies an atmosphere saturated with the life-sustaining gas. All you have to do is rise, open your mouth, and take a deep breath.

Tarpon, and quite a few inshore (and river-dwelling) fishes can do just that. They are mouth-breathers, whose upturned mouths swallow a gulp of air that is shunted to their swim bladder. While the epithet of mouth-breather can be hurled at the slow-witted as an insult, for tarpon it is a stroke of genius. As we learned earlier, most fishes inflate and deflate the bladder thanks to a gas gland and the Bohr effect, but in many species it is connected directly to the throat: air can be swallowed into the bladder, or burped from it. In tarpon and other coastal fishes, the bladder works exceptionally well as an air-breathing lung. Surrounding a tarpon's swim bladder are dense networks of capillaries that absorb oxygen from the inhaled air. Blood vessels are plumbed so the freshly enriched blood flows directly to the heart, which then pumps oxygen to energy-starved muscles throughout the body.[5] Human lungs are remarkably similar, apart from a few differences in plumbing. Thanks to this innovation a tarpon can breathe either water or air—or both. Under normal conditions, a relaxed tarpon may mouth-breathe only twice in an hour; the bladder contributes a paltry 2 percent of the body's oxygen.[6] While swimming, however, more oxygen is needed than gills alone can provide, and that rate kicks up to one breath every four or five minutes.[7] But in oxygen-starved waters, a cruising tarpon must inhale every single minute or it will grind to an asphyxiated halt: the bladder provides half, or more, of the fish's oxygen. Flush with a breath of air, the tarpon's muscles hum despite the energy-sapping water all around.

Warm water holds less oxygen than cold, so a tarpon's mouth-breathing allows the fish to thrive in balmy shallows where few others can survive. Near dawn, when those waters are the most oxygen-starved, a tarpon holds a decided advantage. It can outswim almost any fish it pursues thanks to the turbo effect of breathing air when those around you are breathing water. If attacked by a predator, a tarpon will gulp a mouthful—up to three-quarters of a gallon at once—to supercharge its muscles and dash to safety.[8] Oceanic predators who cannot mouth-breathe are sluggish when hunting in low-oxygen waters, and once a tarpon takes flight, they cannot give chase. Thanks to the swim bladder innovation, a tarpon can readily power a high-speed getaway and enjoy an undisputed advantage, as hunter or hunted, in the safe haven of its shallow water home.

To see their prey, tarpon rely on oversized eyes, rods tuned to shallow-water wavelengths, and a reflective mirror behind the retina. These adaptations allow them in the purplish first light of day to see better than their prey. Juveniles, who prefer murky waters, have rod-dominated eyes that see dark blues and greens best, precisely the colors that suffuse the gloom.[9] When they reach adulthood, however, their sensitivity shifts to purples and blues, which better penetrate clear waters. Adults even develop retina cells that can detect UV light, which is abundant in the crystal shallows found on salt flats. For all these reasons, tarpon fishing must take place at dawn: the fish must mouth-breath and can be spotted rolling through the surface; they can see a fly dropped in front of them; and they fight with all the manic energy that mouth-breathed air can give them.

There is only one compulsion powerful enough to draw tarpon away from the protective safety of their inshore force fields. At around eight years of age, tarpon slip quietly from the flats and enter the deep blue sea: in late spring, the urge to mate beckons irresistibly.[10] Spawning takes place well offshore, where sperm and eggs are released to fertilize, then hatch into tiny larvae. A full-grown female can release as many as 12 million eggs, a prodigious output linked closely to her size: the larger the fish, the more eggs it can carry, hence the largest of tarpons are always females. Eggs hatch into unique larvae, transparent ribbons shaped like a willow leaf, each called a leptocephalus. Larvae rely on invisibility for safety, lack red blood cells that would perilously give them color, and feed on marine snow (like deepwater fishes) that they filter through bristling, fang-like teeth. The exact spawning grounds, somewhere in the open ocean, remain shrouded in mystery, though satellite tags hint at breeding waters off Veracruz, Mexico, nearly a thousand miles from the Florida coast.[11] Crystal-clear leptocephali remain in the deep blue for three or four months until kidneys and gills develop and they morph into tiny tarpon. Once they can tolerate erratic salinity, they swim back to shore and settle in the stagnant safety of brackish lagoons and mangrove backwaters. Already, these juveniles have learned to thrust their heads from the water and gulp a breath of air. Now they are back behind the protective force fields of salinity and oxygen, hunting with aplomb on a steady diet of fingerling guppies, mullets, and the like.

Comfortably ensconced behind the force fields of salt flats and shallow backwaters, a few other species have evolved the physiology needed to live alongside tarpon. Mullet (a family of several dozen species, not the haircut) share a tarpon's propensity to mouth breath, and along with snook (a smaller family) will readily move into freshwater rivers or even hyper-saline lagoons.[12] In tropical regions, Atlantic bonefish (*Albula vulpes*) and their relatives hunt over mudflats and seagrass beds. Bonefish resemble tarpon in their large silvery scales, sleek torpedo-shaped body, and dorsal and pelvic fins set well back toward the tail. Cagey adversaries, a bonefish will put up a blockbuster fight if it can be hooked, and as a sport fish they are widely sought by anglers. Unlike tarpon, however, their mouths are distinctly downturned, better suited for plucking crustaceans from seagrass meadows and unearthing mollusks buried in silt than for gulping smaller fishes from behind as do tarpon. Despite their down-at-the-mouth appearance, bonefish readily roll at the surface to gasp air and use their swim bladder as an ersatz lung. They also share with their shallow water neighbors the habit of spawning in deep seas, and transparent leptocephalus larvae. Bonefish swim in tremendous schools numbering in the hundreds or thousands, though the very largest individuals eschew a crowd, preferring instead a solitary lifestyle.[13] This may be a reflection of one of their hunting techniques: relying on the tides to bring them to rich feeding grounds. When the tide is incoming, bonefish will swim up tidal creeks to places large predators rarely reach, and there they feast on buried and encrusting prey. When the tide turns, rather than risk stranding they ride the outgoing tide down the creek to deeper waters.[14] Thanks to their peripatetic lifestyle, wandering bonefish carry nutrients from rich inshore feeding grounds out to deeper waters, linking these separate ecosystems into a single food network.[15]

TANGLED UP IN BLUE

In the year 1291, the Venetian explorer Marco Polo left the employ of Kublai Khan (son of Genghis) and embarked on a four-year journey to reach his native shores. Having lived in Asia for nearly a quarter-century, his accounts of China and the Far East were some of the first descriptions of those exotic lands ever read by Europeans. During his circuitous return voyage, he explored Indonesia by sea and recorded landing at the city of Palembang, in what is now South Sumatra. There he found mangroves skirting the settlement, providing a protective fringe of forest against the battering served up by monsoon storms. Each tree was rooted on marshy sediment but propped up by dozens of pole-like roots plunging into the sea, like a conference of stilt-walkers jostling together at the beach. A veritable thicket of slender, crisscrossed columns, the roots offer bountiful space for algae, sponges, and mollusks to grow and also create sheltered nurseries for juvenile fishes. Over time, those still waters accumulate sediment, and new land is born from the ocean's sands. If Marco Polo had arrived today, he could never have landed his ship at Palembang, for one simple reason: it is no longer located on the coast. Mangrove trees, tiptoeing seaward on their slender roots, have steadily moved the shoreline over the intervening 700 years and now the city is an astonishing 30 miles inland.[16] At a rate of nearly 250 feet per year, the forest has contrived a land grab from the ocean, building habitat where once there was only water.

Thriving along the world's equatorial shores, mangrove forests fringe nearly a half-million miles of coastline and cover an area larger than the state of Illinois or New York, about 60,000 square miles. They achieve their legendary growth rates thanks to hyperactive photosynthesis fueled by tropical sunshine and lavish inputs of nutrients from rivers draining to the sea. Mangroves expend energy every minute of every day, however, fighting a silent killer: salt water. All mangroves block the uptake of salt with tight membranes between cells, actively pump pure water into their tissues, and shed any salt that breaches those defenses through specialized leaf glands. Despite the energy costs, mangroves are incredibly successful wherever blazing sunshine and salty shoreline are found together: more than thirty tree species have adopted the salt life.[17]

Just as corals build invaluable physical structure in an empty ocean, so too do mangroves weave new architecture on formerly featureless coastlines. Spreading branches, sprawling roots, and captured sand are home to a host of organisms: sloths, monkeys, deer, and even tigers roam these tangled jungles, while seabirds like pelicans and frigates nest in dense colonies amid the leaves. Below the waterline, a thicket of roots is encrusted with algae and all manner of clams, sponges, anemones, and crustaceans, offering a veritable smorgasbord to visitors from stingrays to sea turtles and even manatees. An impressive diversity of fishes—particularly juveniles—tuck themselves into the safe harbor of interlocking roots, like children playing under the dining table, where they feed contentedly while growing into adulthood. Barracudas, snappers, sharks, mullets, parrotfishes, pufferfishes, and even seahorses all enjoy the invaluable nursery of mangrove roots; in the hyper-diverse western Pacific, more than 200 species of fishes can be found in these sheltered waters.[18]

Other fishes visit mangroves intermittently, swimming in with the tide like bonefish, or window-shopping like a great barracuda (*Sphyraena barracuda*) until it surges like a glittering missile into the roots for a morsel. In their daily wanderings, snappers and jacks and others may visit half a dozen mangrove islands, browse a seagrass meadow, and swim out to open water to nab a young mullet or two. Both predators and herbivores connect mangroves, seagrass beds, and coral reefs like marine conveyor belts delivering suitcases of nutrients. Coral reefs tend to occur near mangroves, and seagrass meadows are often found between the two. Each system benefits from the proximity of the other. Corals enjoy a boost to their already impressive native productivity from abundant nutrients that pour out of mangrove forests: some reef corals can thank the forest for as much as a third of the carbon they use to grow.[19]

Seagrass beds hold an intermediary position in the chain of food delivery. As much as half the organic material fertilizing these undersea plants is supplied by mangrove forests; conversely, some 20 percent of the organic carbon found among mangrove roots originates in seagrass meadows.[20] Captivatingly named turtlegrass, eelgrass, manatee grass, and some sixty other species flourish in warm, sunny waters—including seas far from the tropics—and their meadows are some of the most productive ecosystems on Earth (and the most threatened), capturing and storing an estimated 7 trillion (with a t!) tons of carbon.[21] Each year they turn over as much as three-quarters of that productivity, feeding fishes and invertebrates, and shedding bits of leaves and stems.[22] Corals in turn can take up that organic matter, literally munching on the nitrogen-rich grass clippings swept their way by the outgoing tide.[23] Flowing water conveys lunch to the reef, one particle at a time. But the best food delivery comes in larger packages, and fishes make the best nutrient baggage handlers in the sea.

Sharks and jacks, grunts, snappers, and more perform nocturnal migrations between mangroves and coral reefs, feeding among the roots before returning to the reef.[24] Their midnight snacks in the mangroves eventually rain down on the reef in the form of fish feces, providing a boost of fertilizer for the entire coral reef community. In the Gulf of Mexico, gag groupers (*Myctereoperca microlepis*) forage in seagrass meadows before ambling to distant reefs, sometimes crossing 50 miles of open water to get there.[25] As much as a quarter of the grouper's muscle is built from prey harvested in the seagrass: if he is eaten by a shark while visiting the reef, that toothy predator is indirectly eating seagrass as well, even while hunting amid the far-flung corals. On the other side of the planet, off the coast of Queensland, Australia, mangrove feeding subsidizes the mottled spinefoot (*Siganus fuscescens*), a type of rabbitfish. These modest-sized herbivores graze on reefs and help keep algae from overtaking the corals, but mangroves contribute 40 percent or more of their daily harvest of calories. Were it not for mangroves, these little lawnmowers might run out of fuel before fully trimming the reef.

Back in the Caribbean, noisy grunts begin their lives in seagrass beds, but as they grow they seek the safety of mangroves. If none can be found, bluestriped grunts (*Haemulon sciurus*) will move to coral reefs, but they suffer much higher predation in those uncomfortably exposed sites: where mangroves are present, their numbers can be twenty-five times greater than where such hideouts are lacking.[26] Commercially important fishes like yellowtail snappers (*Ocyurus chrysurus*) double in size and abundance when mangroves are located near their reefs, a benefit to the reef and to local fishing villages. Gentle grazing giants like rainbow parrotfishes (*Scarus guacamaia*), the largest parrotfish in the Caribbean, make their homes on coral reefs but are absolutely dependent on mangroves as nurseries.[27] So reliant are they on aquatic kindergartens amid the roots that, in Central American sites where mangroves have been deforested, these marvelous polychromatic fishes disappear altogether.

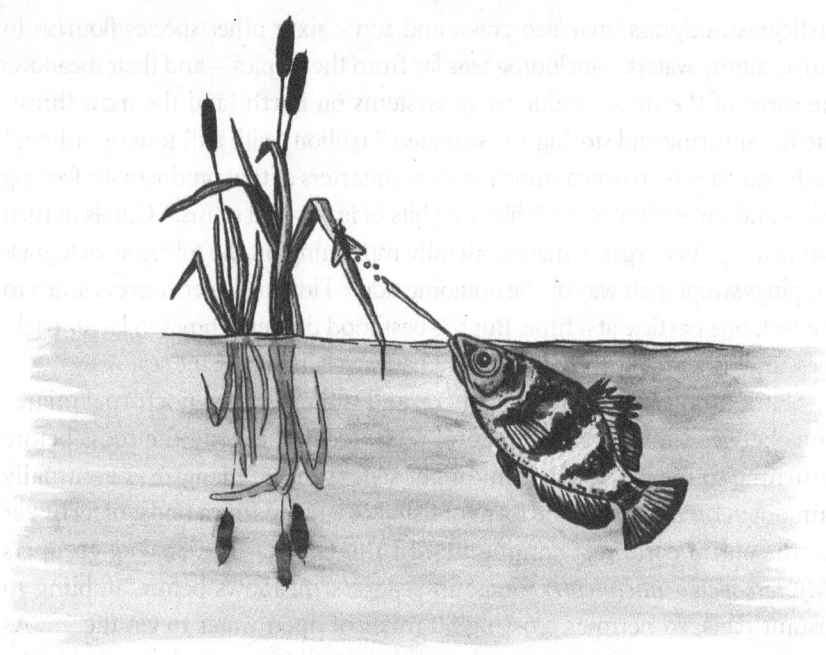

Banded archerfish (*Toxotes jaculatrix*)

Submerged mangrove roots are teeming with life, more than enough for hungry fingerlings to get their fill every day. But just above the waterline, the stilt-like roots are crawling with every imaginable critter the tropics can serve up. Tree frogs clamber and croak, spiders amble and spin, snails glide noiselessly, lizards scuttle and pounce, beetles march about, millipedes, crabs, ants, and hundreds more stroll from pillar to post, just inches above the sea surface. If only a fish could reach out of the water and wield a stick or flyswatter to knock down some of the cornucopia hanging tantalizingly overhead. One small fish, however, has solved this conundrum. It uses neither sticks nor swatters, but rather its own version of a water pistol: it fires an aquatic bow-and-arrow with the unerring accuracy of Robin Hood.

Seven species of archerfishes frequent mangrove-lined estuaries and billabongs from Sri Lanka to northern Australia. They are closely related, all members of the genus *Toxotes*, which fittingly is the Greek word for "bowman." Modest in size, archerfishes rarely exceed a foot in length and resemble small perches with dark brown bars or blotches and conspicuously pointy snouts. Living in the backwaters of mangrove forests, they encounter a lot of floating debris overhead: bits of moss, fallen spiders, the occasional dead beetle. These they pluck from beneath, jabbing their snout to the surface and sucking in the free meal. But somewhere in the distant past, an ancestor had a novel idea, or perhaps just an accidental hiccup, and evolution sprang into action like a geyser. Archerfishes began spitting drops of water at prey and were so successful that their mouth evolved into a precision water pistol. The roof of the mouth has a distinct slot running toward the lips, while the tongue is uniquely shaped to shove water along the groove. Combined, the two mouth parts can blast a jet of water beyond ten times the fish's body length.[28]

As any marksman will tell you, hitting the bullseye does not rely simply on the power of the bow. Archerfishes have developed a suite of skills that improve their odds. They boast exceptional eyesight capable of spotting a target despite confounding camouflage. The retina of each eye is packed with paired color-sensing cones specifically tuned to the brown background afforded by silt-covered mangrove roots. Those cone pairs are known as "offset detectors," a duo that emphasizes contrast: a light-colored bug basking on a slightly darker leaf will be highlighted, appearing almost to glow.[29] In the lower, rearward part of the retina—where light from overhead targets strikes—the cones are so tightly packed that an archerfish's visual acuity is the highest reported for freshwater fishes and compares favorably to terrestrial animals. Even once the prey is seen in high-contrast detail, archerfishes must still compensate for light refraction. Living underwater yet viewing objects above the surface, these clever fishes account for the distortion of light as it crosses from air to water. They achieve this feat by merging estimated distances between themselves, the surface, and the target, and they can accurately separate prey size and distance and even map every detail of its movement in three-dimensions.[30]

All these assessments, a complex set of calculations that would detain all but the sharpest of mathematicians, take place in as little as 100 milliseconds before the archerfish looses its bolt of water.[31] Once the unfortunate insect or spider is super-soaked and knocked from its perch, it tumbles into the water. Even then, the archerfish's trajectory calculator is tracing the complex descent. Gravity pulls downward while the water jet pushes horizontally; combined, these forces drag the insect along what is known as a ballistic path, like a punted football falling beyond goalposts.

To most animals, mapping the landing zone would be nearly impossible, but the diminutive archerfish—computations completed—sprints ahead and intercepts the bug precisely where it crashes into the drink.

If another archerfish observes such ballistic proceedings, it too can puzzle out where the insect will land and surges forward, straight as a laser, to the precise spot where it can catch the prey; triangulation, it seems, is not a feat performed only by cartographers.[32] Experiments meticulously conducted by Stefan Schuster at the University of Erlangen-Nürnberg in Germany unveiled something even more astonishing: captive archerfish, who in the wild hunt in small troops, watch each other's archery efforts and actually learn from them.[33] In Dr. Schuster's words, "observers can 'change their viewpoint,' mapping the perceived shooting characteristics of a distant team member into angles and target distances." Thanks to a bit of studying, subordinate fishes picked up several distinct hunting strategies for nailing bugs on the run. In one, known as "leading," the fish estimated the speed and direction of a moving insect and aimed its shot ahead, like a quarterback throwing a pass to a sprinting wide receiver. But they also learned another approach, "turn and fire," in which the fish rotates its body while blasting a jet, matching the prey's speed and hitting it on the nose (or antennae). Clearly, when aiming for a moving target in a thicket of mangrove roots, it helps to apprentice under an experienced master.

California sheephead (*Semicossyphus pulcher*)

A SUBMARINE FOREST

> To look into such a pool is to behold a dark forest, its
> foliage like the leaves of palm trees, the heavy stalks of
> the kelps also curiously like the trunks of palms.
> —Rachel Carson, *The Edge of the Sea*

Tropical sunshine, throbbing down like a fiery plumb bob onto blistered shores, is transformed into something cool, angular, and demure in the world's temperate latitudes. Move away from the equator, and mangrove forests begin to fade. They simply cannot harvest enough energy to sustain the unrelenting battle against salt. Coastlines, shorn of their protective mangrove roots, are pummeled by waves that strip sediment from shorelines like giant backhoes, gnawing at the base of rocky cliffs. Great boulders tumble into the surf. Yet still a forest survives beneath the waves, one that takes up the duty of mangroves to dampen the ferocity of the sea and provide a safe haven for fishes who need a little tranquility and security.[34] Anchored firmly to the bottom many yards below, a swaying kelp stretches upward to the light. Though not a true plant, these brownish-green algae bear numerous similarities to their distant terrestrial cousins (taxonomically speaking, more like great, great, great, great nephews). A holdfast grips the seafloor, taking the place of roots; a sinuous stipe spirals toward the sun like a trunk; flat and spreading blades, identical to leaves, absorb light to power photosynthesis. One kelp grows next to another, and another, forming tangled forests that cover hundreds of acres.[35] So successful are these algae that more than 130 species of kelps dominate a quarter of the planet's coastlines and provide structure and habitat for an utterly unique marine community.[36]

Bob Steneck, a marine biologist from the University of Maine, has been studying algae and kelp for more than fifty years. He got his aquatic start early in life. "I was always a water baby," he says.[37] "I started scuba diving around ten years old, before there were any certification programs, actually. In 1972 I was asked by a Smithsonian curator who was living on his trimaran sailboat if I wanted to join his research team, and you can imagine that was not a hard choice for me to make! So I spent '73 and '74 living on a 41-foot trimaran doing research." After two years at sea, his knowledge of the marine realm was encyclopedic. "I was a sponge. I could identify virtually all marine organisms by 1973, every angiosperm, alga, fish, invertebrate, I had it all under my belt." Soon he was drawn to the complexities of underwater algae, including kelp. Over his exceptionally productive career he has studied kelp groves around the globe and remains entranced by the experience of diving within them. "I go way back, working on coral reefs all over the world, frankly, but there's nothing like diving through a kelp forest. You are basically swimming through a forest . . . In places like California and Tasmania where I've worked, and in the Aleutians, you're going through these towering kelp forests, and it's very impressive. In some places the water clarity is terrific, and in a *Macrocystis* [giant kelp] bed these things can be 300 feet long. It really is unlike anything else."

Of course, meandering underwater through tangled ropes of giant algae carries its risks. Unflappable while scuba diving, Dr. Steneck nonetheless is keenly aware of the challenges posed by kelp. "I was working in the Aleutians with two German colleagues. I'm working away, and one of the younger guys came and he told me to come, that he needed a hand. The senior German scientist had started getting tangled in the kelp. He started rotating and he got tangled even more. When I came over to him there was this tower of kelp with two fins sticking out of the bottom. You couldn't even see him. So all I could do was take out my dive knife to cut off all the fronds. I found where his head was and I just gave him the thumbs-up, you're going up now."

For the kelp, growing in water has distinct benefits over a life on land. Freed from gravity, kelp can reach towering heights with very little in the way of mechanical support (though their stalks are loaded with unique, elongated cells that spiral upward, giving them an elasticity that terrestrial trees would envy). Many species suspend themselves from hollow bladders pumped full of buoyant gases that stud the upper branches and leaves. Thanks to these inflatable bladders, the accurately named giant kelp can reach heights exceeding all but the tallest of redwood forests. Swaying in cool temperate waters, kelp enjoy a continuous bath of rich nutrients that supercharges their growth. Around the globe, these oversized marine algae capture a staggering 150 teragrams of so-called "blue carbon" every year (1 teragram is 1 trillion grams, or 1 million tons, or a tad more than the weight of San Francisco's Golden Gate Bridge).[38] That harvest enables them to sustain an entire ecosystem, pinched between the cobalt-colored depths and the pounding surf, with a nutritious algae salad. Steneck reveals an unseen side of the forest's bounty: "A lot of photosynthesis leaks out of kelp. It's actually very common for all algae. Maybe as much as 40 percent of the photosynthates leak out of the kelp, and heterotrophic [food-eating] bacteria take advantage of that . . . and mussel growth rates in the vicinity of kelp are much greater than where there are no kelp."[39] So lavish is kelp productivity that each year literally tons of nutrients are shipped by currents to the continental slope, and even the deep sea.[40] Kelp can sustain extraordinary abundance and diversity, Steneck summarizes, because "you've got structure, you've got particulate food, you've got algal detritus, and so you actually have quite an advantage."

Within these dense and tangled jungles, coastal fishes find a sanctuary that protects them from the vicissitudes of wave and tide, offers safe refuge and protected breeding grounds, and teems with delectable life. The first morsels to set up residence on holdfast, stipe, and blade are diverse colonies of invertebrates. Towing their protective homes with them like pop-up campers to a national park, snails, clams, crabs, and amphipods (marine versions of a garden roly-poly) soon constellate the towering alga. In Norwegian waters more than 80,000 individuals were found on a single giant kelp, within a forest that harbored nearly 250 different species.[41] "All of the hitchhikers on the sides of the kelp," Steneck explains, "they too get out of the benthic [bottom] boundary layer; that means there's greater water flow—water flow is one of the bigger drivers of nutrient availability." Where there are invertebrates in abundance, there are invertebrate-eaters, and fishes of all sorts are drawn to kelp like hungry students to free pizza. In Ireland, over 300 species of fishes and invertebrates were found in surveys of just four kelp forests.[42] Worldwide, thousands of fish, invertebrate, and marine mammal species call kelp home.[43] While traversing Tierra del Fuego in Chile, Charles Darwin marveled at the diversity of fish that could be found in kelp, and its resilience in the face of oceanic adversity. "I know few things more surprising than to see this plant [actually an alga, even Darwin gets it wrong now and then] growing and flourishing amidst those great breakers of the western ocean, which no mass of rock, let it be ever so hard, can long resist . . . The number of living creatures of all Orders, whose existence intimately depends on the kelp, is wonderful. Amidst the leaves of this plant [!] numerous species of fish live, which nowhere else could find food or shelter."[44]

Just as fishes depend on kelp, so too do kelp groves rely on fishes. The seemingly robust forest actually teeters on a delicate ecological see-saw, and fishes lean heavily on the scales. Off America's west coast, kelp forests are cruised by California sheephead (*Semicossyphus pulcher*), bulky wrasses painted attractively with broad bands of midnight blue and sunset orange (the scientific name pays homage to their pulchritude, while their blunt head and white chin inspire their ovine common name). These burly carnivores hunt urchins, lobsters, crabs, and other critters that scuttle between the dense holdfasts. Like many fishes of cold waters, they grow slowly, breed late (at age seven or eight), and enjoy long lives of fifty years or more.[45] Powerful jaws are equipped with strong canines honed by evolution to pierce the tough shells of their prey. If the urchin or lobster is too well armored, sheephead may resort to a method blacksmiths might use to crack walnuts: they choose a rock as an anvil and slam their prey upon it until it shatters, a rare example of tool use by a fish.[46] Confusingly, a similarly named species inhabits the opposite coast: Atlantic sheepshead (*Archosargus probatocephalus*, occasionally dubbed "the fish with human teeth" for their broad front incisors) are robust fish who also crush invertebrates found while prowling nearshore waters, though they live far from kelp forests and are as yet unacquainted with tools.

California sheephead are particularly fond of urchins, and this predilection is a boon to kelp, because a mob of urchins is the alga's worst nightmare. When sheephead and other urchin-eaters disappear—from either overfishing or changing oceanic conditions—urchin populations usually erupt. Gangs of urchins gnaw ravenously at kelp holdfasts, chewing through the trunk until the entire kelp floats free, and perishes. If unchecked, these spiny buzzsaws can deforest acre after acre of kelp, leaving behind an urchin barren: bare, lifeless sand devoid of structure and food where once there was abundance. A famous example of such cascading effects occurred in Alaska at the turn of the last century, when hunters virtually exterminated sea otters for their luxurious pelts. Otters feast extravagantly on urchins, and when the fur trade decimated their populations in the late 1800s, the urchins exploded and thousands of square miles of kelp were wiped out.[47] Only after decades of intensive conservation efforts, particularly the Marine Mammal Protection Act of 1972, have otter colonies been restored in parts of their former range, where urchins now are kept in check and kelp can once more thrive.

On the coast of Maine, Bob Steneck's early studies documented wild swings in the enduring tussle between fish, kelp, and urchins. Where once only a narrow fringe of kelp persisted after heavy fishing unleashed years of unbridled urchin deforestation, intense harvest of sea urchins for Japanese culinary markets shifted the balance dramatically. "Without a doubt, kelp were the first to respond, and remarkably fast. I have videos that I took at my [barren] study site in 1993, and by 1995 it was 100 percent kelp. You know, for people who think about natural history broadly, the very possibility of going from a completely deforested to a completely forested state in just two years is just mind-boggling."[48] These days, however, urchins seem to have the upper spine, as their populations are booming worldwide.[49] "The bottom line is that the global rise of sea urchins, I think, follows on the tail of the global collapse of coastal groundfish, that is, fish that feed off the benthos."

In Californian waters, this saga played itself out all too clearly. In the early 1980s a commercial fishery began targeting sheephead. A decade later sport fishers added their hooks to the siege, and combined catches soared to more than a half-million pounds per year.[50] When numbers declined, size-based regulations were put in place, causing a perverse consequence. Sheephead, like most wrasses (portrayed in the coral reef chapters) change sex from female to male when large, harem-holding males disappear. Imposing a minimum catch size meant that anglers took mostly males, who then were replaced by smaller females. The remaining pint-sized females produced far fewer eggs, and the population's ability to rebuild after fishing was hampered even further.[51] Today, the sheephead is officially listed as a declining and vulnerable species. Urchins, as their fishy predator's abundance slipped, were unleashed and began decimating kelp forests. To make matters worse, the biggest sheephead are the only ones who can chew through large urchins, and large urchins—like large fishes—release more eggs, a biological rule that placed sheephead and urchin populations on opposite trajectories: the former plummeting while the latter boomed.[52] Kelp forests were savaged as a result, clear-cut into urchin barrens at an alarming rate. Fortunately, the state has since established a number of Marine Protected Areas where sheephead can grow safely to their accustomed larger size and greater reproductive potential. Thanks to the spillover effect, ecosystems outside the reserves benefit from reproduction inside. Protected and plentiful larvae, juveniles, and adults drift and swim across boundaries, boosting their own populations and knocking back those of kelp-nipping urchins.

Smaller grazers also contribute to keeping kelp healthy, by nibbling encrusting invertebrates and calcified algae off the branches and blades. Among these helpful cleaners is the kelp perch (*Brachyistius frenatus*), a diminutive and rhombus-shaped fish swathed in glittering, brass-colored scales. They pluck all manner of encumbrances from the kelp blades, akin to cleaning moss from solar panels, a scouring that rejuvenates the growth of the algae.[53] Clinging snails, gnawing isopods and amphipods, blanketing hydroids, burrowing worms, these nuisance critters are favored entrées on the kelp perch's menu. In addition to keeping the kelp clean, these little perches also serve as fish groomers, an activity more commonly seen in coral reef cleaning stations. Garibaldis and kelp basses, stunning opaleyes, brawny halfmoons, and many more kelp fishes rise to grooming salons near the surface and solicit a de-parasitizing session from these nimble valets. Not a true perch, kelp perches belong to the curious surfperch family (aka seaperch), a group of fishy oddballs because they give birth to live young. Males internally impregnate females, using a swollen part of their anal fin to transfer sperm. Embryos develop wholly within the female, after some delay, and feed on some of mom's tissues as they develop, emerging as perfect miniatures of their parent.

Kelp rockfish (*Sebastes atrovirens*) is another pint-sized denizen of the topmost blades, who sway with the waves and conceal themselves among the leaves as youngsters. Irregular, dark brown blotches on a background of dingy yellow only add to their camouflage. They also adopt curious postures while lolling amid the kelp. Marine biologist and self-styled humorist Milton S. Love bemusedly reported, "among the rockfishes, kelps seem to be among the least concerned about remaining upright, as you can see them lying sideways or even upside down, often with their lacy pectoral fins extended and lazily undulating in the current."[54] Presumably this attitude, or lack thereof, allows them to appear more kelp-like—and less fish-like—as they grow out of their awkward (and risky) teenage years.

Yet another disguise artist, giant kelpfish (*Heterostichus rostratus*) relies even more heavily on superbly matched and shifting colors (though the marketing department must have been on vacation when their dowdy name was handed out). More elongate than the previous rockfish and surfperch, they sport a pointy nose and extended, fringe-like dorsal and anal fins that superficially resemble an eel's. Green, red, and brown morphs, each splotched or barred or striped, can all be found. Colors can be changed quickly, to match the background of their particular patch of kelp, but the different shades also hint at gender and age. Males tend to be brown (and live among brown-hued algae), while females mostly favor green or red (and reside in like-colored districts); juveniles are green or greenish-brown.[55] Patterning can enhance any camouflage by breaking up outlines, so the bedecked fish fades into kelp fronds rather than presenting a recognizable silhouette. Among giant kelpfish, females typically are barred (lines running vertically, from back to belly), or plain-colored, while males, in contrast, seem to prefer stripes (head to tail) or blotchy mottling. But both sexes can shift at will: when males or females find themselves in wide-bladed giant kelp, they adopt stripes; in finer algae they shift to barring or mottling. Camouflage begins, however, long before adulthood. When female giant kelp lay eggs (unlike surfperch, they are oviparous, not viviparous), even the clutch of sticky little eggs is pre-colored brown or red to match the alga that serves as crib and cradle.

Why does this diverse community of fishes work so hard to breed, and remain invisible, within stands of kelp? Because the forest is also prowled by fearsome predators, hungry and sharp-eyed, always on the lookout for inattentive or unconcealed prey. Giant sea bass (*Stereolepis gigas*) and kelp bass (*Paralabrax clathratus*) inhabit kelp groves as juveniles, plucking crustaceans and mollusks from the branches and the seafloor below, but they mature into fierce, large-bodied, big-mouthed hunters of fish. Kelp bass are speckled with pale blotches on top of greenish brown (marketing team still on holiday, though some people supportively call them calico bass), but they can change colors quickly to match their surroundings. They spawn in aggregations of several hundred individuals, where white-checkered males court females with a variety of dance moves and fin bumps. Their ample reproductive output (spawning continues throughout summer) allows them to withstand modest fishing pressure, and today they are targets of a thriving artisanal fishery in southern California. As adults, they eat just about anything they can suck into their gaping maws, including kelp perch, rockfish, and others trying to hide among the fronds.

Giant sea bass are named adequately, if uncreatively, since they frequently reach 500 pounds, with record-setters tilting the scales beyond even 800 pounds.[56] Large males make powerful booming sounds, especially when disturbed, adding to the crackling of shrimp and the rest of the undersea orchestra. And for a time, they were disturbed incessantly, by a tenacious fishery intent on catching every last one of these giants: if you are a fish, it is most unfortunate to be both large and delicious. Fortunately, after it tumbled onto lists of critically endangered species, capturing this giant has been banned, and the population has made a remarkable recovery, particularly in the protected waters of Catalina Island.[57] When not being fished, they pursue a life that takes them in and out of coastal kelp forests. There, although they unmenacingly accept routine cleanings by kelp perch and even juvenile sheephead and kelp bass, they are formidable predators when appetite overtakes hygiene. Over their seventy-plus-year life span they will gulp lobsters, squid, crabs, skates, and just about any kind of fish they can find.[58] Their unmatched size makes them apex predators in their environment, and nearly every species that scuttles or swims is on the menu. Except one, that is, a fish who wields a unique and inflammatory defense against being swallowed.

Swell sharks (*Cephaloscyllium ventriosum*) are compact fish, just a yard in length, whose tan bodies are beautifully draped with irregular blotches of ochre sprinkled with speckles of white and black. By night they swim amid ropes of kelp, hunting for fishes and the occasional crab or mussel. Some are even bioluminescent, a curious phenomenon that may aid their camouflage or foster communication during nightly raids. By day they snooze in social clusters while camping around the bases of the kelp stalks, where their unique patterning helps hide them from predators. But predators do find them, larger sharks and terrifyingly oversized bass, and suddenly the swell sharks' lack of size and speed become distinct disadvantages. Discovered, they dart into the kelp forest, desperately trying to outdistance their hunters. But if caught, they have one more trick up their sleeve. As a last resort a swell shark will guzzle mouthfuls of seawater (or air, if near the surface) into a distensible stomach. As far back as 1947, the renowned marine biologist Eugenia Clark was cutting her teeth on shark dissections and noted that swell sharks actually have two stomachs. "The stomach of sharks is divided into two parts—the cardiac stomach, which follows the oesophagus, and the pyloric stomach, which follows the cardiac stomach."[59] Filled like a party balloon, the foremost stomach bulges until the swell shark nearly doubles in size, rendering it too big to be swallowed. "As in the puffer fishes, air or water is gulped into the stomach, inflating the belly region of the fish . . . [and] the cardiac stomach swells out in all directions." To make even less of a meal of themselves, a threatened swell shark will bite its own tail, latching on tightly and curling its body into a U-shape, and a rather bloated U at that. Confronted with a swollen balloon that can no longer be swallowed, the perplexed predator swims away for other less-inflated prey.

Thanks to all these specialists, kelp forests are incredibly diverse (especially those in the Pacific) and packed with uniquely adapted species found nowhere else. Like rainforests, in their delicate balance they cultivate biodiversity. Perhaps it is the kelp forest's ability to buffer the changing conditions that batter coastlines—erratic temperature, light, salinity, and wave force—that makes them most attractive to fishes. It can be challenging to thrive near the shore, where change is the only constant. Still, a few fishes who swim in the sea are ready to confront variety, and even use wildly changing conditions to their advantage. To do so, they will swim halfway across an ocean and confound the world's brightest minds for centuries.

10 SWEET AND SALTY

Eels, Salmon, and Alewives

A SALTY BREEZE STIRRED THE LEAVES of a fig tree casting dappled shade on two Greek scholars sipping from cups of wine. Wave tops glinted under a Mediterranean sun that rose 2000 years ago. Young Theophrastus turned to Aristotle, a dozen years his senior, and posed a query that was simple yet impishly challenging: "master, from where do eels arise?"

"Well," replied Aristotle with a modicum of grandiosity, "the mullet goes up from the sea to marshes and rivers; the eels, on the contrary, make their way down from the marshes and rivers to the sea."

"That's all fine and good, master," said Theophrastus, "but I wasn't inquiring about mullet, and you haven't answered the question. As your eminent studies revealed, fishes reproduce by admixing their sperm and eggs. But what of these mysterious eels?"

Aristotle trotted out discoveries about the slender, slippery fish he had published in his masterpiece *The History of Animals*: "The eel is neither male nor female, and can engender nothing, as no eel was ever yet seen with an egg. Furthermore, while all male fishes are supplied with milt, the eel is an exception: with the eel, the male is devoid of milt and the female of spawn. In point of fact, this entire species of blooded animals proceeds neither from pair nor from the egg."

Theophrastus' own brilliant mind was quick to spot the conundrum. "But master, if they have neither males nor females, and yield neither eggs nor milt, then how is it that every year a multitude of eels visit our waters? What could be their birthplace?"

The elder man painstakingly outlined a sequence of logic that would make Plato, his own teacher, proud. "There is a species of mullet that grows spontaneously out of mud and sand," he began.

"Oh my," sighed Theophrastus, "we're back on the mullets again."

Unperturbed, Aristotle forged ahead. "This occurs in a pond in the neighborhood of Cnidos. This pond at one time ran dry, about the rising of the Dog Star, and the mud all dried up. At the first fall of the rains there was a show of water in the pond, and on the first appearance of the water, shoals of tiny fish were found in the pond."

His student would not relent. "That may very well be true of mullets, for of them I know but little. But I doubt it has much to do with eels."

"There can be no doubt that the case is so," thundered Aristotle. "For in some standing pools, after the water has been drained off and the mud dredged away, the eels appear again after a fall of rain. In time of drought they do not appear even in stagnant ponds, for the simple reason that their existence and sustenance is derived from rain water."

"From rain water? I thought you said from mud and sand."

With a bombastic flourish, the celebrated Greek overruled this churlish objection. "From the facts enumerated it is quite proved that certain fishes come spontaneously into existence, not being derived from eggs or from copulation."

"But how is this possible, master," spluttered Theophrastus, "and what could trigger these fishes to suddenly appear?"

Aristotle batted aside his pupil's misgivings. "As a proof that these fish occasionally come out of the ground, we have the fact that in cold weather they are not caught, and that they are caught in warm weather. Obviously they come up out of the ground to catch the heat."

"To catch the heat," repeated Theophrastus, wanly. "You have got to be kidding."

"I never kid," retorted the great philosopher. "Eels are derived from the so-called 'earth's guts' that grow spontaneously in mud and in humid ground; in fact, eels have at times been seen to emerge out of such earthworms, and on other occasions have been rendered visible when the earthworms were laid open by either scraping or cutting."

Open mouthed, Theophrastus slowly blinked twice. "Earthworms," he enunciated. "So let me see if I've got this straight. According to you, eels appear magically from mud and sand, or possibly from rain water, but

European eel (*Anguilla anguilla*)

occasionally from inside of earthworms, all because they are somewhat chilled and need a bit of sunbathing." He snickered, "I think you have drunk too much wine this day, my dear master."

Never one to concede the last word, Aristotle brought the conversation to an abrupt end with one of his favorite catchphrases: "so much for the generation of the eel!" With that, he promptly rolled over for a nap.

Theophrastus chuckled, stretched under the fig tree, and mused, "I wonder if anyone will unravel this riddle. Certainly it has stumped even the great Aristotle."[1]

SLIPPERY CUSTOMER

> Throughout the Middle Ages and even in our modern
> times, there has been a veritable frenzy to find a male eel.
> —**Sigmund Freud**, *Letters to Eduard Silberstein*

Aristotle's mystery would go unsolved for at least two millennia. Nobody in all that time could divine how eels reproduced, nor even whether there were distinct males and females. Finally, in the year 1707, a suspiciously plump female eel sold in the Italian port of Comacchio was found to be swollen with eggs. Eels, it would appear, did indeed have gender. But where were the males? This new puzzle would vex scientists for 200 more years, until the arrival of another towering intellect. Sigmund Freud, just nineteen years old and newly enrolled in the University of Vienna, was dispatched to Italy in 1876 to search for eel testes. One year earlier, a Polish zoologist named Syrski had written of mysterious "lobed organs" in larger eels netted near Trieste (some over 12 feet in length!) that he suspected were the long-lost gonads.[2] Young Sigmund was given more than 400 specimens to dissect and examine under a microscope, which he described in meticulous detail. He was stymied, however, by the simple fact that he did not know exactly what he was looking for. After weeks rummaging for testes, he was becoming testy himself. "Recently, a zoologist in Trieste claimed to have found testicles and thus to have discovered the male eel," Sigmund groused in a letter to a classmate, "but since he apparently didn't know what a microscope is, he failed to provide an exact description of them."[3] In the end, however, he convinced himself that he had indeed identified the elusive testes, using a bit of logical double-negativity that would have done Aristotle proud. "The microscopic examination of the lobe organ does not contradict the view that the lobe organ is the testicle of the eel," he wrote in his very first scientific publication.[4]

Freud seemed to know instinctively why his eel examinations were inconclusive. "I found the most advanced state of the lobe organ only in the larger eels, from about 400 to 430 cm, more frequently in September and the following months than in March. During the whole period of my investigations, however, I came across forms of the lobe organ in smaller eels which I must regard as underdeveloped."[5] Demonstrating an impressive grasp of the eel life cycle, having only handled immature specimens, Sigmund hypothesized that the testes were not yet fully formed because "the eels are not in the least prepared for the reproduction business when they go out into the sea." Freud made his way back to Vienna, perhaps to prepare for the reproduction business himself, as he would soon propose to his wife Martha. Twenty years after he left Italy, a mature male eel with fully formed testes was finally caught near Sicily, dispelling Aristotle's mystery once and for all.[6]

There was one very good reason why the reproduction of eels had stumped Aristotle, Freud, and the long line of scientists that stretched between them. That reason is that eels mate very far from Europe. Eels belong to a select group of fishes that are diadromous, traveling between salt and fresh water to breed. Specifically, eels are catadromous, a cross-word vowel bonanza describing fish who reproduce in salt water and later swim into freshwater streams and ponds where they spend the majority of their lives. European eels, triggered by a complex set of environmental and hormonal cues, swim downstream from their freshwater home until they reach the ocean, and only then begin to develop the sexual organs that Freud sought. Autumn marks the onset of the breeding season, explaining nicely why young Sigmund found more advanced organs in the largest (and oldest) eels caught in September.

Johannes Schmidt, a Danish biologist who started his career at the dawn of the twentieth century, was already aware that eels reproduce at sea. Less celebrated than the scientists who came before him, he would outwork both Aristotle and Freud to solve the last outstanding puzzle: where do eels mate? Having completed his doctoral research on mangroves, Schmidt stumbled across a single eel larva in the North Atlantic, a transparent ribbon just three inches long. Eels share the same larval form as tarpon and bonefish, a leptocephalus. From that serendipitous beginning in 1904, it would take him seventeen years, three research vessels, twenty-three sampling freighters, and the whole of World War I before he had his answer. He began finding smaller larvae the further southwest he traveled from Europe: first two inches off the coast of Africa, then just an inch and a quarter in the middle Atlantic, under an inch as he approached the Caribbean, then nearly microscopic. After capturing and examining more than 6000 leptocephali, and painstakingly documenting their locations, he had his map: the smallest larvae all came from an elliptical area now known as the Sargasso Sea. "Hosts of eels from the most distant corners of our continent," he summarized with a certain flair, "shape their course south-west across the ocean, as their ancestors for unnumbered generations have done before them. How long the journey lasts we cannot say, but we know now the destination sought: a certain area situated in the western Atlantic, N.E. and N. of the West Indies. Here lie the breeding grounds of the eel."[7] The coda of this epic detective story was contributed by a young biologist working for the New York Zoological Society, appropriately named Marie Fish. In 1926 she discovered actual eggs of the American eel in samples collected from the Sargasso Sea, proof positive that the Western Atlantic was indeed the cradle of eel reproduction. Triumphantly, she laid to rest a scientific riddle that had persisted for 2300 years: "the sea has given up the last secret concerning the life history of the American eel which it has jealously guarded for so many centuries."[8]

At the heart of these concentric mysteries are several species of freshwater eels, cousins to morays and their coral reef ilk, but residing primarily in freshwater lakes and rivers across the globe. Aristotle and Freud were bamboozled by the European eel (*Anguilla anguilla*), but in the Sargasso their eggs and larvae mix with those of American eels (*A. rostrata*). As the larvae develop, they enter the Gulf Stream and make their way north and east. American eels (distinguished only by their number of vertebrae and muscle segments) develop quickly, and after a single year are strong enough to fight the current, turn west, and hightail it for the coast of North America. Their European counterparts grow more slowly, requiring two to four years to attain the same degree of independence, and consequently are swept by east-flowing currents to the opposite side of the Atlantic. Amassing near their respective coasts, in their respective graduation years, great shoals of juvenile eels gather. By now they have transformed out of their ribbon-like leptocephalus form into slender versions of their adult selves. Still they are nearly transparent, earning them the moniker of "glass eels." By the millions, glass eels wait for slack river flows and rising tides before plunging into river mouths and battling their way upstream, onto the continental land mass where they will spend their entire adult life. At least, that is, until the siren call of mating summons them back to the sea.

Of the remaining eighteen species of eels, some glide through Australian streams (short-finned eel, *A. australis*), slip amid ponds in Japan and China (Japanese eel, *A. japonica*), or meander the rivers of Southeast Asia (giant mottled eel, *A. marmorata*). Despite their far-flung distribution, most eels are quite similar in form and behavior. Highly elongated fishes, with long and supple backbones that permit them to swim with a distinctive side-winding motion (appropriately dubbed "anguilliform"), they can even swim backwards. Adult eels are draped with minute scales, so small they barely can be detected, but their bodies also are covered with a thick mucus that renders them slippery, slimy, and probably better protected against parasites. Forward-facing eyes and prominent nostrils allow them to see and smell prey, which they pursue voraciously. Along their muscular back a ribbon-like dorsal fin extends all the way down the body, wraps around the pointed tail tip, and joins with a similarly fringing anal fin, like a low fence that outlines two-thirds of their body. European eels have slightly underslung mouths giving them a faintly bemused appearance; their American cousins are distinguished by a more pointed snout (their scientific name *rostrata* means "beaked"). Most eels are a dark brownish-yellow as adults, though populations inhabiting clearer waters tend to be lighter in color. At this stage they are sometimes referred to as "yellow eels," and they may live in their freshwater homes for thirty years or more. They are biding their time, waiting for a signal, and they are exceptionally patient. Captive eels have been known to survive fifty or even eighty years in aquariums. In Sweden, where it once was not uncommon to toss an eel into a freshly dug well so as to keep it free of insects and other non-potable creatures, one well-dweller is alleged to have survived for 150 years.[9]

While eels pass the time in ponds and rivers, they grow steadily from the glassy immature phase into large, brawny predators. Formidable hunters, they stalk prey by night, taking crustaceans, insects, and all manner of fishes, occasionally even surging from the water's edge to seize an unwitting frog or baby bird. During the day, eels retire beneath stones or sediment. If a hard winter should strike, they survive by burrowing into the mud for a few months. Should their pond or stream dry up, they are capable of traveling significant distances overland, winding through wet herbs and grass, even climbing low walls and other obstacles by braiding themselves together into a pillar. If moved from its home, an adult eel will migrate many miles to return, showing tremendous navigation abilities. After several decades spent in upland waters, however, subtle signals prompt the eel to embark on a final, and massively ambitious migration. Rachel Carson, in many ways the poet laureate of conservation, dramatized how a female eel might sense these cues: "A strange restiveness was growing . . . For the first time in her adult life, the food hunger was forgotten. In its place was a strange, new hunger, formless and ill-defined. Its dimly perceived object was a place of warmth and darkness."[10] That strange new hunger is the song of the sea, summoning eels from Europe and America to enter the ocean and make their way to the balmy Sargasso Sea.

The magnitude of this transoceanic undertaking is staggering. An eel must navigate downstream and evade all manner of dams, sluices, dikes, mills, power plants, and other obstacles.[11] Its wall-climbing and grass-clambering skills become invaluable, even life saving. Along the way predators from bears to hawks and more seize the chance to dip a paw or talon into a stream positively boiling with tasty flesh. Should the eel survive its overland journey and reach the ocean, its migration will have barely begun. For this voyage, however, the eel's old body no longer serves. Prompted by entry into salt water, its fins grow longer, for better propulsion. Eyes enlarge, and turn blue, the better to see in the open ocean. Its sense of olfaction sharpens, allowing the eel to smell its way toward the Sargasso.[12] The yellowish color is replaced by silver on the belly, and stripes on the flanks. Reproductive organs begin to develop; at the same time its stomach and intestines cease functioning and nearly disappear. From this moment on, the eel will eat no more, subsisting only on fat stores built up during years of freshwater hunting.[13] In readiness for deep diving, the eel's swim bladder prepares for frequent use: the gas gland ramps up its pumping ability, and guanine crystals are deposited in the bladder walls to decrease its gas permeability.[14] Swim bladders in silver eels are five times as effective at generating and retaining gas as they were just a few weeks earlier, in their yellow-bodied selves. But its very entry into the ocean, where the eel is surrounded by salty water, represents the most fundamental challenge of all. The eel must adapt, and quickly, or risk dehydration and death.

All living organisms are composed mostly of water, and if immersed in an ocean, the salt inexorably draws out that internal moisture. This is why fish are cured (i.e., dried) with salt, and why you get thirsty after a swim in the sea. It is not, however, why your fingers get "pruney" when swimming, a reaction actually driven by the nervous system; as evidence, people with nerve damage to their hands never get wrinkled fingertips. Fishes who spend almost their entire lives in freshwater begin to dehydrate as soon as they enter the sea. In response eels drink seawater, and now a desalination battle begins. Proteins like prolactin and hormones like cortisol and angiotensin course through the body, initiating a wave of physiological changes to fight salt.[15] When the first mouthful hits the esophagus, salts are absorbed so the water reaching the stomach is a third less salty.[16] Next, the swallowed water passes into the intestine, where both water and salt are absorbed into the bloodstream by newly stimulated cells lining the gut. Those intestinal wall cells spend energy to actively pump salt out of the gut, and water is drawn along passively: the eel actually gains some internal water from the sea. But the blood is now higher in salt, and as it reaches the gills, that excess salt must be eliminated. Salt is actively driven across the delicate folds of the gills, shed back into the ocean while precious water is retained. Eel gills are six times more water-impermeable in the ocean than in a lake, a vital adaptation that blocks internal water from diffusing into the sea.[17] Fish kidneys, unlike those of land animals, are not capable of making concentrated urine, so they are of little use in this transition; however, in a few pages they will help salmon survive entry into fresh water. Thanks to the sweeping changes from mouth to gills, eels are able to survive in their new briny environment within a matter of days. Now they are ready to venture across an ocean.

From the shores of Europe, a migrating female eel sets out toward the southwest, along with tens or hundreds of thousands of kinfolk, and follows the seaboard as far as the coast of Africa. There she turns to the right and hitches a ride on a powerful west-flowing current that carries her across the Atlantic. The round-about route is actually easier to swim, thanks to the current, than if she had set a straight course to the Sargasso Sea. Since migrating eels have ceased eating, she must power her peregrination entirely with stored energy, converting fat and protein into miles as she swims methodically toward the New World. Now her highly efficient side-winding style comes to the fore, as she has just embarked on a journey of some 3500 miles: anguilliform swimming burns only one-sixth the energy of a typical fish to cross the same number of miles.[18] Still, upon arrival at the breeding grounds she must court a mate and lay eggs—from 2 million in smaller eels to 5 million in larger individuals— so energy saved by hitching a ride on a cross-Atlantic current will become invaluable.[19]

As she swims, she adopts an odd habit of changing depths. During the night she will swim at depths of around 500 feet, but by day she descends to 1500 or even 3000 feet. Predator avoidance may partly drive this yo-yo profile, but the leading hypothesis suggests thermoregulation is the cause. Swimming in deep, cold water (around 50 °F) may delay development of her reproductive organs, stalling their depletion of precious energy until absolutely necessary; ascending by night into warmer waters boosts her metabolism and swimming speed, giving her the best of both climates.[20] Along the journey, she will rely on her newly honed olfactory system, and an internal compass, to navigate. Eels can detect the intensity and bearing of Earth's magnetic fields, allowing them to maintain a westerly course while crossing an ocean largely devoid of landmarks.[21] Even with all these advantages, when this eel reaches the Sargasso she will have melted 40 percent of her fat stores to fuel the voyage;[22] the remaining 60 percent will be mobilized to produce millions of eggs.[23] After six months or more at sea, this eel has reached the end of her journey and the culmination of her long story arc. Soon after releasing her eggs to float to the surface of the Sargasso, she will perish, her life's work—and a round-trip crossing of the Atlantic Ocean—complete.

Most species of eels are semelparous: they reproduce just once in their lifetime, then they expire, making every baby eel an orphan. When the microscopic, leaf-like larvae hatch they drift into currents bearing them to Europe and North America. A young eel does not return to the stream where its parents lived, a homing instinct found in other fishes who migrate between fresh and salt waters, such as salmon. Across Europe the entire eel population is genetically similar, since parents from thousands of freshwater sources all mix together in the Sargasso, and their offspring upon returning select rivers nearly at random.[24] Still, for reasons not yet understood, some streams are more appealing than others, and glass eels assemble in enormous numbers at the mouths of these rivers in preparation for the uphill march. Gathered there, they make easy targets.

Fishes strike from below, birds plunge from above, all manner of hungry animals feast on the defenseless glass eels, also known as elvers. People also stalk them, stretching nets across river mouths and scooping up juvenile eels by the millions. Since Aristotle's time and before, people around the world have captured eels for food, but today in the state of Maine a pound of elvers can fetch more than $2000. Mostly the elvers are shipped to Asia, where aquaculture programs raise them into adults for sale to fish restaurants and sushi shops. In part due to this booming demand, eel populations worldwide have plummeted. European eels are now critically endangered, their numbers having fallen by 99 percent since the early 1980s.[25] Once they made up as much as half the fish biomass in freshwater ponds and streams, an abundance almost unimaginable today.[26] American eels—who, like cod, helped the *Mayflower* pilgrims survive in their new land—are faring little better, now listed as endangered.[27] Fishery catches of adult eels are sternly managed around the world in an effort to rebuild their populations, and glass eel harvests have been banned in all but a select few streams. For eels, fishing is not the only threat. Widespread installation of dams, diversions, dikes, power stations, and hundreds of other structures that interrupt river flows also block downstream eel migrations.[28] Industrial chemicals like PCBs leach into freshwater systems and are gradually absorbed by eels, coming to rest in fatty tissues. During their oceanic migration, fat reserves are mobilized to power the journey, and those toxic chemicals exact a heavy toll.[29] And like most fishes, eels suffer the attentions of parasites, including one worm that invades the swim bladder, impairing the eel's buoyancy and ability to swim properly across the wide ocean. For animals who spend so much time in both fresh and salt water, it is no surprise they are affected by the hazards of both worlds.

Little wonder, then, that the mysterious eel baffled Aristotle. He was stumped because it seemed impossible that a fish could live for so long without mating, could travel far from home to do so, and could show no signs of reproductive organs until the very end of its life. That they can swim halfway around the world to court, mate, and die is almost beyond comprehension. One theory postulates that the warm seas sought by eels were much closer a few tens of millions of years ago when these remarkable fishes evolved, because the Atlantic Ocean was much narrower. As the continents drifted, the Sargasso Sea slid further and further away but eels kept adapting to cross the ever-widening gulf, like ultra-marathon runners gradually upping their distance. Despite more than two millennia of study, to this day there is much about the life history of eels we do not know. Consider this: an adult eel has never been found in the Sargasso Sea, nor has any living person ever observed two eels mating. Slippery in action and origin, eels yet guard a few secrets; like an enchantment, mystery shrouds them still.

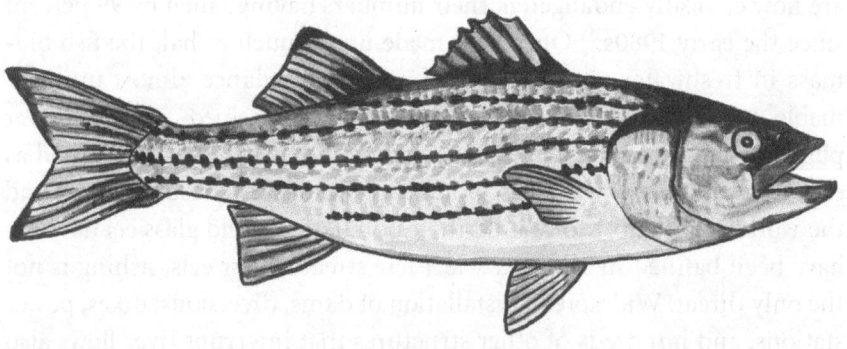

Striped bass (*Morone saxatilis*)

WHAT COMES DOWN MUST GO UP

Although the habit of migrating in and out of fresh water is rare (just 1 percent of all fishes do so), over 440 species in fifty-eight different families have adopted the diadromous lifestyle. Anadromy—breeding in lakes and rivers, like salmon—is more common than the eel's catadromy, with one-third of species choosing this route to reproduction. Of no little significance is the fact that more than half of diadromous fishes are commercially important species, such as salmon and eels, shad and sturgeon.[30] This unique life history must provide substantial benefits to those species able to cope with the transition between two very different environments, but there is disagreement over the origin of diadromy. The leading hypothesis posits that resource availability enticed fishes to expand their watery horizons.[31] Near the equator, ocean productivity is relatively poor while freshwater streams boast considerable resources. In these places, catadromy would have arisen as saltwater fishes invaded the land in search of more food. Conversely, near the poles, crystal clear rivers and lakes on land offer scant nutrition, but the seas are booming with productivity, tempting freshwater species to descend toward the ocean and the promise of a ready meal.

This explanation has a certain elegance, though not all diadromous species fit the pattern: both giant mottled eel and African longfin eel (*A. marmorata* and *A. mossambica*), for example, live near the equator.[32] An alternative theory postulates that in cold latitudes fresh water would be invaded less frequently since ponds fluctuate dramatically in tempera ture and can even freeze solid. Others contend the abundance of saltwater predators make streams a safer place to rear young. Combinations of all the above likely apply, in varying permutations, to different fishes: if the cost of migration outweighs the cost of staying put, then adopting a diadromous lifestyle becomes advantageous. And, if the ocean is a pathway rather than a barrier, an additional set of benefits kick in. Take New Zealand, where half that isolated nation's fishes (including eels) are diadromous.[33] Genetic evidence reveals that sea-swimming allowed species to colonize new streams around the main islands, bolster established populations that were declining, and re-colonize waters where local extinctions had taken place. Crossing the ocean even permitted mainland fishes, over geological time, to settle on remote offshore islands. The Australian longfin eel (*A. reinhardtii*), for example, appears to have reached New Zealand within only the last fifty years, having successfully swum over from the continent of koalas and kangaroos to reach the land of kiwis.[34] Where most freshwater fishes treat the ocean as a salt-ridden, poisonous barrier, diadromous species know an opportunity when they see it.

What may be considered a colonist on one continent, however, can be an invader on another. Take the alewife (many in the Great Lakes would add an emphatic "please!"). This innocuous-looking member of the herring family is a foot-long, silver-sided fish with a dark and shallowly forked tail, and a propensity to swim upstream in huge numbers. The alewife (*Alosa pseudoharengus*) is native to the northwestern Atlantic, where it routinely enters rivers of eastern North America for breeding: like salmon, it is anadromous. Unfortunately, one inquisitive alewife eventually discovered Canada's Welland Canal, built in the early 1900s for shipping traffic to bypass Niagara Falls. Alewives seized an opportunity to circumvent the falls, which heretofore had been an insurmountable barrier to these underpowered little fish, and by the 1930s they had reached the Great Lakes.[35] Their population soared in this new freshwater universe, and soon they were the most abundant fish in Lake Michigan, exceeding half of all fish biomass.[36] Municipal drinking water intakes were routinely clogged by dense knots of alewives.[37] By 1974, commercial fishers were harvesting in excess of 20,000 tons annually, without so much as denting the population.[38] Ballooning schools savaged the larvae of resident fishes such as lake trout, a popular sport fish, whose numbers plummeted like a barrel tumbling over Niagara.[39] Another anadromous pest, the sea lamprey (*Petromyzon marinus*), also had entered the great lakes with devastating effect. Known as the vampire fish, these eel-like parasites attach a round mouth bristling fearsomely with teeth to the flanks of larger fishes and proceed to gnaw, tear, and slurp the life from their unfortunate victims. Lake trout, hit with a double whammy of predator and parasite, were headed for extinction.

But alewives and lampreys decimated their own food supplies, swiftly eating themselves out of house and home. Seasonal fluctuations in prey abundance, water temperature, and nutrient levels led to mass alewife die-offs: millions upon millions of fish corpses piled high on lake beaches where they rotted, to the considerable visual and olfactory distress of would-be sunbathers. Here was an invasive species in desperate need of control. Fisheries managers would eventually find a savior in two non-native species, chinook and coho salmon (*Oncorhynchus tshawytscha* and *O. kisutch*). Voracious predators in their own right but native to the Pacific Ocean, chinooks and cohos were stocked into Lake Michigan in the 1960s, and the other Great Lakes soon followed suit. Over the following forty years, nearly 1 billion salmon were poured into alewife-dominated lakes in the United States and Canada.[40] They soon became flagships of a resurgent sport fishing boom and took pressure off lake trout by making significant inroads in alewife numbers. Chinooks in particular devour alewives the way moviegoers plow through popcorn: in some years alewives made up more than 90 percent of chinook stomach contents.[41] Bottom trawl surveys in Lake Huron showed alewife numbers sank by 50 percent in the 1980s, and by the early 2000s had nosedived by 99 percent.[42] Eventually, the popularity of salmon sport fishing encouraged lake managers to maintain alewife populations as food to sustain valuable coho and chinook stocks. Today these Pacific and Atlantic newcomers have established permanent populations that oscillate in response to environmental factors and each other's abundance. A new equilibrium has been reached, in a novel ecosystem that North Americans accidentally created but learned to manage for environmental stability and food production.

When European colonists first reached the New World, they nearly starved to death; arriving in winter, without agricultural know-how, or even farming tools, will tend to do that. But they were saved by the bounty of the sea. When the *Mayflower* pilgrims could no longer reach nearshore cod, they would rely on anadromous species that came to them. "In the year 1623 the Plymouth colonists had but one boat left," wrote a historian in the early 1800s, "which then was the principal support of their lives, for that year it helped them for to improve a net wherewith they took a multitude of bass, which was their livelihood all that summer."[43] Fortune, and migratory pathways, smiled on the pilgrims, as they had set up their colony near spectacularly rich runs of striped bass (*Morone saxatilis*). Also known as stripers, or rockfish for their habit of congregating and cavorting around river rocks, some robust individuals can weigh over 100 pounds and live up to thirty years or more.[44] Stripers are handsomely shaped, with a pointed nose and a deep belly tapering gracefully to a shallowly forked tail fin. Silver flanks are boldly striped from head to tail with eight or nine dark lines. Like sports fans wearing team jerseys, striped bass can herald their origin with colors: the dorsal side of coastal individuals is tinted olive green, while offshore residents exhibit more a bluish cast.[45]

After reaching maturity (females in eight years, males in just four), they migrate into estuaries and up freshwater rivers to mate. Striped bass are impressively fecund, with the largest females releasing as many as 5 million eggs. Juveniles frolic in fresh water for one or two years, then make their way back down to the sea where they devour smelt, herring, alewives, and lots of crustaceans.[46]

Reports of their abundance while running upriver are awe-inspiring, and it is little wonder that their bounty was once a mainstay of coastal dinner tables. Residents of Plymouth in 1637 gushed, "there are such multitudes, that I have seene stopped into the river close adjoining to my house with a sande [seine net] at one tide, so many as will loade a ship of 100 tonnes."[47] Even in the 1960s researchers recorded "an aggregation of large striped bass forming a 4- to 8-foot-wide band along the east bank of the river. This band, which was formed by several thousand fish, was about 1,000 yards long."[48] That this latter report was from California's Sacramento River, where only 435 fish had been introduced a century earlier and 2500 miles from their native Atlantic, shows just how adaptable and prolific are striped bass.

As has been seen with cod, however, seemingly inexhaustible populations of fish can collapse after only a few years of overfishing. In the case of New England, the colonists soon conceded that striped bass was on the decline. In probably the first wildlife conservation statute enacted in North America, the General Court of the Massachusetts Bay Colony decreed in May of 1639 that "it is forbidden to all men, after the 20th of next month, to imploy any codd or basse fish for manuring the ground."[49] In other words, no more grinding up stripers into fertilizer, something the pilgrims had also done with menhaden. Instead, striped bass would be protected solely as a food fish, and a delicious one at that. "The Basse is an excellent Fish, both fresh and Salte. They are so large, the head of one will give a good eater a dinner."[50] Commercial fishing in the twentieth century unfortunately overwhelmed even their immense reproductive output, and numerous striped bass fisheries were closed outright in the 1970s. Strict catch limits were applied, and later modified to implement a "harvest slot": only medium-sized fishes can be caught, a policy that protects both developing juveniles and large, reproductively important females.[51] Populations today are rebounding to the delight of fisheries biologists as well as sport fishers.[52] As far back as 1849, the exhilaration of hooking a striper was trumpeted, when guide books sang the praises of "the boldest, bravest, strongest, and most active fish that visits the waters of the midland States."[53] It seems only fitting that anglers, whose annual licenses provide substantial funding for striped bass management, today play a key role in conserving a fish that has sustained Americans for centuries.

AND MILES TO GO BEFORE I SLEEP

> The Waters of this river is Clear, and a Salmon may be Seen at the deabth of 15 or 20 feet. Passed three large lodges on the Stard Side near which great number of Salmon was drying on Scaffolds.
> —William Clark, *Journals of the Lewis and Clark Expedition*

They called the place Celilo, meaning "echo of falling water" in several tongues, and the thunderous roar of the cataracts could be heard for miles. Indigenous peoples had gathered at this site for some 11,000 years, making it the oldest continuously occupied settlement in North America.[54] Wishram, Wasco, Wyam, Chinooka, and many others were drawn to the broad river and its roaring waterfalls by a fish; more precisely, by the yearly arrival of millions upon millions of salmon. Nets the size of dinner tables were deployed on long handles from steep banks, dipped into roiling waters to catch salmon disoriented by the raw power of the falls. Chinook salmon (*Oncorhynchus tshawytscha*) can attain weights exceeding 120 pounds, and annual harvests at Celilo were turned into thousands of tons of pressed and dried fish.[55] A portable and eminently tradeable source of protein that could be stored for two years or more, preserved fish from winter salmon runs helped build a sprawling economy and some of the richest cultures in the world. When Lewis and Clark passed the area in 1806, on their return journey from the Pacific coast, they were astonished by the sheer size of Indigenous populations. Contrary to popular belief, theirs was not an exploration of a vast, empty continent; instead, the expedition found itself leapfrogging from one native settlement to the next, through a densely populated and intricately managed landscape

Sockeye salmon (*Oncorhynchus nerka*)

where they were repeatedly rescued by the kindness and generosity of local tribes. At Celilo, they traded for fish and other goods side by side with people from Indigenous nations across the West. From ornamented shirts and deerskin moccasins to onions, horses, and tools, "all of those articles they precure [sic] from other nations who visit them for the purpose of exchanging those articles for their pounded fish of which they prepare great quantities."[56] A born marketer, William Clark dubbed the place, "the Great Mart of all this Country. Ten different tribes who reside on Taptate and Catteract Rivers visit those people for the purpose of purchaseing their fish . . . and Such articles as they have not."[57]

Today, the bustling mart—and indeed the waterfall itself—is no more. Celilo Falls (also called Horseshoe Falls, or Columbia Falls for the name given its river by white settlers), plus an impressive series of downstream cataracts and rapids, were drowned by the construction of the Dalles Dam in 1957. Part of an extensive system of hydroelectric power production in the West, the dam is 200 feet high, nearly 9000 feet wide, and takes its name from a French voyageur term for rapids. On the eve of construction more than 5000 Indigenous fishers still plied their nets from nearly 500 sites along the steep banks.[58] But the salmon runs were soon finished, blocked by an insurmountable wall of concrete. Cold waters that drained from high mountain snowpacks, ideal for rearing juvenile salmon, would no longer flow unimpeded to the sea. Native communities on both sides of the Columbia were displaced as rising waters flooded fishing sites and market grounds. Although fish ladders were built for salmon migrating upstream, and a spillway added in 2010 for juveniles heading downstream, the dam remains a colossal impediment. For the salmon, not to mention Pacific Northwest native peoples, life would never be the same.

Most salmon grow to adulthood in the richly productive waters of the northern Pacific, devouring crustaceans and small fishes who are slurping a rich stew of plankton. But those planktivores also devour eggs, so salmon—like striped bass—enter the mouths of freshwater rivers and swim furiously upstream to reach mating grounds where their eggs can be deposited safely. There are seven species of Pacific salmon (all in the *Oncorhynchus* group like the chinook), separated by a continent from their cousin the Atlantic salmon (*Salmo salar*). In both oceans, these fishes grow from juveniles to brawny, powerful adults: some 95 percent of their weight gain occurs in the marine realm. Most salmon are shaped like muscular torpedoes, with dorsal and anal fins affixed fairly far aft, as is characteristic of strong, swift swimmers. The mouth is set slightly low on a pointed snout, and jaws are studded with sharp teeth befitting a carnivore. Body color ranges from greenish-yellow to silvery, depending on the species, as do diverse patterns of ornate spots, speckles, and stripes. Thick flanks drive powerful thrashes of a salmon's shallowly forked tail, pink-orange muscles that provided abundant food to Indigenous Americans and now to diners around the globe. The exact color of salmon muscles reflects their diet and geography, seasoned with a dash of genetics. Astaxanthin, a carotenoid pigment (named after carrots), cannot be synthesized by salmon but is abundant in their diet of shrimp, krill, and other marine crustaceans.[59] Sockeye (*Oncorhynchus nerka*) and coho muscles are the darkest red, while some chinook harbor a recessive gene that programs for ivory-colored flesh; a few oddballs even show white on just one flank while the other is pink.[60]

Assigning common names to living organisms can be fraught with confusion, and the words "salmon" and "trout" offer befuddling examples. There are salmon who skip migration entirely, remaining land-locked in fresh water their whole lives like some sockeye, but still they are called salmon.[61] Meanwhile among trout, typically seen as permanent freshwater residents, there are a few species who undertake migrations to salt water, such as the rainbow trout (*Oncorhynchus mykiss*). Even worse, salt-inhabiting rainbow trout often are labeled steelhead salmon. The unfortunate reality is that there is no concrete definition of the word trout, a term applied with distressing inconsistency to various species of *Oncorhynchus* and *Salmo*. "Char," another moniker for freshwater members of the salmon family (in the group *Salvelinus*), muddies the waters further since some char are bewilderingly referred to as "sea trout." All these fishes, however, are similar in appearance, streamlined and sporting an extra dorsal fin near the tail (the adipose fin); all are predatory in both fresh and salt waters, but only true salmon will migrate from oceans up into streams and rivers.

After three to five years of growth, a seafaring salmon reaches full adult weight and is ready to reproduce. By the millions, passionate salmon males and females gather along coastlines, readying themselves for an upstream journey that can range from a brief, casual jaunt to an arduous migration. Some king salmon (a regal synonym for chinook, reflecting their status as the largest of all salmon) will travel up the Yukon River from Alaska to British Columbia, traversing nearly 3000 miles.[62] Around the Pacific thousands of rivers once hosted salmon migrations, although dams and other obstacles have reduced that list to a fraction of its historical number. Regardless of the count, when salmon depart the vastness of the sea and enter river mouths, they are funneled from a dispersed life in an open environment into dense parades up a narrow freshwater boulevard. That concentration of living animals, passing by river mouths and rapids in immense quantities at predictable times is what made salmon the largest Indigenous fishery in the Americas. The Pacific Northwest was home to some of the most advanced native civilizations on the continent, largely because a year's supply of salmon could be caught and dried in a few hard-working weeks, leaving abundant leisure time during which wood carving, weaving, bead art, dancing, dramatic storytelling, and other forms of culture could flourish.

Before oceanic salmon test themselves against a gauntlet of fishing nets and human-imposed barriers, they undergo significant physiological changes. They have spent years at sea, fattening and preparing for migration, and now they are ready to return home. Sockeye salmon may travel 8000 miles or more through ocean waters as they develop.[63] King salmon spend four years at sea traversing more than 10,000 miles, but all these fishes unerringly return to the precise stream in which they were born.[64] Magnetic sensory cells have been found in salmon skulls, a trait they share with eels, that enable them to follow compass bearings. But it is this fish's sense of smell that enables them to find a freshwater needle in a saltwater haystack. Salmon possess exceptionally acute olfaction and just need a whiff of a few scent molecules from their native stream to find their way. Experiments comparing blinded salmon (wearing dark sunglasses actually) to those with their noses plugged by cotton confirmed that only those fish with unblocked nostrils were able to find the stream of their birth.[65] Other ingenious experiments with juvenile salmon revealed they "imprint" on their home stream, meaning they can sense lower concentrations of odors to which they were exposed as youngsters.[66] A superior sense of smell may not be the only course-plotting tool salmon have at their disposal: juvenile salmon can even detect the sun's position, and thus celestial navigation likely plays a role.[67] Despite these tools and skills, some salmon will inevitably swim up a river that is not their own. Known as "straying," these migratory wrong-turns actually provide considerable benefits, since salmon could never colonize new rivers nor restock declining runs if they exhibited only error-free navigation.

Once an adult salmon reaches the mouth of its natal river, it stops eating and henceforth relies wholly on stored energy (though this is not true of Atlantic salmon, who do return to the ocean after inland breeding forays). Just offshore, they wait for spring floods to scour away tidal sandbars, clearing a path from the sea. Upon entering the river, however, they are exposed to fresh water, a liquid as dangerous to them as salt water is to eels. To fight bloating and the loss of salt, salmon rely on special cells in the gills to actively take up salt; membrane proteins that pump salt from those cells surge in abundance as a salmon readies for migration. Meanwhile, kidneys work overtime to draw excess water from the bloodstream and manufacture a very dilute urine. Even in the bladder, additional salt is pulled from the urine before it is jettisoned back into the river.[68] Adapting to fresh water uses valuable energy, and a salmon will need every volt in its battery banks to power the arduous journey through currents, rapids, and waterfalls. For this reason, salmon entering rivers are at the pinnacle of physical condition, tuned for a marathon swim upstream, dramatic physiological alterations, and protracted battles on the mating grounds. Fishers who target salmon runs—whether modern-day commercial fleets, Indigenous netters, or hosts of bears, herons, and more—are catching majestic fish at their muscular best.

Longer and more strenuous migrations require more energy and are undertaken only by the largest species: chinook, chum (*Oncorhynchus keta*), Atlantic, and coho, all of whom can bulk up to 30 pounds or more. On their way upstream, all salmon fight against river currents that range from feeble to firehose. Their streamlined shape, which permits speed and acceleration during ocean hunting forays, now allows them to slip through the onrushing waters. Powerful contractions of thick flank muscles drive vigorous tail thrusts, propelling them upstream and occasionally launching them into the air. Salmon are famous for leaping up rapids and waterfalls, circus stunts engrained in their DNA by some 50 million years of evolution. In Idaho, chinook will climb to over 7000 feet of altitude, so powerful is the drive to reach the cold, fast-moving, highly oxygenated waters ideal for rearing eggs and larvae. Along the way they may hurdle falls 10 or 12 feet high in a single bound, the equivalent of a pole-vaulter clearing a bar 30 feet above the mat. Some fish make the jump in a single attempt, others fail at first then circle to regroup and try again. Many shrewdly take advantage of a rapid's standing wave, just below where a waterfall strikes the river, a counter-rotating eddy that thrusts water upward into a wavelike fountain and serves as a springboard for the salmon's vault. After days or weeks of surmounting such obstacles, the strongest and most fortunate salmon reach the end of their migration: the spawning grounds.

In cold upland streams and ponds, males arrive first and await females. Once on the scene, a female sets about building a spawning nest called a redd, a shallow trench dug in gravel (though sockeye redds can be 5 feet long and a foot deep). Females compete vigorously for the best nesting sites, where pebbles are just the right size, and current sluices just enough cold, oxygenated water overhead but not so much as to wash away the eggs.

Meanwhile, males have adopted dramatic costume changes in preparation for the big dance. Sockeye and coho exchange their mantles of silver and gray—excellent camouflage in the open ocean—for cloaks of the brightest carmine. In sockeye, the head is repainted a contrasting greenish-yellow; in coho (known in the ocean as a "silver"), olive-green wraps the head and back; chum salmon turn mostly olive, with large maroon blotches on their flanks. Skin pigments are repurposed from pink muscle tissue, at some energetic cost, explaining why the meat of late-run salmon becomes deathly pale. Male sockeye and pink salmon (*Oncorhynchus gorbuscha*) develop a pronounced hump on their backs, responsible for the latter's nickname of "humpy." The jaws of many species become grotesquely prolonged and bent, with the upper maxilla twisted into a hook and overlapping a curled lower mandible (Pacific salmon species share the scientific name *Oncorhynchus*, Greek for "hook snout"). Transformed males begin to battle for the attention of egg-bearing females: bright colors serve as a proxy for an individual's physical health, while the hooknose gives an advantage when fighting rival suitors.

After digging her nest, a female selects a mate, as best as one can in waters roiling with fish, then deposits eggs into the redd. Her chosen male rushes in to fertilize the eggs with milt. Moving upstream, the female digs another redd, the current carrying some of the excavated gravel downstream to cover the preceding nest. Meanwhile, some males who might not win by virtue of color or fisticuffs try a different strategy. Chum salmon can adopt female coloration in mere minutes, a disguise they use to slip past rival males and quickly fertilize eggs while the macho boys still strut and wrestle. Over the course of several days, females will lay hundreds or thousands of eggs (depending on the species) and spend their final hours protecting their redds until the eggs settle safely into the gravel. Then, her work complete, she dies. All around her, vigorous scenes of life's origin mingle with tableaux of death as the cadavers of spent fish fill the breeding grounds, and then are swept downstream. Still, in death there is life, as the bodies of dead salmon provide important fertilizer to aquatic environments below, and a valuable food source to scavengers like eagles. In Alaska's Wood River basin, salmon even sustain a flower, a petite herb related to parsley and known as kneeling angelica. Umbrella-shaped clusters of white flowers are pollinated by a carrion fly that lays its eggs on salmon corpses; each year, the plant times its flowering precisely to coincide with the salmon run and blooms within ten days of their arrival.[69]

In the redd, half the salmon's eggs will be eaten or die of their own accord, the remainder hatching into tiny larvae after a month or three. Of those, just a quarter will survive to become parr, juvenile salmon.[70] Despite these heavy losses, they are nowhere near what ocean-breeding fishes experience (as much as 99.9 percent of larvae)[71]. Upland lakes and streams have fewer predators than the sea, and the farther uphill one travels, the more scarce egg-eaters become. Parr remain in their birth waters for anywhere from one to three years before succumbing, like adult eels, to the call of the ocean. A few depart early, like pink salmon who quit fresh water nearly as soon as they hatch. Whenever they begin, all embark on the reverse journey of their parents, over rapids and fish ladders, until they reach the river's mouth and taste salt water for the first time. Now silvery in color, the youngsters are called smolt, a word sharing roots with "smelt," and referencing the extraction of equally shiny metal from ore. Some will gather in nearshore eelgrass for a time, relying on those protective meadows for safety and food until they are ready to venture into the open ocean. Curiously, many smolt enter the sea tail-first, as they find it easier to face the down-rushing river current and swim with slightly less speed, traveling gently backwards.[72] It is perhaps fitting that these juveniles, in culmination of a two-generation journey covering hundreds of river miles, rejoin the ocean facing the same direction as their parents: upstream, toward their origin.

These days, regrettably, many parr never smell the sea, and many adults never reach the breeding grounds. Around the world, rivers have been managed for priorities other than fish migrations, and countless salmon stocks have vanished. Dams like the Dalles have blocked upstream and downstream movement. Logging dumped tons of sawdust into rivers before the practice was banned, and winter-cut logs were floated downriver in the spring, just in time to barricade the annual upstream salmon migration. DDT, before it too was banned, was sprayed widely over forests to control pest outbreaks, but it savaged aquatic insect populations on which the parr rely for food. Mine tailings blocked rivers. Livestock to this day disturb river bottoms and trample streamside vegetation. Irrigation returns water to rivers warmer, saltier, and with less oxygen. Outflow from power plants also heats rivers. Global climate change is doing much the same, raising the temperature of lakes and streams and lowering their oxygen levels, changes that alter the timing of migration,[73] raise rates of disease and mortality,[74] and inhibit juvenile development.[75] Pitted against this parade of iniquities, it is astonishing that salmon have survived at all. But tenacity and adaptability are salmon's strongest traits, and survive they do. Along the way they have received a little help from their friends in fisheries departments, who annually rear hundreds of millions of salmon fry in hatcheries then release them into streams to boost wild populations.

Yearly salmon runs are still strong on the Copper River in Alaska, where shoals of the regal chinook have been stable for years, under the watchful eye of fisheries managers and commercial fishers. Sockeye in Alaska's Bristol Bay also continue to provide bountiful but sustainable harvests, as do coho in the southeastern waters of that state. Alaska in particular has done well to conserve its salmon, codifying their protection in the state constitution and preemptively stocking rivers with hatchery-reared larvae. Whereas overfishing has decimated populations in some regions— the Connecticut River, for example, and countless others in eastern North America—careful management in a few rivers has ensured consistent stock sizes and stable catches.

In some ways, the solutions are easy. Easy to identify, that is, but far more difficult to put into practice. As far back as the early 1800s, what was required to save salmon was widely known. Oregon Territory's 1848 constitution mandated that all dams across rivers known to host salmon must include a bypass for both upstream adults and downstream parr. Earlier still, in 1824, a review of the declining state of salmon in Britain outlined a remedy that applies just as well today: "Remove the obstructions and the fish locks; keep the fish to the natural stream; prevent all unsizeable and unseasonable fish from being taken; protect them during the fence days and let no fish be taken but by fair and legal nets . . . and we soon will have no reason to complain of the scarcity of salmon."[76] Sage advice that unfortunately was largely ignored for 200 years. Today, despite renewed interest in removal of obstructions (nearly 2000 dams have been eliminated from US rivers), we are prophetically facing a scarcity of salmon, and many other marine fishes too, and we all have reason to complain.

Part IV
TIDE TO TABLE

"A NORMAL DAY STARTS AROUND 4:00 a.m. I have three alarm clocks, and I turn each of them off and I stuff them deep in my sleeping bag," says salmon boat captain Megan Corazza with a chuckle, "and then I finally wake up."[1] Her great-grandparents moved to Alaska in the 1930s, her grandfather fished for the Snug Harbor cannery, and she was raised aboard family boats in the Prince William Sound fishery. "I've been on salmon boats since I was one year old," Corazza relates, "So for me that was life, there was no other, we didn't get summers, we were on the boat all the time and that was completely normal to me." At the age of twenty she became one of the youngest skippers in the state. "I was at college, and I was so busy with what I was studying, and I just looked at this piece of paper, and I had hardly any money in my bank account, and I signed it," she laughs again, warmly. "And all of the sudden I owned a boat, and a big debt."

Though the art of fishing came naturally to her, being a skipper presented new challenges. "My mom took me out on my dad's boat before we got my boat in the harbor, and she said, 'I'm gonna teach you how to drive a boat.' I was driving through the harbor, and she pointed at this dock and said, 'Okay, I want you to very slowly nose up to that.' So I came at it pretty slow, and I put it in reverse, and I had no idea there was such a big delay on the shifter . . . and it actually shifted in reverse about the time that I rammed into the dock." She chortles at the memory, "I just watched the whole dock undulate like a wave all the way to the shore. So I knew nothing about these big boats."

Today, Corazza is a successful and experienced captain who divides her time between a house in Homer and hardworking months aboard the FV *Centurion*, her boat. "The whole crew lives together all summer long, and some summers we don't come back to town at all." After her three alarm clocks roust her from bed, she and three crew members get straight to work. "I spend the morning looking for fish, I'm listening to my radio group talking about what they're seeing for fish," she describes. Meanwhile, "the crew is getting the deck ready . . . and then my skiff man goes out and checks the oil and everything in the skiff, and gets the skiff started." When salmon have been found and the boat is in position, captain and crew get down to the business of fishing. "I turn around and give a thumbs-up, my one crew 'pulls the pin'—meaning they release the skiff—and the skiff spins off the back of the boat, connected to the net, and starts pulling the net off my boat, and I drive in the opposite direction."

The *Centurion* is a purse seiner, dragging an enormous net with a drawstring. "[We] do a quick round haul, which means you can see where the fish are and you just do a circle around them, and then close your net up." Working in tandem, "the skiff and the boat come together and the purse line, which runs through rings on the bottom of the net, you pull the purse line up. Then you're stacking the net on the boat and you're pulling the purse line up until all the purse rings come up beside the boat and you can put them on this big metal hook. And now you've got all the fish trapped, they can't go out and dive under the net, because the net now is like a purse, the whole bottom gathered and right beside the boat." With the fish safely netted, the crew turns to getting them on board and into a hold filled with refrigerated seawater. "So then I haul gear and the crew stacks the net back on the boat, until we get to the very end. At the end, all the fish are just right beside the boat in this big bag, and we roll the bag on board."

In one day of hard work, skippers can bring as much as 40,000 pounds of fish into port, mostly pink and coho salmon. But fishing is fraught with risk, and dangers lurk everywhere. "On deck you're dealing with so much weight, and so much strain. You've got a skiff pulling on the side of the boat," she describes, "and if the skiff man hits the end of the rope too hard, it can snap the line and come back and have a snap-back and take your crew out at the knees." Hauling up the net, when it is weighed down by tons of fish, is particularly hazardous. "You've got a huge amount of strain on the purse line, and it can snap and hit people in the chest. I've known a couple guys who had it snap into their abdomens, and kind of wreck their guts." Crew members have fallen into the fish hold, and skippers have stumbled on heaving decks and broken legs or hips. "You've got a deck winch, which is a big capstan that's used to pull up the purse line," she continues, in quiet tones. "We call it 'the mutilator' because it wrecks people. If you get a piece of clothing caught, and there's four wraps of lines going around the deck winch, it'll suck you in and it will break your arm and rip your arms off, and pop your eyeballs out of your head." She speaks from experience. "I'm not making these things up, this has happened to many friends of mine." Even once the fish are loaded on board, threats lurk. "In our engine rooms we have Freon, because we're refrigerating our salt water, so you've got a whole system with a deadly chemical going through it, where if you get a pinhole in your copper tube you're gonna die." As skipper, these dangers are uppermost on her mind. "When I run my boat, getting the most fish is not my priority. Having a healthy, safe, and happy crew at the end of every day is my priority."

This captain's own well-being is sustained in part by her catch. "Salmon is my most favorite thing to eat," Corazza exults. "I feel super healthy when I eat it, and I think it has amazing health benefits, especially for our brains. So my pantry is always full of salmon, and my freezer is full of salmon." Comfortingly, her pantry has been well-stocked for decades. Over forty years of fishing Prince William Sound, she is reassured by the fishery's stability. "I do think in my short—my career feels long, but it's such a pin drop in the world—I don't see a reason to question this environment's ability to keep supporting salmon runs." Those runs are annually bolstered by juvenile salmon fry released from hatcheries. "In Prince William Sound I think that the success is in part due to the hatchery system," she explains. "What I have noticed as a fisherman is that having a hatchery in place takes a lot of pressure off the wild stocks . . . If a hatchery stock comes in, and the wild stock that year is weak, the fishermen can fish on the hatchery stock, and they never have to get anywhere near the wild stocks."

Corazza distinguishes such hatcheries from closed-cycle aquaculture and admits, "I get kind of grossed out by thinking of genetically modified salmon that are maturing in a matter of months, and being fed weird dyes to get the natural color in their flesh . . . it doesn't seem like good food." But she is quick to point out what should be an obvious truth: "there's nobody that's more concerned about the conservation of salmon than salmon fishermen, who need the salmon to be there every year." Whereas aquaculture can produce large amounts of salmon on a low budget, she believes in supporting the trade of fishing. "We want to be out fishing our coasts rather than raising our fish in pens." When she says "we," she means her family, and her local community. "The majority of fishing that I've ever known in my life happens on small, family-owned boats . . . our salmon fisheries, and most of the halibut fishermen I know, are all small boats that have been in families for years and years." Painting a poignant portrait of a profession passed from generation to generation, Corazza says of the Alaskan salmon boats, "we raise our kids on them, and we fish by our parents."

11 | FISH TO THE RESCUE

Feeding a Hungry Planet

DESPITE HARDWORKING MONTHS AT SEA BY FISHING CAPTAINS like Megan Corazza, the oceans can no longer supply as much salmon as the world consumes. Today, more than three-quarters of the salmon we eat comes from aquaculture farms, primarily in Norway and Chile, where over 3 million tons of these diadromous fishes are produced every year.[1] People around the globe relish seafood, and they eat more of it with each passing year. Fishes, crustaceans, mollusks, and algae—collectively known as blue foods—provide 17 percent of the protein eaten on the planet. Some peoples are more blue than others: Icelanders set the world record, according to the United Nations, each year pickling, frying, baking, and grilling more than 180 pounds of seafood per person.[2] By comparison, US residents eat a modest 50 pounds per year, close to the global average. Seafood has always been a staple for coastal peoples, making up half or more of their traditional diets, but its popularity has boomed worldwide in places as far from the shore as Kansas City or Alice Springs. Since the 1960s, global consumption has doubled, and today the value of exported seafood exceeds that of beef, chicken, and pork combined.[3]

For centuries, the immense productivity of our oceans was capable of feeding the world's hunger for seafood. Salmon, tuna, cod, anchovy, and more delivered ever-larger catches in the decades following World War II. Improvements to boats, fishing gear, navigational equipment, and high-seas fish processing allowed us to draw from the ocean's resources with greater effectiveness. Catches swelled like the rising tide; so too did fishing fleets, coastal economies, and the range of options available to diners. But like any other marine predator, people eventually reached the fundamental limits of how much predation the seas could withstand, and the tide began to ebb. In the words of University of British Columbia fisheries biologist Daniel Pauly, "European and North American fisheries in the North Atlantic, and those of North Asia in the Pacific, peaked in the 1970s with all major fish populations exploited to the fullest."[4] In response, fisheries began to expand in three directions: geographic (new regions), bathymetric (deeper waters), and taxonomic (stranger fish). Technology, petroleum, and the dedicated labor of fishing crews forestalled the inevitable, but eventually the high-water mark was reached: marine catches topped 90 million metric tonnes [99 million US tons] by the 1990s and have been declining since.[5] Just take a moment, however, to marvel at that number. Earth's oceans are capable of producing 200 billion pounds of seafood, every year, more than double the yield of all terrestrial livestock in the United States.

Nevertheless, it seems clear we have reached a limit. Marine scientists warn that many commercial fisheries could collapse by the middle of the century.[6] One simply cannot take more from the seas than they are capable of replacing. Clearly, we must live within our means, and a fundamental first step is to understand the nature of each fish.

Atlantic salmon (*Salmo salar*)

ONE FISH, TWO FISH, DEEP FISH, NEW FISH

No fear can stand up to hunger, no patience can
wear it out.
—Joseph Conrad, *Heart of Darkness*

One of the fastest-growing seafood markets is for salmon, and sales of their appetizing, carrot-colored fillets are at an all-time high. Their routine of migrating between fresh and salt water, however, makes these fishes challenging to manage. Wild salmon are caught mostly at the onset of their spawning run, either by seines drawn behind boats or in gillnets deployed near shore. Once the fishes enter fresh water they begin to bloat, gradually spoiling the quality of their meat; to some Indigenous communities, a spawning salmon is categorically inedible once it has quit the ocean. Fisheries managers count fish entering the river mouth and wait until enough have passed—a number based on detailed estimations of current stock size and likely spawning success—to ensure the population will be sustained by their upstream reproduction. When enough salmon have crossed this freshwater starting line, boats and netters in the sea are given the green light. Even then, only a pre-ordained amount of fish can be harvested, approximately equal to the number who will enter the adult population that year. Much uncertainty exists, however, since it is impossible to count all the fish in the sea or determine the exact proportion of juveniles who graduate to adulthood. Estimation is a tricky business, halfway between art and science, but getting the numbers right is critical: catch too many fish, and the population could go bust. Since the collapse of a stock can be catastrophic, catch levels are usually set conservatively. But after the harvest target is decided, one whale-sized question remains: who gets to catch the fish?

For years, fisheries managers would set a simple quota—the total weight of salmon that could be caught in a season—then weigh the landings of each boat as they came to port. When the combined catches equaled the annual allocation, the season was promptly closed. This "total allowable catch" approach was intuitive, but deeply flawed. Fishing captains realized that the faster they landed fish, the greater share of the total catch they could claim. The race to fish was on. Each year, after fisheries scientists set the catch limit and starting date, boats would crew up, gear up, and dash onto the high seas for a "derby" of nonstop fishing. Weighed down by a hold overflowing with fish, seiners wallowed dangerously in northern Pacific waters that could turn ferocious, or deadly, in a matter of minutes. Even today, commercial fishing is one of the most hazardous occupations on Earth: in the first fifteen years of this century more than 700 fishers in the United States lost their lives, a fatality rate nearly thirty times the national average.[7]

Beyond the human tragedy, economic and biological catastrophes also resulted from the race to fish. While some boats, through skill or good fortune, might land hefty catches and earn huge paydays, others would limp into port with holds half empty. Meager paychecks might not cover the immense costs of fishing, from crew salaries and gear repair to insurance and loan payments; bankruptcy and unemployment beleaguered the industry. Meanwhile, overfishing routinely plagued derby fisheries because landings are measured only at port, while boats still at sea might have tons of fish on board, boosting the total catch over the limit. By 1974, nine-tenths of the world's commercial fish stocks, salmon and otherwise, were drastically overfished, exploited "beyond biologically sustainable levels."[8] It was as plain as the horn on a unicornfish's face. Our system for managing catches was failing, and fishes were disappearing.

To the rescue came a "tentative suggestion" published a year earlier by conservationist and adroit fly fisherman Frank Christy.[9] Recognizing that "drastic steps" were required to remedy overfishing, he proposed that fishers or boat owners "be allocated a share of the yield from the fishery, expressed as a percent of the catch." Christy's tradable "fisherman quotas" were adopted by several Canadian and Icelandic fisheries in the 1970s, including Atlantic herring, and as a national policy by New Zealand in 1986. Soon, individual transferrable quotas (ITQs) would become the most widely used fishery management system in the world, but not without controversy.

At the outset, an initial allocation of shares is established, each representing a fixed percentage of the total catch. Shares are awarded to current boat owners, typically on the basis of their historical catch levels, and give them an exclusive right to land their quota. If you hold a one percent share, and the fisheries department sets the total allowable catch for salmon at 10 million fish, then your boat can catch 100,000. Operators who do not acquire a quota are denied access to the fishery, a painful but necessary side effect of trimming an oversized fishing fleet. The race to fish is now suspended, because each boat can land only its own quota and no more. Captains who come to port with too many fish cannot sell the excess catch and face stiff penalties. Share owners do not own the resource itself—the population of fish—only a right to land a proportion of the annual harvest. Their quota ownership, however, promotes stewardship of the fish and curbs competition with other fishers. Quota shares can be sold between boat owners (the "transfer" part of the ITQ acronym), their value set by the market between buyers and sellers. Fishers whose boats are too expensive to operate will eventually sell their share to other, more efficient operators. Whereas previously an unrestricted number of boats fished furiously for a few days, then sat idle after the annual harvest was exceeded, under an ITQ system a reduced fleet can fish steadily and safely until each boat's share is landed and sold.

ITQs have generated success and controversy in equal measure. Broadly speaking, those who can hold onto quotas do well. Boats no longer over-invest in gear and crew, fish are landed in better quality, and harvests are more stable, helping push prices higher. Icelandic fisheries of demersal species like cod and haddock saw revenues jump by US$15 million in the first year after ITQs were implemented.[10] But only shareholders are allowed to fish, which limits the number of boats that enjoy the economic revival. In the British Columbia fishery for bottom-dwelling sablefish (*Anoplopoma fimbria*), for example, a program initiated in 1990 led to the retirement of a third of the fleet. Such limitations on fleet size and fishing effort, plus improvements to efficiency and quality, have helped rebuild fisheries that adopted catch share systems. In one review of early ITQ adopters, twelve of twenty fisheries showed increases in the population size of the target fishes.[11] Today more than 100 fish species in the United States are managed through catch shares, like Gulf red snapper (*Lutjanus campechanus*), whose populations rebounded by 300 percent.[12] In Namibia, an ITQ system was implemented to revive an utterly collapsed hake fishery (*Merluccius capensis* and *M. paradox*); after ten years the population had grown by 30 percent, and profits climbed even more.[13] Globally, the proportion of commercial fish populations that are harvested sustainably, once a distressing 10 percent of all fisheries, has swelled to 65 percent.[14] Much more remains to be done, but ITQs have clearly made a positive impact on the health of wild fish populations.

As with any system of haves and have-nots, it is essential to listen to those who are left behind. In an ITQ system this happens in two ways:

either you are unable to obtain an initial share, or you are later forced to sell it. Because quotas usually are applied to fisheries already in decline, there may be fishers who have temporarily retired their boats to wait out the downturn; unfortunately, they can be denied quotas, especially if the initial allocation emphasizes currently active fishers. In some cases, a few in-the-know fishers buy up boats and fishing permits prior to an ITQ's launch, akin to insider trading, and snare for themselves an outsized number of shares. Such developments have disenfranchised Indigenous people and small-scale fishers, who lack the resources and connections to place a wager on the upcoming system. Still, some ITQ systems do emphasize enrolling Indigenous and artisanal fishers at the outset, to maintain ancestral livelihoods.

Once catch shares are distributed, however, they can be bought by other boat owners and even by people or corporations with no connection whatsoever to the region. Economic downturns that might be shrugged off by a large corporation could be calamitous for a small-scale fisher. When economic hardship strikes, often the most valuable possession in a captain's household is their catch share. A medical emergency, a newborn son, or a daughter going to college may tempt, or compel, a lower-income fisher to sell their quota. The problem is, prices keep going up, and an artisanal fisher who sells their share is unlikely to be able to buy it back. In Australian waters, when an ITQ system was implemented for southern bluefin tuna (*Thunnus maccoyii*), an initial 143 shareholders shrank to just 63 after only four years.[15] Little by little, fishery access has consolidated as catch shares of the minnows fatten the portfolios of market whales. Dishearteningly, shares are increasingly amassed by hedge fund managers, rather than boat owners, who view them only as "positive investment vehicles" in byzantine investment strategies, rather than as access rights to a noble profession.

Consolidation of shares into outside hands, especially when fisheries were originally dominated by artisanal or Indigenous fishers, has rightly been met with mistrust, frustration, and fury. In Alaskan halibut fisheries where ITQs and Indigenous rights ran headlong into one another, the native fishers lost. Across forty-four villages on southeastern islands, some two-thirds of boats were decommissioned, 82 percent of quotas vanished, and more than half the people who used to hold shares no longer have the right to fish.[16] Poignantly, this troubling side effect was foreseen by Frank Christy, who had already counseled that "no fisherman could acquire through lease more than a certain percentage of the total yield. This provision would prevent monopoly controls."[17] Consolidation and loss of Indigenous fisher rights is not inevitable, it should be noted. In New Zealand, where Maori people retain rights to a mere fraction of their native lands (less than 6 percent), they control far more of the sea's bounty, holding 40 percent of ITQ shares.[18]

Despite the gamut of arguments against quotas, and numerous case studies in which rights-based management has fallen short of expectations, one point cannot be denied. Fisheries from the 1970s to the 1990s were failing in their fundamental charge, to harvest fishes while ensuring that future generations would also find fish in the sea. The utter collapse of Atlantic cod fisheries off the eastern shore of North America offers a sobering example of just how badly a fishery can implode, putting thousands out of work, forever. While ITQs are by no means a faultless system, and many improvements are still needed, open-access fisheries are no longer feasible. There simply are too many people buying fish, and too many fishers catching fish, to permit unfettered access to the sea's fragile bounty.

Today, more than twenty countries rely on ITQs for management of hundreds of fish species. Christopher Costello, an economist at the University of California in Santa Barbara, reviewed more than 11,000 fisheries and confirmed that "implementation of catch shares halts, and even reverses, the global trend toward widespread collapse."[19] Another study, led by Duke University's Anna Birkenbach, found that "catch shares slow the race to fish," and she forcefully underscored a crucial point: "the alternative is no management, which leads to both overfishing and economic waste."[20] Firm supervision of a fishery, via catch shares or other approaches, is essential. University of Washington marine biologist Ray Hilborn evaluated over 1000 fisheries and found that "in regions where fisheries are intensively managed, stock abundance is generally improving or remaining near fisheries management target levels," and that those fishes are twice as abundant as in weakly managed areas.[21]

Salmon fisheries enjoy management benefits not shared by most fishes, thanks to a simple advantage: they can be counted. When spawning runs begin, salmon pass through narrow, shallow waters at the river's mouth, where biologists can accurately assess their numbers. Salmon were once fantastically abundant on both coasts of North America, and along Europe's western shores. But by the late twentieth century, most of those wild populations had collapsed, doomed by inland obstructions and overfishing on the high seas. Today, there are virtually no thriving Atlantic salmon runs in the eastern United States. Even after dam removals and other measures improved Maine's Penobscot River, for example, just 1000 fish were counted there in 2019.[22] Only in Alaska have US salmon runs remained strong, but as those stocks also began to decay, alarm bells tolled. Several decades later, Alaskan ITQ programs have been successful in restoring populations of chinook, sockeye, and coho. Some of the societal outcomes are unquestionably nettlesome, but for now salmon in the eastern Pacific may be spared a repeat performance of their Atlantic cousins' demise.

While ITQs are capable of saving fish populations from overexploitation and extinction, one thing they cannot do is make more fish magically appear in the sea. To accomplish this, we are moving beyond the harvest of wild animals and turning instead to fish farming and ranching, to the realm of aquaculture.

FARMED FINS

> We must plant the sea and herd its animals using the sea as farmers instead of hunters. That is what civilization is all about—farming replacing hunting.
>
> —Jacques Cousteau, *The Ocean World of Jacques Cousteau*

At the dawn of human history, hunter-gatherer lifestyles gave way to farming when wild lands could no longer support our ancestors' expanding populations. So, too, have fisheries turned to aquatic farming, to fill the gap between what oceans can deliver up and what people want to gobble down. Aquaculture, which includes saltwater fishes like tuna and salmon, freshwater species like carp and tilapia, as well as shrimp and other crustaceans, oysters and other mollusks, and a lot of edible algae, has seen explosive growth in the last few decades. Aquaculture farms now produce blue foods exceeding all wild-capture fisheries combined, an estimated 135 million tons in 2020 alone.[23] Even after luxuriant harvests of algae are shorn from the total, farmed marine and freshwater species (including shellfish) account for half of global aquatic animal production. Some 600 million people also rely on aquaculture and fisheries for their livelihoods. Blue foods, from ponds and seas, constitute a significant chunk of the world's agricultural trade, generating annual revenues in excess of US$280 billion.

As far back as 4000 years ago, Chinese aquaculturists were raising fishes in ponds to meet the needs of a settled population far from the sea. Species like carp were, and remain, the most commonly farmed fish: freshwater pens are easier to maintain, carp reproduce obligingly in captivity, and they eat a vegetarian diet readily grown nearby. The first how-to manual on the subject recommended building a large pond with nine islands, stocking it with female and male carp, adding aquatic vegetation, and sprinkling in a few turtles for good measure. "The turtles are heavenly guards, guarding against the invasion of flying predators," the booklet explains. "When the fish swim round and round the nine islands without finding the end, they would feel as if they are in natural rivers and lakes."[24] Penned more than 2500 years ago, this brilliant handbook was authored by Fan Lee, a former politician whose surname fittingly means "carp."

A couple hundred years later the Romans were rearing fishes in captivity too, mostly trout and carp, but they also experimented with farming saltwater fishes in lagoons connected to the Mediterranean. Variations on the two approaches, freshwater ponds and saltwater pens, spread across Europe into Africa, and soon the whole world was engaged in some form of aquaculture. By 1987, an estimated 11 million tons of fish and shellfish were being produced.[25] A decade later that amount had tripled, and by 2020 reached nearly 100 million tons.[26] From ancient beginnings in a pond with nine islands, today's tidal wave of aquaculture has swept the planet, making commercial fish production the fastest growing food sector on Earth.[27] Blue farming has arrived, and it is here to stay.

Some three-quarters of all aquaculture fishes are freshwater species like tilapia, carp, and trout. Of the saltwater one-quarter, however, salmon are undeniably the kings: one-third of all mariculture fishes are Atlantic salmon, some 3 million tons of them in 2020. Also in the top ten mariculture species are coho salmon and rainbow trout (*Oncorhynchus kisutch* and *O. mykiss*), each at 200,000 tons annually. Captive salmon are born in laboratories and reared in cold freshwater tanks to mimic their upland nurseries. After a year or so, the youngsters are moved to pens floating in sheltered ocean bays or fjords. Norway, with its abundance of fjords, leads the world in salmon mariculture, raising a third of the global total. Fashioned from heavy netting, a salmon pen is a huge cylinder or cube that hangs from floats; its bottom, also a net, is anchored by cables to the seafloor below. Growing salmon are fed pellets of fish meal blended with soybean and other vegetable proteins. They are, after all, predatory fish and need plenty of protein to flourish.

There is great efficiency in the sea, however, when compared with the land. Because fishes live in a gravity-free medium, their bones need not support as much weight as terrestrial livestock, and their muscle-to-bone ratio is much higher than for a pig or cow. Most fishes also do not maintain a warm body temperature, something that forces livestock to burn additional fuel. Thanks to buoyancy and metabolic advantages, raising 1 pound of farmed salmon requires only around 2 pounds of feed,[28] whereas that same pound of ranched cattle will chew through 6 to 10 pounds of grass and feed.[29] Salmon aquaculture also generates fewer greenhouse gases than a comparable beef or pork operation, about on par with chicken farming.

As with all agriculture, fish farming must confront an array of challenges if it is to improve sustainability, environmental responsibility, and social equity. Mariculture corrals fishes in high densities, sometimes in the same waters as their free-swimming relatives. Disease can sweep through such crowded pens with terrifying speed and mortality, like the bacterial outbreaks of *Piscirickettsia salmonis* in Chile (the world's second-largest salmon producer after Norway). Parasites also invade salmon pens, including blood-sucking copepods known as sea lice that attach to flanks, cheeks, gills, and even eyes. Wild fishes captured near fish pens are often afflicted by the same diseases and parasites. Recently, the government of Canada committed to banning open-pen salmon operations in British Columbia to reduce the spread of these maladies into wild populations. A more creative solution has emerged in the form of cleaner wrasses, who adore plucking parasites from larger fishes. Several wrasse species are being evaluated in Norwegian mariculture operations, and in early trials the 4-inch goldsinny wrasse (*Ctenolabrus rupestris*) was found to devour more than fifty large sea lice per day.[30] Even adding as few as one wrasse per every hundred salmon in a pen was effective at controlling infestations, significantly improving the health of the corral's silvery livestock.[31]

Little by little we are refining our management of fish farming, in part by better understanding the fishes themselves. New York University biologist Becca Franks has studied salmon captive-breeding operations for years and found the fish can be "highly aggressive towards each other, taking chunks out of each other's flesh when they're in their recirculating systems on land."[32] Nobody would have guessed the cause. "At certain phases in their life they're typically housed in these blue tanks," Dr. Franks elaborates. "Perhaps not for any reason other than it's just what they've done in the past couple of decades." But when researchers finally challenged the color scheme, an invaluable discovery was made. "When you give them an option for a different background, their aggression drops to near zero." During experimental trials in British Columbia, coho salmon strongly preferred black backgrounds, and in these tanks their aggressiveness fell by a factor of four.[33]

In another aquaculture project, this time of a freshwater fish in Europe, managers could not induce females to release eggs, no matter how hard they tried. It turns out that all they needed to do was listen to the fish, literally. "They realized at some point that in this species the males vocalize to the females to lead to the spawning release," Franks explains. "And if you allow the males to vocalize, and you allow the females to hear the vocalizations, they readily would release their eggs . . . and you could have fertilization within a captive environment." The staff had been keeping males and females in separate ponds, thus preventing the gentlemen from serenading the ladies. These anecdotes reveal a truth about aquaculture: fish biology, natural history, and welfare all matter. In short, to grow a fish you have to know a fish.

Other pitfalls in aquaculture are more imperceptible than tank color or fish song. Farmed salmon are painstakingly bred for rapid growth in captivity, at the expense of all other concerns. Speed and maneuverability are of secondary interest, as is eyesight, fecundity, and the determination to struggle upstream and mate. But ocean storms occasionally rupture pens, releasing lab-reared fish into the open sea. Once they breed with their wild cousins, the genes of captivity may be passed on, carrying those same shortcomings with unknown consequences for wild populations who very much need speed, maneuverability, good eyesight, and a healthy libido. In Alaska's Prince William Sound, an estimated 25–30 percent of wild-caught salmon carry genes bred in captivity.[34] Farmed salmon also tend to produce fillets with higher concentrations of toxins, such as PCBs and dioxins, than their wild relatives.[35] While levels generally remain within European and US safety standards, such findings are prompting concern over the wisdom of eating farmed salmon, the sources of their feeds, and the management of their water and wastes.[36]

Whether aquaculture is carried out on land or at sea, fish waste poses a significant dilemma. Fish ponds accumulate enormous quantities of feces which are eventually pumped to the ocean when pond water is exchanged. Excessive outflow of nutrients leads to local algae blooms, smothering of bottom habitats, even red tides and marine dead zones. When pens are sited in the ocean, waste accumulates beneath them, as one might expect from a corral holding 200,000 fish or more. Raining from above, fish feces radically alter seafloor habitat, changing the benthic fauna and harming wild fishes who rely on a healthy bottom environment. But the most significant challenge plaguing salmon aquaculture, and farms of tuna, seabass, and other predatory species, is not the pollution that rains out of their cages. It is the food that goes in.

Sardines, anchoveta, herring, and other small schoolers are caught in astounding numbers by encircling seine nets. Every year, more than 10 billion pounds of anchoveta are pulled from South American waters alone.[37] The majority of that catch is ground and pressed into fish meal and oil, to be used in animal feeds and fertilizers (continuing an agricultural tradition that began with bird guano 200 years ago). Peruvian anchoveta alone account for about half the worldwide production of fish meal, and a third of its fish oil.[38] Much of the resulting animal feed is eaten by terrestrial livestock, household pets, and so forth, but almost half the planet's fish meal, and three-quarters of its fish oil, are destined for aquaculture.[39] Where anchoveta once fertilized the fields of Europe, now they nourish the farms of the sea. Caged carnivores are accustomed to eating fish for dinner, but the use of captured fishes to feed farmed salmon is problematic. Rather than reducing fishing pressure on wild ocean populations, mariculture may simply be shifting overfishing from salmon stocks to anchoveta schools.

Today, anchoveta harvests are only about a third of their 1970s peak: overfishing has taken its toll as demand for fish meal soared in the 1980s and 1990s. Fisheries that target small schoolers are also tested by the deeply cyclical nature of their populations: when upwellings go slack, fewer nutrients are delivered to the surface, and fish abundance plummets. If catch allocations are not adroitly reduced, the fleet can gravely depress the population. In response, Peru imposed rights-based management in 2009, allocating catch shares and compensating or retraining those fishers who left the fishery, about a quarter of the fleet. With fewer boats in the water, overfishing has been reined in, incomes have gone up, and fish meal quality has improved. Furthermore, the fishery has adopted principles of adaptive management, in which real-time assessments of anchoveta abundance are used to set the total catch.[40] This management system is flexible, allowing catch limits to expand in good seasons, or shrink when waters are less productive. Despite the hardship such fluctuations can impose on fishers, who have to get by with less in lean years, it is the only option for highly variable populations like the Peruvian anchoveta and may ultimately safeguard the fishery against climate change and erratic shifts in life-giving upwellings.

While better management can improve the sustainability of small schooler fisheries, better husbandry of farmed fish can reduce the demand for fish meal. Urged by vocal environmentalists and squeezed by rising feed prices, aquaculture has responded. Today, farmed fishes grow faster with less feed, their food is made with alternative oil and protein ingredients, and fish meal is increasingly manufactured from formerly discarded bycatch and fish-processing wastes. Salmon feeds now use wild-caught fishes for only half their fish meal, as little as 10 percent in some operations, and trimmings (the cast-aside scraps of seafood processing operations) supply a third of the world's fish meal production.[41] Globally, one pound of farmed fish now requires just a quarter-pound of wild fish in its feed. That fish-in:fish-out metric is higher for salmon and eels but has improved substantially. In 1997 a captive eel guzzled nearly 5 pounds of wild-caught fish meal to produce a pound of farmed eel, today just 3 pounds; meanwhile salmon once gulped more than 3 pounds, nearly double their intake today.[42]

"Insects to the rescue!" is not a slogan trumpeted by summertime mosquito-slappers, but aquaculturists have been singing the praises of bugs for years. Predatory fishes need protein to grow, and they prefer the taste of fish. But just as alternative nonmeats like Impossible Burger and Tofurky have shown that the public's palate can be fooled, fish farmers too have turned to alternate protein sources to feed their marine livestock. Formerly an obscure insect of tropical forests, the pint-sized black soldier fly (*Hermetia illucens*) has grown into a colossal hero for aquaculture. Black soldier fly eggs hatch into larvae called maggots, which feed in the wild on decaying leaves, fruits, and all manner of rotting vegetation. But in commercial insect-rearing facilities they can be served virtually any kind of vegetable food, including compost. In as little as a month, a mountain of compost collected from restaurants, farms, and households transmutes into a seething pile of inch-long larvae which can convert up to 70 percent of the compost into maggot meat.[43] And fortunately for aquaculture, farmed salmon love maggot meat. Tests carried out in Norway's Institute of Marine Research (home to cod-trusting Bjarte Bogstad) proved that salmon grew just as swiftly on feed prepared from black soldier fly larvae as from fish meal, and the resulting fillets were identical in nutrition and flavor.[44] If scaled up, insect factories could digest much of the world's organic waste into feed for farmed predatory fishes, while also relieving the overfishing pressure on wild small schoolers like anchoveta.

CORRALLING THE TIGER

Mediterranean sardines, like their South American anchoveta cousins, swim in immense schools and are seined in vast numbers, frequently destined to become fish food. So heavy is their overfishing that so-called common dolphins (*Delphinus delphis*), who rely on wild sardines as a prime food source, are now downright rare. All those sardines are being harvested for a single purpose: to feed captive tuna as they are fattened for market. Tuna are so staggeringly valuable, to diners and to fisheries, that many attempts to rescue their dwindling populations have been defeated by the power of the dollar and demand of the dinner fork. Wild tuna are often caught using baited longlines, despite the appalling levels of bycatch (a polite word for collateral damage) snagged by this indiscriminate method, including sea turtles and toothed whales. Advances have been made, however, by using weights to sink the longline hooks below typical turtle depths and limit their bycatch.[45] Nets also encircle schools of tuna when they can be found, by spotter planes or by taking advantage of tuna's penchant for swimming under pods of spinner and spotted dolphins. In the 1960s, San Diego tuna boats began setting nets on dolphin pods and killed an estimated 4.8 million dolphin between 1959 and 1970.[46] In one of the earliest public movements to protect oceanic wildlife, outraged citizens rose up in protest, and in 1972 the Marine Mammal Protection Act made it illegal to harm whales and dolphins in US waters. Expanding their reach, environmental groups and consumers made a global push for dolphin-safe tuna and have succeeded in changing an industry. Today more than 800 tuna companies in the United States and beyond, including 95 percent of canned tuna, adhere to dolphin-safe standards.[47] Among other advances that have been won, purse-seining boats can no longer intentionally set on dolphins, and they are required to "back down" their nets, leaving an explicit escape route. Public protest remains a powerful tool to compel change.

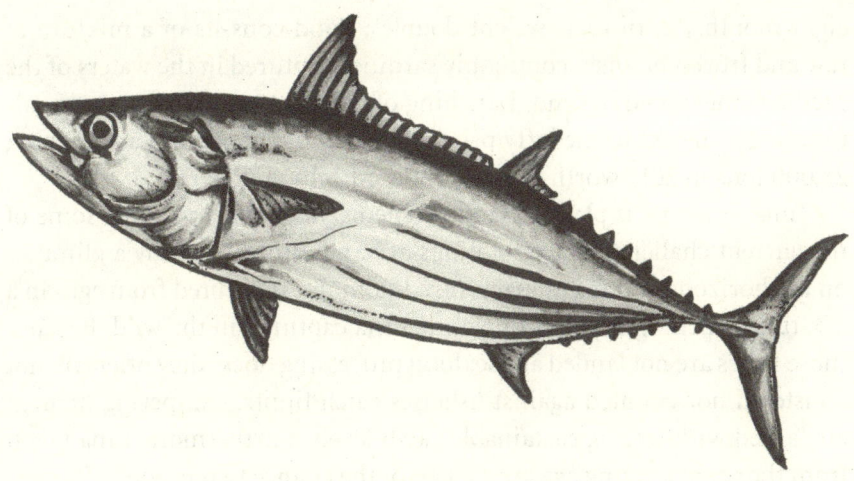

Skipjack tuna (*Katsuwonus pelamis*)

As a result of increasingly sophisticated fishing operations, however, and swelling desire for top-dollar seafood, catches of most species of tuna have been declining for years. In 2019 alone, fishing fleets landed over 6 million tons of tuna; mackerel and other tuna-like relatives raised that total to 9 million.[48] Sadly, more than 40 percent of commercial tuna stocks are being harvested beyond biologically sustainable levels. The largest tuna species are so heavily hunted that all are listed as threatened or endangered. In the case of southern bluefin tuna, an estimated 95 percent of its population has already been fished out.[49] Still, there is cause for optimism: about 10 percent of tuna stocks are actually increasing, due to smarter management and firmer control of fishing pressure.[50] Skipjack, boasting the highest reproductive rate of any tuna, has yielded stable catches around 3 billion pounds annually, making it the third-most harvested fish species for eleven straight years.[51]

In response to shrinking tuna populations and rising consumption (or at least rising prices), tuna ranching has galloped onto the scene. In a typical ranching operation, juvenile tuna are caught in the open ocean, transferred to a cage of netting, then towed to the mainland. When the cage reaches nearshore waters sheltered from storms and swells, they are transferred to a permanent ranching pen, a hockey puck–shaped cage some 150 feet across and 30 feet deep. Inside, the tuna are fed twice a day for up to eight months, until their weight doubles. Food consists of a mixture of raw and frozen baitfish, commonly sardines captured in the waters of the Mediterranean and beyond. Ranching of bluefin tuna alone, an aquaculture target thanks to the lofty prices they command, yielded a whopping 28,000 tons in 2018 worth a jaw-dropping 1 billion dollars.[52]

Tuna ranching is plagued with problems, however, and while some of the current challenges have solutions at hand, others are only a glimmer on the horizon. Whereas aquacultured salmon can be bred from eggs in a lab, tuna ranching depends on young tuna captured in the wild. Because those fishes are not landed at a seafood processing dock, they often are not registered nor counted against fisheries catch limits, hampering managers tasked with setting sustainable catch levels. Furthermore, tuna taken from the sea at a young age are robbed of the chance to reproduce. Targeting youngsters therefore is a double hammer blow, removing current fish from the population and blocking future fish from hatching.

To address this dilemma, researchers in Japan and Europe are laboriously studying every aspect of tuna reproduction. By controlling water temperature and light levels while simultaneously manipulating hormones, they have succeeded in rearing tuna from eggs entirely in captivity.[53] If their experimental technique becomes standard across tuna ranching, the possibilities are staggering: a single bluefin female weighing about 200 pounds is estimated to produce 9 million eggs in just one spawning event.[54] If just one in a hundred of those eggs could be fattened up to the size of the mother, the resulting 9000 tons of tuna would exceed the annual bluefin catch of the nation of Japan.

Still, fattening a captive tuna requires prodigious amounts of food. Tuna's hot-blooded bodies burn a lot of energy just to stay warm, far more than a cooler-blooded salmon: a bluefin tuna requires about 15 pounds of feed to put on a single pound of flesh, seven times more than a salmon.[55] Unfortunately, switching to insect larvae–based food is not an option, as tuna turn up their snouts at any meal but fish. Much will have to be done if the growing tuna ranching industry is to avoid starving the seas to fatten the livestock. In 1995, one southern Australia sardine fishery collapsed precisely because of overharvesting for tuna ranches: in a single year of overfishing, sardine numbers plummeted to a mere quarter of their former abundance.[56]

Like salmon, tuna held in dense captivity also suffer high rates of parasitism and disease. Sea lice are perhaps the most troublesome pests. In captive tuna, it is not uncommon to find one type of sea louse (*Caligus chiastos*) attached to the eyes where they gnaw at the cornea, damaging the keen vision on which a tuna relies for hunting.[57] Infestations can be extreme: Craig Hayward and his Australian colleagues studied southern bluefin enclosures and found three-quarters of the penned fish were infected, with a skin-crawling average of 265 lice spiked to each fish.[58] Heavy infestations, if left unchecked, can cause tuna to waste and die after just a couple months in a ranching pen. This particular louse has an alternate host, a scavenger fish called Degen's leatherjacket (*Thamnaconus degeni*) who feeds on the scraps and waste that accumulate beneath the crowded pen.[59] The lice are thus perfectly poised to leap into action as soon as a new herd of tuna are corralled and infect the entire group. Infection levels can be reduced, though, by shifting the pens into deeper waters and limiting overfeeding, actions that reduce the excess food accumulating under the pen so leatherjackets (aka lice launching pads) no longer congregate.

Cramming hundreds of tuna into a ranching pen can damage not only their well-being, but also the health of the environment nearby. Beneath and around the cages, dense snowstorms of falling food, scales, and feces have significant impacts. Even studies commissioned by the Tuna Boat Owners Association in Australia revealed harmful impacts on the seafloor up to 500 feet away from the pens.[60] As nitrogen and phosphorus-rich wastes fall from pens, they settle on the bottom at rates fourteen times higher than are found in nature.[61] Dense fish farming operations risk triggering eruptions of algae, disease, and dead zones around their cages. These effects are made worse by the relative calm of inshore waters; farther out to sea, currents carry away more of the waste and dilute the cage's effect. On the other hand, moving pens to deeper waters leaves them vulnerable to lashing storms, and at greater risk of catastrophic failure.

Seafood buyers and sushi lovers may just have to accept some fundamental limits of eating the largest of the ocean's top predators. Tuna are fast moving, far ranging, have a searing metabolism, and require vast amounts of prey fishes to reach adulthood. History has shown that they endure only modest annual harvests. While smaller species like skipjack and albacore can tolerate the predations of a robust fishery, the true kings of the sea—bluefin and bigeye—may never withstand more than the lightest of fishing pressure. Though the supply may be augmented by production from aquaculture operations, restrictions on the capture of food fishes must, for the moment, limit the scale of those ranches. Big tuna, like lions and tigers and bears, may just be animals that are better admired in their native waters than dispatched to a dinner plate.

TOO MUCH OF A GOOD THING

Atlantic cod (*Gadus morhua*) was for many years a mainstay of Indigenous peoples in North America, and new arrivals from Europe, but the bounty did not last long. A few centuries after English settlers established themselves on the shore, Canadian and American cod stocks had wholly collapsed. Why did this happen? The answer is the same around the world: too many boats and nets, buoyed by government incentives, are chasing too few fish. Economists have a word for this imbalance, overcapitalization, and its effects have been devastating. A global analysis led by renowned fisheries biologist Boris Worm (and man most likely to lend his name to the next James Bond villain) estimated in 2012 that two-thirds of marine stocks were being fished beyond their capacity to recover. Incredibly, more than 100 of them faced fishing pressure ten times higher than their sustainable target.[62] His frequent collaborator Daniel Pauly asserts that the global fleet is double or triple the size that fish stocks can bear. Government support of this fleet is unbending, despite fisheries making "a miniscule contribution to the GDP of advanced economies—in the United States, even less than that of the hair salon industry."[63]

   In part to sustain the employment historically provided by fishing, nations continue to prop up fisheries. "The world's governments provide over US$30 billion in subsidies each year," writes Pauly, "about one-third the value of the global catch."[64] That is a shocking statistic, and it is difficult to imagine another industry that could rely on governmental support equal to a third of its revenue. But fishing has always been a noble occupation, and it holds a soft spot in our collective heart—otherwise known as a blind spot that obscures the judgment of industry and government. In the case of cod in the western Atlantic, it was a blind spot that led to permanent closure of a once-booming fishery. Boat owners, crew members, fish buyers, and government representatives refused to accept that the cod fleet was far too large, and nobody had the courage to take the first step back from the precipice. So, over the cliff went the entire industry, never to return.

Overcapitalization is rampant in world fisheries. Subsidies to the North Atlantic fishing industry alone add up to over US$2.5 billion per year. Those free dollars encourage new boats to enter a fishery, and ageing boats to remain active, even after fish populations are overtaxed. Profligate subsidies also undermine cooperation, as fishers in the same fleet view one another as rivals to be outspent rather than as partners with whom fishes must be justly shared. In 2021, a group of nearly 300 scientists from forty-six countries published a challenge to the World Trade Organization, urging that "WTO members must prohibit fisheries subsidies."[65] The authors highlighted the threat posed by foreign fleets, which exploit weaknesses in international treaties governing the high seas to flaunt national management regulations. "Subsidies to distant-water fishing fleets," they railed, "threaten lower-income countries that rely on fish for food sovereignty." Even worse, some of those fleets subject their crews to abysmal wages, appalling work conditions, forced labor, and shocking abuse; they have rightly been accused of modern-day slavery.[66]

Auspiciously, the WTO passed a historic agreement in 2022, with all 164 members agreeing to prohibit "fisheries subsidies that contribute to overcapacity and overfishing, and eliminate subsidies that contribute to IUU [illegal, unreported, unregulated] fishing."[67] The WTO manifested commendable empathy to developing nations, noting that "improving the sustainability of fisheries is critical . . . especially to the millions of mostly poor people who make their living by fishing." Hard work remains: two-thirds of member nations must ratify the agreement, negotiations to protect fisheries in developing countries must be concluded fairly, and subsidy-free fleets across the globe must adapt to new financial realities. Shrinking a fishing fleet puts captains and crew out of work, along with a network of associated laborers, while retraining programs and boat buy-backs only partially offset the hardship. But if the bitter prescription of eliminating subsidies can be swallowed, it could finally cure a syndrome of overcapitalization that has plagued fish and fisheries for more than fifty years.

If fishing pressure is restrained, cod and other demersals can provide a sustainable source of food for the world, and a steady source of income for coldwater fishers. Countries like Norway have solved many of the core problems, and their carefully managed fisheries appear healthy. Today, cod and haddock and hake constitute 10 percent of the world's aquatic export market, a robust share that can be maintained for years as long as appropriate measures are taken.[68] Management of northern Pacific schools of haddock, halibut, and others has been admirably successful, according to the Food and Agriculture Organization, although salmon are singled out for some declines in Canadian and Californian waters. "Most species except salmon stocks in this region are healthy and well managed, primarily due to science-based advice . . . and to good governance that has helped reduce fishing pressure from distant water fishing nations."[69]

Even when fishing pressure is reined in to match the productivity of slow-growing fishes like cod, the gear used to catch them can cause long-term damage. For demersal fishes, the invention of the otter trawl was once a brilliant idea. Dragging a net along the seafloor was a surefire way to capture fishes that hugged the bottom. But cameras on remote submersibles have shown the extent of the devastation caused by bottom trawling. In many regions, such trawlers rake a single patch of seabed dozens of times each year. There is no chance for the essential elements of healthy habitat to recover, a protracted process that can take years, and soon there is nowhere for fish to forage, dodge predation, court, and breed.[70] All too quickly, life-sustaining sea floors can be converted into biological deserts, evicting demersal fishes from their seafloor homes. The ecological collapse also releases approximately 1 billion tons of planet-warming carbon dioxide every year, equivalent to the entire aviation industry.[71]

Bottom trawling is particularly problematic for artisanal fishers, and nowhere is this more evident than in tropical waters. There, small-scale fishers have for years earned an honorable living from their boats, nets, and hooks. But when industrial trawlers begin assaulting the ecosystem, dozens of species suffer, and artisanal boats come to port empty. One reason is that marine productivity in tropical waters drops off steeply with depth: near to shore, the sea is full of life, but sail just a few miles out and the waters are largely empty. This abrupt gradient encourages trawlers to ply their nets very close to the coast, precisely in the waters that artisanal fishers have used for centuries. Clashes inevitably occur, of the David versus Goliath variety. In Indonesia, an uproar of violence and protest prompted the federal government to ban nearshore trawling throughout the archipelago. Although the trawlers yielded huge catches of shrimp, invaluable to a developing nation hungry for seafood export income, they were catastrophic for artisanal fishers. Shortly after the ban went into effect, small-scale fisheries rebounded dramatically, and lost fishing employment was revived.[72]

In warm, shallow waters around the world's tropics, coral reefs, sea-grass beds, and mangrove fringes have sustained small-scale fishers for millennia. But rapid growth of coastal communities is now weighing on tropical seas. Displaced farmers from interior lands often migrate to coastlines, where fish ostensibly are easily caught for dinner. Entry of large numbers of untrained fishers into a small fishery, however, can be disastrous. While the responsibility does not fall to the landless individual, but rather to the global market forces that displaced them, the effect on the ocean is the same: too many hooks for too few fish. And not only hooks are employed. Whether due to desperation or to avarice, tropical waters have been assaulted by scandalous forms of fishing, including lobbing dynamite or even cyanide into the water. Such methods are indiscriminate, killing large numbers of nontarget species, and they have been responsible for the near-total destruction of reefs around Jamaica and other tropical sites.

Unfortunately, in developing nations we often do not know how many hooks, nor how many fish, are involved in a fishery. For a wide range of systemic reasons, reporting of fish harvests is spotty in many tropical countries, which handcuffs proper fisheries management. Dirk Zeller of the University of British Columbia studied the artisanal harvests of American Samoa in the 1980s and 1990s and estimated that true catches were seven times larger than reported in official statistics. Furthermore, their contribution to the local economy was an astonishing nine times larger than reflected in government ledgers.[73] With this degree of underreporting, accurate monitoring and management of tropical fishes, and their benefits to local fishers, can be as murky as squid ink. It is nearly impossible to administer a fishery in the dark, especially when those who most need the livelihood—artisanal and small-scale fishers—are underrepresented or absent from official records.

Even with accurate assessments, applying fisheries management tools developed in the temperate zone to the peculiarities (and delights) of tropical waters can be challenging. ITQs, for example, are predominantly designed to manage single-species fisheries. But tropical waters are gloriously rich in species, the very reason they are revered by snorkelers and scuba divers. Management systems must be adapted to handle fisheries that catch many species at once, and even the loudest of cheerleaders admit ITQs are a tool that does not apply well to tropical multispecies fisheries. When poorly managed artisanal and unregulated commercial fishing collide, it is the ocean's creatures who suffer most. By some estimates, fish abundance in tropical Asian countries is now less than 10 percent of historical levels, thanks to mismanagement and overfishing.[74]

The solution to overfishing, according to fishers, community members, and conservationists, will involve a combination of gear restrictions, catch limit enforcement, but also meaningful stakeholder participation. Old models of top-down fisheries management, with remote scientists setting catch allocations and governments handing out quotas, evoke disturbing reminders of painful colonial legacies. Moreover, such models often do not help sustain fish populations nor support the livelihoods of fishers. Instead, advocates of a bottom-up approach, like Vandick Batista of the Federal University of Alagoas in Brazil, recommend "engaging stakeholders, especially those that are involved in the day-to-day use of a resource, such as artisanal fishers and their families."[75] Approaches to participatory governance as diverse as the fishes they manage are being trialed in countries throughout the tropics. Central to their success is acknowledgment that input from local fishers must be sought long before gear and catch limits are implemented, and that cooperation must be emphasized throughout the process. What is referred to as an ecosystem management approach, something that pairs well with the flexibility of adaptive management, works best when it accepts that ecological, social, and economic fluctuations are bound to occur. In Batista's words these advanced systems, "more fully embrace the inherent uncertainty and complexity of both the fisheries and the communities that exploit them."

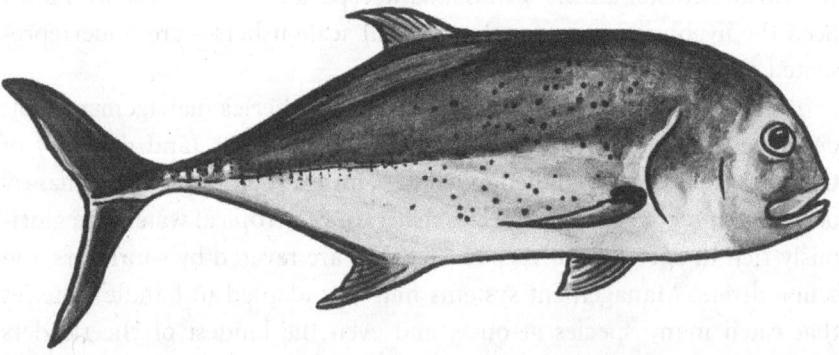

Giant trevally (*Caranx ignobilis*)

SANCTUARY IN THE SEA

One touch of nature makes the whole world kin.
—John Muir, *Our National Parks*

As developed and developing nations make demonstrable progress toward improved fisheries management, it remains evident that achieving sustainable ocean stewardship on a global scale will require a mosaic of solutions. One key piece of the marine conservation puzzle is a protective measure dating back to John Muir and the establishment of the US National Park System. Setting aside land—or water—is critical to preserving diversity and abundance in sensitive habitats. The most effective tools we have in the struggle to save the seas are marine protected areas, and the United Nations recently set an ambitious goal of placing 30 percent of the marine realm into protected areas. Currently the United States has enrolled a quarter of its seas,[76] but the standout leader is Australia, where nearly half of national waters fall within marine protected areas.[77] The planet, however, has some work ahead of it: just 8 percent of the world's oceans are currently protected, and assigning seas to a marine reserve does not automatically constitute a guarantee that those waters are fully safeguarded.

Marine protected areas (or MPAs) bestow a diversity of benefits so broad that resistance to their establishment is receding like an outgoing tide. Currently, the main opposition comes from what Daniel Pauly refers to scornfully as the "fishing-industrial complex," an echo of outgoing US President Dwight D. Eisenhower's warning about the outsized power of the military-industrial complex.[78] The former president's words, uttered in 1961, were prescient and apply just as well to the fisheries of today: "we—you and I, and our government—must avoid the impulse to live only for today, plundering, for our own ease and convenience, the precious resources of tomorrow."[79] Industrial fishers often oppose marine protected areas since they view them as closures of former fishing grounds. This is usually true, because protected areas tend to be enacted in locations with high fish abundance, diversity, and productivity: precisely the spots where boat captains would like to fish. But the benefits of closing down those sites outweigh the lost harvests. Marine protected areas offer vital refuges and usually enhance fisheries and local economies by ensuring fish live longer and grow larger, amplifying reproduction

(often immensely), exporting juveniles who recharge harvested populations nearby, and establishing tourism destinations that yield beneficial
new sources of revenue.

Marine protected areas have a long enough history to assess their contribution to maintaining healthy oceans. More than 1600 protected areas
have been inaugurated in US waters alone, including the tongue-twistingly
named Papahānaumokuākea National Monument north of Hawaii which
encompasses nearly 600,000 square miles. Within its safeguarded waters,
fishes like bird-nabbing giant trevally, graceful silky sharks, and hundreds
more enjoy a respite from fishing vessels that pursue them elsewhere on
the high seas. Around the globe, a flood of MPAs have been declared since
the 1970s, more than 11,000 sites and counting, including ten massive
reserves each covering over 100,000 square miles.[80]

Marine protected areas, particularly those that include no-take zones
where all extractive activities are prohibited, are good for both fishes
and their habitat. Within reserves, seafloor environments recover more
swiftly, even after trawling.[81] Fishes and other marine life increase in
number and size, as does their diversity. A review of 124 reserves around
the globe, coordinated by Sarah Lester from the University of California,
revealed that biomass within reserve borders swelled to five times that
of pre-establishment levels.[82] In some cases the gains were extraordinary.
Large predatory fishes including snappers and groupers were a remarkable twenty-eight times more abundant in the Philippines' Bongalonan
Reserve. In the Governor's Island reserve of Australia, rock lobsters (made
famous by the B-52s hit song) were twenty-three times more plentiful.
Perhaps of greatest importance was Lester's confirmation of the so-called
"spillover" effect, whereby animals that reproduce within the shelter of
a marine reserve eventually disperse, revitalizing populations in nearby
waters and sustaining fishery harvests. Biomass in sites lying beyond
reserve boundaries nearly tripled after protections were put into place, and
animal densities doubled, benefiting both fish and fishers. In the Caribbean nation of St. Lucia, even modest-sized marine reserves promptly
lifted catches by artisanal fishers in nearby waters by 46–90 percent.[83]
Back in Hawaiian waters around the world's largest no-take reserve, spillover has boosted yellowfin tuna catches by 54 percent, showing that even
highly mobile predators benefit from marine reserves.[84]

One of the most significant contributions of marine protected areas
is made by females who call them home, or more specifically, by their

eggs. Fishes of both genders grow significantly bigger within a reserve, where the abundance of large fishes can increase tenfold after establishment.[85] Outside the reserve, the heftiest individuals are usually prime targets for fishers because they command the highest prices, and because fisheries management is often structured around minimum-size limits. But catching large fishes, especially large females, can drastically impair a population's capacity to reproduce. Big females produce more eggs, so landing the largest individuals removes the best breeders. The relationship between female size and egg production can be mind-boggling: a single 24-inch red snapper (*Lutjanus campecheanus*) carries over 9 million eggs, more than could be deposited by a team of 200 females each 16 inches in length.[86] The vast productivity of large fishes is one of the key shortcomings of size-based management, but the bounty of eggs they release into the sea is an outsized advantage of marine protected areas.

Like many solutions, marine reserves are not a perfect remedy for all problems. They can be costly to implement, their establishment may be met by opposition, and their placement may inadvertently favor certain fishers over others. As climate change warms the oceans, fishes are shifting their distributions and may permanently migrate beyond a fixed boundary. Variations in reserve size, degree of protection, pre-establishment degradation, nearby fishing pressure, and other circumstances can also temper their effectiveness. In the northwest Mediterranean, for example, the Medes Islands MPA has shown mixed outcomes. Established in 1983, six of the reserve's fish populations have been surveyed annually for nearly twenty years.[87] Five of them, all heavily targeted by fishers, have steadily increased in abundance, though it may take up to thirty years for their numbers to fully recover. Gilthead seabream (*Sparus aurata*), a delectable porgy who favors seagrass beds, sadly has not recuperated. An initial recovery stalled, likely because fishers were targeting spawning areas unfortunately located near the reserve's border. Judicious expansion of the reserve would seem to be called for. Marine reserves that are too small, or that suffer from inadequate monitoring and enforcement—a direct result of underfunding—often show fewer and weaker benefits than larger and better protected areas.

In the Caribbean nation of Belize, Glovers Reef Marine Reserve spans more than 30,000 acres (about 50 square miles) of turquoise waters and rich coral reefs. By the mid-1990s a large no-take zone had been established, but monitoring is chronically underfunded. While the resident population of Caribbean reef shark (*Carcharhinus perezi*) remained stable from 2000 to 2013,[88] parrotfishes gradually became more scarce.[89] Recovery of these colorful coral-eaters may have been hampered by algal overgrowth, which built up during years when overfishing decimated herbivorous fishes who normally keep algae under control. Researchers from the Smithsonian Institute who conduct regular studies in the reserve suspect both legal and illegal fishing are hampering the reef's revival (artisanal fishing is still permitted in broad zones).[90] Shark conservation studies in Australia underscored that no-take zones, where fishing bans can be challenging to enforce, are outperformed by no-entry zones that strictly prohibit boat traffic of any kind.[91] Despite parrotfish declines in Belize, however, tourism to Glovers Reef has soared, bringing much-needed revenue to a country with the smallest economy in Central America.

The world-famous Galápagos Islands are encircled by an enormous marine reserve covering some 50,000 square miles of Pacific Ocean, a watery expanse nearly impossible to adequately police. National and distant-water fishing vessels sit on the reserve's border waiting for tuna to cross, a (legal) tactic known as "fishing the line." According to Jorge Ramírez, fisheries biologist with the Charles Darwin Foundation, fishers from mainland Ecuador (to which the islands belong) are increasingly supportive of the reserve. "Right now the tuna fishers are more aware about the importance of the Galápagos Marine Reserve," he explains, "they understand the reserve is like a nursery for tuna . . . and they declared publicly the importance of the Galápagos reserve."[92] Distant-water fleets, however, are another story. When a school nears the border, disreputable boats may switch off their satellite tracking systems, meant to control incursions, and sneak into the reserve. In 2017 a Chinese vessel emblazoned with the overly long name of *Fu Yuan Yu Leng 999* was nabbed by the Ecuadorian coast guard. Inside they found more than 300 tons of fish taken illegally from the reserve's waters.[93] Worse, critically endangered scalloped hammerhead sharks (*Sphyrna lewini*), who travel to the waters of Galápagos to breed, made up most of the illicit catch.

For organisms that tend to stay put, a marine reserve can be a fabulous place to live. Highly mobile creatures like sharks, however, are more difficult to conserve. To find food, especially in an increasingly depleted ocean, they must travel great distances, wandering in and out of reserve boundaries. Sharks are particularly at risk, since a vigorous market in Asia for shark fins has painted a target on the backs of these magnificent animals. Vast smuggling networks have sprung up to move dried shark fins illegally across borders, often the same networks that traffic illicit tropical timber, weapons, drugs, and even human beings. So valuable are the fins that they are frequently sliced from the shark while still on the high seas, the body then dumped overboard to hide the evidence. Fortunately, a strong alliance to denounce shark finning has emerged between marine conservationists and Asian business leaders. No longer is shark fin soup considered standard fare for weddings, especially among younger couples, and the purchase of fins is increasingly reviled. Despite the exceptional mobility of sharks, marine protected areas located sufficiently far from fin marketplaces (800 miles or more) have been shown to effectively protect them from slaughter.[94]

One nation has gone so far as to ban shark fishing entirely. As one of the world's foremost destinations for snorkelers and scuba divers, the Republic of Palau believed there was greater value in viewing undersea wildlife than in harvesting it. Economic analyses in 2010 revealed that shark diving generated an estimated US$18 million every year, nearly one-tenth of the country's gross domestic product.[95] Much of the proceeds go directly into local salaries and domestic tax revenues. Conversely, if all those sharks were harvested and sold, the earnings would total a paltry $10,800. Palau has long practiced a fishing management strategy, rooted in Indigenous wisdom, that sets aside large areas for spawning. The shark fishing ban was viewed as an extension of that philosophy, known as "bul." Now a no-take zone covers 80 percent of Palau's waters, nearly 200,000 square miles of ocean, an area larger than France; local fishers retain access to the remaining 20 percent. President Tommy Remengesau, Jr., explained the true meaning of bul: "It is prohibition in the sense that you're doing this to benefit your children, because you have to think about tomorrow and the day after and the years coming."[96]

Comparably huge marine reserves like Australia's Great Barrier Reef have also shown that tourism revenue can compete with the sums generated by fishing. Visitors to the Great Barrier Reef support some 64,000 jobs, making the sector one of Australia's largest employers.[97] Around the planet, coral reef tourism contributes some US$36 million to the world's economy.[98] That income may be just a blenny-sized fraction of the revenue of global fisheries ($151 billion in 2020), but when all types of coastal and marine sites are included the figure jumps like a marlin.[99] In Europe alone, this blue tourism generates an estimated $180 billion per year, and employs more than 3 million people in meaningful, safe, and rewarding careers.[100] Whether large or small, marine protected areas have shown undeniable success at achieving a trident of goals. They can protect wild fishes and their habitats, provide solid sources of local employment, and guarantee the future of fisheries, for tomorrow and the day after and years to come.

SEA OF CONFUSION

How do you know what you are eating? On land, this is a pretty simple question: chicken looks like chicken, beef tastes like beef. But when it comes to seafood, it is discouraging how frequently fraud and impersonation are perpetrated. The label on a fish you buy at a seafood counter or seaside market may bear little relationship to the fish you are about to eat. By some accounts, as much as one-third of seafood sold in the United States was mislabeled.[101] Another study suggested this number may be as low as 8 percent,[102] while still a third investigation tested sushi sold in Los Angeles restaurants and found nearly half of the fish was mislabeled.[103] In some cases, mistakes in handling and shipping can lead to honest labeling errors; in others downright fraud is being committed to dodge import bans, mislead investigators, or overcharge buyers. Regardless of the cause, the result is the same: consumers may inadvertently finance disreputable fisheries while trying to support sustainable sources. Mislabeling can be particularly damaging to wild populations because the fishes substituted for well-known species most often come from fisheries whose cheap prices reflect shoddy, short-sighted management that perpetrates overfishing, habitat damage, and high bycatch.

Sharks and deepwater species are mislabeled with disturbing frequency. Shark meat is often sold in South Africa under the vague moniker of "ocean fillets,"[104] and in Greece more than half the shark meat samples identified by DNA barcoding were labeled as other, less objectionable fish species.[105] Orange roughy was the recipient of a deliberate rebranding campaign (changing the name from slimehead), but other fishes of seamounts and the deep sea are fictitiously renamed in an effort to boost their marketability. This practice is fraught, since such fishes are often exceptionally slow to reproduce and cannot withstand more than very light fishing pressure. Patagonian toothfish (*Dissostichus eleginoides*) is now listed on menus as Chilean seabass, though it is not a bass but rather inhabits deep waters near the Antarctic ice shelf. Another popular restaurant item is marketed as monkfish but properly known as a goosefish (*Lophius americanus*) for its homely, even terrifying appearance.

If diners do not know what they are eating, they do not know what fishery they are financing. Absent this information, consumers cannot vote with their dollars for fisheries that are sustainable, whose fish stocks are not overexploited, whose gear does not damage habitat or ensnare bycatch, and whose fishers enjoy stable income and decent working conditions. To provide such data, independent certification systems now evaluate fisheries on behalf of buyers. A number of options exist, but the most prevalent—and sometimes the most controversial—is the blue checkmark label of the Marine Stewardship Council (MSC).

Brian Perkins spent six years as the MSC's regional director for the Americas, retiring in 2021. A native Mainer, his voice betrays occasional traces of New England as he relates his personal introduction to the seafood industry, on the Icelandic island of Grímsey. "I landed as a hitchhiker backpacker, then twenty-one years old, and in the summertime went up to see the midnight sun."[106] Remote and tiny, Grímsey is home to a hardworking fishing community, which Perkins joined. "One of the boat owners asked me if I wanted to work on his fishing boat that winter, and I agreed." About his lack of preparation, he confesses, "I don't come from a fishing background at all." With endearing understatement, he recaps the challenge of working in Iceland's stormiest months: "it was an interesting experience to fish on a 35-foot wooden boat on the ocean in the wintertime."

Founded in 1996, the Marine Stewardship Council is, in Perkins' words, "the gold standard" of sustainability certification. Fisheries seeking the blue checkmark label undergo a rigorous assessment process that lasts a year or more, testing their performance against twenty-eight indicators of stock status, habitat impacts, and fishery management. Evaluations are conducted by third-party assessors, paid by the fishery itself. The MSC may grant immediate certification, as was the case with well-managed fisheries in Norway and Iceland. "Think of us like the Good Housekeeping seal of approval . . . of the good job that the fishers and the management are doing." Or a conditional certification may be offered if the fishery is not yet up to par but shows improvement on the indicators. "One of the things that does is rather than only being able to certify the very best fisheries, we are able to certify fisheries that are doing a very good job but have a few areas that need improvement, and this provides the incentive and the encouragement to improve."

An example comes from South Africa, where a small fleet fishing for shallow-water hake (*Merluccius capensis*) applied for certification. "There were five vessels that wanted to sell into the European marketplace," Perkins explains, "and in order to do that they needed to get certified because the European market is quite particular about only selling MSC certified fish." Bycatch, however, was a serious problem. "That fishery, when it went into assessment, was killing about 20,000 seabirds a year in bycatch. And that is obviously not a sustainable practice. A chunk of those birds were albatross, which is particularly endangered." At the MSC's urging, the fleet consulted with the nonprofit BirdLife International, which recommended adoption of tori lines, "basically long nylon streamers that stream off the stern of the boat and prevent the birds from getting to the hooks while they're still on the surface." The streamers worked and spread through the industry. "They reduced the bird bycatch from 20,000 to 200 by using that system," Perkins proudly reports. "South African fisheries management, looking at those five boats that were using the tori streamers, made it mandatory on all longline vessels."

MSC certification opens the door to new and larger markets and allows seafood buyers to support a sustainable fishery which can then charge more for its fish. In this they resemble the Forest Stewardship Council, whose certification of responsible timber operations gives consumers of wood and paper the opportunity to put their money where their hearts are and reward well-managed suppliers. In the United States, companies like Kroger, Whole Foods, Walmart, and Target promote the sale of sustainable fish; in the Netherlands, more than 85 percent of all fish sold is certified.[107] Ideally, the certification should allow fish buyers—from individual shoppers to restaurant buyers to commercial distributors—to check whether a fleet, no matter where they operate, is fishing conscientiously and then decide whether to pay a higher per-pound price for that sustainability. Currently more than 80 percent of US fishery landings are MSC certified,[108] while 15 percent of the globe's wild marine catch has netted the seal of approval.[109] US numbers can be a little misleading, though, since the majority of fish caught in the United States is shipped overseas,[110] while paradoxically the preponderance of the seafood eaten by Americans is imported.[111]

Despite the MSC's noble aims and significant achievements, facets of the certification process have rankled marine conservationists. The non-profit is funded by donations, but also from licensing fees paid by fisheries to display the coveted blue label. Some argue this arrangement creates an incentive to certify substandard fisheries so the MSC reels in more income. Others have objected to the organization's certification of fisheries using damaging gear, like bottom trawlers. Perkins defends the practice, suggesting that such gear can be employed under strict conditions. "We have a number of trawl fisheries that are certified. Some of them take place on sandy bottoms where the impact is minimal. Some of them are in places if they get into a sensitive environment they have 'move on' rules. So if the trawl pulls up a piece of deep water coral . . . they agree to avoid that area."[112] Even sandy sea floors, however, can be home to diverse and vulnerable invertebrate communities, and remote submersibles have documented damage where an MSC-certified fishery trawls for Greenland halibut (*Reinhardtius hippoglossoides*).[113]

Other fishing techniques with poor track records have also been certified. The organization awarded their checkmark label to several tuna fisheries that rely on floating aggregation devices, despite the distressing bycatch of this tactic, estimated at five times greater than in the open sea.[114] When it comes to deepwater fishes, like the Chilean seabass (née Patagonian toothfish), marine biologists argue that we lack data fundamental to assessing the fishery. Jennifer Jacquet, professor of environmental studies at New York University points out, "nothing is known about this fish: no eggs or larvae have ever been collected."[115] In this case, market opportunity rather than demand seems to be steering the ship. "One guy starts catching them, starts experimenting with cooking them," Dr. Jacquet narrates, before swatting aside the fallacy that all fisheries are helping feed the hungry. "[He] introduces this fillet into New York high-end markets where you could sell fish for thirty or forty dollars a pound. There's no argument that you could make that this is a food security item."[116] Despite the paucity of scientific data, in 2009 certification was granted. Recently, MSC also awarded its label to the highly controversial Antarctic krill fishery, which has been accused of robbing the sea of a resource critical to polar communities of seals, penguins, and great whales.

More broadly, Jacquet argues that "the incentives of the market have led the MSC certification scheme away from its original goal, toward promoting the certification of ever-larger, capital-intensive operations. Small fisheries that use highly selective, low-impact techniques, such as hook-and-line fishing or hand picking, are often sustainable, but make up only a tiny fraction of MSC-certified fisheries." Given that around half of all fishes landed worldwide are caught by small-scale fishers, who lack the resources needed to apply for certification, the process effectively shuts the door on a giant fraction of the global catch. Brian Perkins openly admits, "it's one of the things that we're struggling with as an organization . . . to try to figure out how can we adapt to better serve the small-scale fisheries."[117] In the past few years improvements have been made, and fisheries in the developing world now comprise about 7 percent of certified stocks.[118] Policy experts have suggested that small-scale fishers could achieve certification more readily by sharing expensive assessment costs between different fisheries targeting the same species, certifying multi-species fisheries as a single unit, and by promoting domestic support so certified fishes can also be sold locally.[119]

Another objection leveled at the MSC is that the certification process does not always lead to healthier fish stocks. An independent analysis of MSC-certified stocks in the northeast Atlantic found more than half of them still were "exploited above the maximum sustainable level" in their first year of certification.[120] An earlier review had found just 31 percent of globally certified stocks were overfished.[121] Lead author Rainer Froese, of Germany's Helmholtz Centre for Ocean Research, underscored that countries need not rely solely on the MSC and can independently ban sales of overexploited fishes. Certainly customers can also consult other seafood sustainability rating programs, such as those led by the World Wide Fund for Nature, the Marine Conservation Society, and Seafood Watch. And it would be unfair to place the burden of rescuing all declining fishery stocks, a planetwide problem, on the shoulders of any one of these organizations. Froese emphasized that certified stocks, despite their problems, remain a better option for the consumer: "it is still reasonable to buy certified seafood because the percentage of moderately exploited, healthy stocks is three to four times higher in certified than in noncertified seafood."[122]

In spite of all the challenges to comprehensively assessing a fishery, the MSC courageously waded into a shadowy labyrinth of mislabeled fishes and misleading claims and is trying to shed a little light. Certainly not all MSC certified fisheries are perfectly sustainable. But there are success stories. Alaska pollock (*Gadus chalcogrammus*) is fished throughout the Bering Sea and constitutes the second-largest fishery in the world. The Gulf of Alaska and Aleutian Islands fleet was awarded MSC certification in 2005 and recertified several times since. Worldwide catches had fallen from a peak of 9 million tons per year in the 1990s but stabilized by 2002 and have sustainably yielded some 5 million tons every year since.[123] The fishery, one of the first to be managed with transferrable catch shares, accounts for some 30,000 jobs in the United States alone. Sold in fish sticks, fast food, and countless other products, more often than not pollock is the first seafood a person tries as a child. Employing midwater trawls—nets pulled below the surface, but high above the bottom—the fishery boasts a laudably low bycatch rate estimated by NOAA Fisheries at less than 1 percent.[124] Because there are no agricultural inputs, pollock fishing also has an enviably small carbon footprint, on par with tofu; meanwhile, beef's carbon hoofprint is a hundred times larger.[125] Alaska pollock seems to be just the kind of fishery that deserves to be certified by organizations like the Marine Stewardship Council, and purchased by consumers intent on supporting sustainability, employment, and the environment.

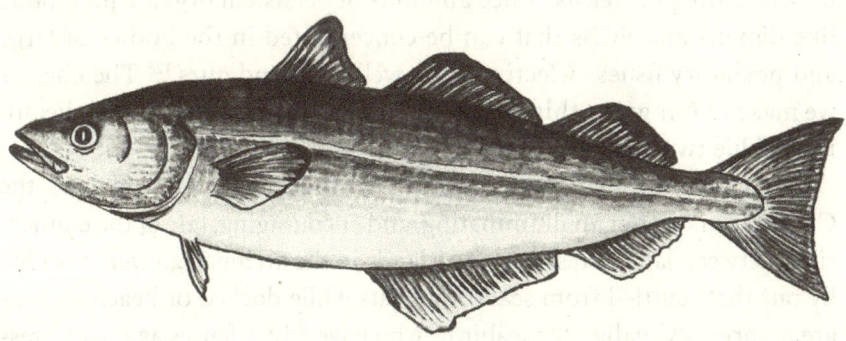

Alaska pollock (*Gadus chalcogrammus*)

INVISIBLE, EXTRAORDINARY, INDISPENSABLE

When you finally see what goes on underwater, you
realize that you've been missing the whole point of the
ocean. Staying on the surface all the time is like going to
the circus and staring at the outside of the tent.
—Dave Barry, *Miami Herald*

While the oceans may seem invisible beneath the waves, many of
our deeds on land are felt all too keenly by the sea. Diadromous and
coastal fishes, whose habitats are directly connected to continents,
are particularly at risk. Although the sight of oil spills on the ocean
is heart-wrenching, as is the harm done when petroleum smears bird
rookeries and sea otters, their direct impact pales in comparison to the
indirect effects of runoff from land. The Mississippi River alone drains
nearly half the continental United States, delivering a brew of silt, fertilizers, pesticides, and industrial chemicals into the Gulf of Mexico. Algal
blooms are one inevitable result of too much runoff, and their decay can
lead to anoxic dead zones, particularly when overfishing has decimated
seafloor scavengers who would otherwise digest the excess algae instead
of leaving it to rot. Tarpon and menhaden and bonefish are affected
by runoff plumes and dead zones, as are Gulf shrimp and the fishers
who toil to deliver them to our tables. Shoreline development and the
elimination of key breeding habitats like salt marshes and mangroves,
whether for high-rise hotels or aquaculture ponds, is kneecapping the
ability of many nearshore fishes to reproduce. Inland factories, even
backyard fire pits, release trace amounts of persistent organic pollutants
like dioxins and PCBs that can be concentrated in the bodies of large
and predatory fishes, affecting their wellbeing and ours.[126] The choices
we make in our green third of the planet increasingly impact fish health
in the blue two-thirds.

Far out in the Indian Ocean, more than a thousand miles from land, the
Chagos Islands offer an illuminating, and encouraging, tale of the connections between land and sea. Many islands in the archipelago were invaded
by rats that scuttled from seafaring boats while docked or beached. Rats
are notoriously malign for seabirds, who have few defenses against terrestrial predators (the main reason they nest on remote islands). But not all

islands in the archipelago were infested, and intensive removal programs have eradicated rats from several of the isles, setting in motion a natural experiment that reveals how readily we can restore the oceans through better land stewardship.

The differences between the rat-free and rat-infested islands could not be more stark. Where nest-raiding rats are present, seabird populations plummet. With them, a key source of nutrition vanishes: seabirds feed bountifully at sea, then defecate copiously on land, transporting marine nutrients to the islands. But as those nutrients disappear, a complex web of interactions begins to unravel. Trees and native vegetation decline by a factor of four, then nitrogen in soil and leaf litter fades.[127] Less natural nutrition reaches the sea, with nitrogen inputs falling nearly twentyfold, and nearshore algae are deprived of fertilizer. Algae crops suffer, and damselfish farmers decline in health and diminish in number. The abundance of zooplankton, which also depend on nitrogen inputs, sinks by two-thirds. When plankton is scarce, giant manta rays who normally feed near the islands are forced to quit the coastlines in search of richer pastures: manta sightings near affected islands tumbled from one every 14 minutes to precisely zero.[128]

On islands where rats are absent, however, studies by Nicholas Graham from the UK's Lancaster University revealed an ecosystem that was vastly healthier. Seabirds were 750 times more abundant, and soil nitrogen was 250 times higher than on rat-infested islands.[129] Damselfish and other fishes, thanks to more luxurious algal pastures, grew faster and saw their combined biomass increase by almost 50 percent. Algae grazing rates, a key determinant of coral recovery, were also boosted, improving the reef's ability to withstand storms and climate change. Dr. Graham fingered the culprit and underscored what we can do to support the sea: "rat removal should be a conservation priority for coral reef islands. The return of seabirds would benefit not only the island, but also adjacent nearshore marine ecosystems." Protecting damselfish, seabirds, coral reefs, and manta rays is well within our grasp, but it is a job that starts on land.

Meanwhile, out at sea, a future of stable fish populations and sustainable seafood is now visible on the horizon. Many of the solutions have been identified already and are supported by years of scientific evidence. What remains is the heavy lift of achieving publicly acceptable remedies, ones that policy makers and fishers and consumers alike can embrace. Our planet is not going to stop eating salmon, cod, snapper, and tuna.

Fisheries already harvest food from an area four times larger than terrestrial agriculture.[130] But the world can reduce overfishing by shrinking catch quotas, modifying gear, and limiting subsidies that prop up today's overcrowded fishing fleets. Despite some resistance from fisheries, it is clear that healthy and well-managed seafood stocks will actually be more, not less, productive in the long term. Some estimates suggest that if all the world's fisheries were managed properly, the rebuilt stocks would yield an astonishing 18 million tons more seafood than we are producing right now.[131] As it is, more than 80 percent of all seafood landings today come from stable stocks, including two-thirds of the ten most landed species, so the dream of sustainability is within reach.

Much of the work of repairing fisheries will need to be done by governments. Daniel Pauly, 2023 co-recipient of the distinguished Tyler Prize for Environmental Achievement (with fisheries economist Rashid Sumaila), lays the responsibility for saving the fishing industry from itself directly at the doorstep of the world's legislative bodies. "Governments are the only entities that can prevent the end of fish. For one thing, once freed from their allegiance to the fishing-industrial complex, they are the ones with the research infrastructure capable of prudently managing fisheries. For another, it is they who provide billions of dollars in annual subsidies that allow fisheries to persist despite the lousy economics of the industry."[132] The sea can bounce back from overfishing, and will, but only after fisheries are compelled to slacken their grip on the sea. Rashid Sumaila coined the term "infinity fish" to describe what the oceans could then provide: "If we manage them well, if we use them wisely—they can continue to deliver fish to us for food, for income, for jobs—forever."[133]

Individual people can also contribute much to sustaining wild fish populations, today and into the future. Adopting a vegetarian diet is a healthy option and immediately alleviates pressure on wild fishes. Those who do choose to eat seafood can carefully select which species and fisheries they support. Households can become better informed and purchase fish from well-managed fisheries, following guidelines from auditors like the Marine Stewardship Council and Seafood Watch. These organizations should be pressed, like all organizations, to improve their methods, their accuracy, and their equity. But such votes by pocketbook nourish those fishers who are sustaining their stocks, while starving unlicensed and unregulated operators who decimate theirs. Catch-share systems can be improved, so artisanal fishers are treated fairly, and so quotas are no longer concentrated in too few hands. Government agencies can require fisheries and distributors to faithfully record the chain of custody of every fish caught in the sea. Novel approaches, like the block-chain system piloted in Mauritius, show new technologies can provide that traceability from net to plate. Consumers can patronize fish farming and ranching while at the same time urging the reduction of wild-capture fish meal, the adoption of creative new feed stocks, and a stronger emphasis on farming nonpredatory species. These improvements can alleviate the strain on wild fish populations and supply people with healthy seafood, a source of protein better in many ways for the planet than terrestrial livestock.

And no matter what you eat, you can enjoy blue tourism, visiting some of the majestic and awe-inspiring sites the sea has to offer. Snorkeling and scuba diving offer a ticket to the world beneath the waves, making visible what was formerly invisible; so do glass-bottom boat tours, sea kayaking, paddle boarding, and trips to an aquarium. Blue tourism provides invaluable support for marine protected areas and sustains the local economies that surround and depend on them. Sharing the rapture of swimming with whale sharks, snorkeling over coral reefs, or watching fishes behind glass will foster that delight in others, from naturalist guides to government ministers, and even school groups you might bump into at the aquarium.

By assembling a stained-glass window of these myriad solutions, we can complete our evolution from hunter-gatherers, who hauled wild animals from the sea, to farmers and ranchers who responsibly cultivate food for a hungry planet. The oceans are magnificently resilient, they have rebounded many times before, and they will recover again when given a chance. Above all, we must understand and respect the nature of fishes, lest we ask of them more than they can give and test them more than they can bear. Our vast blue seas, which harbor some of the most extraordinary and indispensable animals on Earth, deserve nothing less.

EPILOGUE

AMID THE JUMBLED CORALS, sponges, and rocks of an Australian reef, something remarkable is taking place. There, a blackspot tuskfish (*Choerodon schoenleinii*) is eating cockles. In and of itself, this is not peculiar; fishes munch on shellfish all the time. But this tuskfish is particularly notable: it uses a tool to open its lunch.[1] After nabbing a cockle on the reef, the fish carries the robust shell in its mouth to a nearby rock shaped like an anvil. There, the industrious tuskfish slams the shell against the boulder, repeatedly, until it cracks open and the delectable bits are exposed. A pile of shells strewn around its base reveals that the anvil has been used many times. Improbably, this assiduous tuskfish has finned its way into the world of tool use, joining human beings, a few other primates, and a handful of clever birds in taking a great behavioral step forward. Several other fishes also have taken to wielding tools: yellowhead wrasse,[2] orange-dotted tuskfish,[3] tripletail wrasse,[4] and graphic tuskfish have all picked up the trick.[5] Even the California sheephead, that lovable denizen of coastal kelp forests, has been known to dash sea urchins against boulders to crack them open.[6]

What these fishes have in common is progress, and perhaps a dose of curiosity. Somewhere in their dim ancestral past, a tuskfish must have gripped a particularly hardy cockle shell and wondered, "how shall I open this?" Another may have spied a large rock and thought, "what would happen if I bash this shell against that stone?" Now fishes may not have the literal power of speaking to themselves, but one tuskfish must have tried the technique, and presumably tried it quite a few more times before

perfecting the trick. That animal discovered, through curiosity and ingenuity, how to resolve a dilemma and feed itself in a novel way. And the discovery spread, perhaps through observation and imitation, so that anvil
use is now common among blackspot tuskfish. Like carpenters, blacksmiths, plumbers, and other noble tradespeople, these humble fish have
managed to solve old problems by applying new tools.

> One can remain alive long past the usual date of
> disintegration if one is unafraid of change, insatiable in
> intellectual curiosity, interested in big things, and happy in
> small ways.
> —Edith Wharton, *A Backward Glance*

Fish, in their small ways, are endlessly surprising. They inhabit a world
deeply unfamiliar to us, so their most basic adaptations seem novel, even
entrancing. We are far more familiar with the adaptations of terrestrial
wildlife. Giraffes use long necks to browse high vegetation. Porcupines
rely on stiff spines to defend themselves. Parrots crack seeds with their
heavy bills. But because the realm of fish is largely concealed from us,
their behaviors, forms, and modifications provide a rich opportunity for
discovery. The challenges they face are unique, as are the opportunities
they enjoy. Seawater is dense, requiring brawn to push through, but it is
buoyant and can support creatures who grow to immense proportions.
Oceans are cold, sapping strength and slowing growth, but they can be
fantastically productive. While much of the sea's depths are too dark
for vision, water transports chemical signals and sound communication across meters, miles, and longitudes. The upper layers of the ocean,
seven-tenths of the planet bathed in sunlight, offer unparalleled possibilities for photosynthesis, but its bounty inevitably sinks to the bottom of the
sea. On land, the equivalent would be if berries from every shrub floated
slowly into the sky, where terrestrial animals could not reach them. The
marine world poses unique challenges, and it is little wonder that fishes
have evolved so many fascinating and creative responses.

On our underwater safari, we have toured many of these extraordinary adaptations. A blue-chinned parrotfish can transform from female
to male. Billfish and tuna traverse thousands of miles to find sufficient
food, and many species are warm-blooded to boost their muscle power

and efficiency. Prey often hide in the dim and chilly depths, but sword-fish and marlins can rely on heated eyeballs to improve their vision while hunting in the near-dark. Anchoveta reach astonishing abundance, when upwellings are strong, and have adopted complicated schooling patterns to defend against ravenous predators. On the other end of the size spectrum, giants like mantas and whale sharks eat the tiniest of foods, plankton, and rely on enormous bulk to sustain themselves between patches of richness. Sharks hit upon a successful formula millions of years ago and have thrived ever since. On coral reefs, diversity spawns diversity, thanks to the endless number of interactions between denizens, and thanks to the nutrient-accumulating power of corals and sponges, two unheralded animal groups on which the very existence of reefs is founded. In more temperate and polar seas, great productivity clashes with slow growth, gifting the planet with some of its most important food fishes, but also challenging fisheries to manage them cautiously and intelligently. Responses to the already novel environment of the ocean become truly bizarre in the extremes of the deep abyss, where light, pressure, chemistry, and energy combine to produce some of the strangest beings on the planet. And along coastlines, the immense opportunities presented by mangroves and kelp, as well as the vast freshwater resources of the continents, have prompted coastal and migratory fishes to hone their own unique adaptations: magnetic navigation, salt-tolerance, mouth-breathing, even precision archery.

The proverbial bull shark in the ocean's exquisite china shop is us. Like it or not, people have acquired stewardship over our planet and its oceans, and we bear a responsibility for managing and mitigating our widespread and tangible impacts. Plastics are accumulating in the seas, and in fish tissues. Climate change is shifting ranges, migration routes, and timing of spawning. Acidification slows shell growth and threatens coral reefs. A cocktail of chemicals slips into the sea, elevating disease rates and endangering fish reproduction. Nutrient runoff triggers dead zones in coastal waters important to both fish and fishers. Bottom trawlers damage habitat, and factory-sized fishing vessels turn fishes into corporate profits. Overfishing, simply taking fishes from the sea faster than they can reproduce, threatens populations of many wild animals whose behaviors and adaptations inspire and amaze us.

But if we are unafraid of change, as Edith Wharton encourages, and if we accept the challenge of stewardship, we can improve our treatment of the oceans and build on progress already underway. New South Wales,

Australia and the state of California have banned single-use plastic bags, to slow the flow of plastic into the sea. In 2022 the US Congress moved to bar all trade in shark fins.[7] Seagrass beds in Tampa Bay, Florida, have recovered to lushness not seen since the 1950s.[8] Consumer awareness, at restaurants and fish counters, is helping promote sustainable fisheries. Marine tourism is surging and generating much-needed revenue for coastal communities. And considerable progress has been made, around the world, in management of our fisheries.

A recent review of wild-capture fisheries, by environmental economist Christopher Costello and fisheries biologist Daniel Ovando, found cause for optimism.[9] Many of the world's most valuable fisheries, comprising a third of all fishes caught, are in good health thanks to intelligent management and strong governance. Their review also suggests that much of the damaging influence of climate change can be successfully mitigated by better monitoring and administration of fish harvests. The pair specifically highlight the value of adopting rights-based approaches like ITQs and conclude, "the future of fisheries can be bright with effective fishery management interventions."

Augmenting the catches from wild-capture fisheries are the harvests from aquaculture. These operations have exploded in productivity, geographical range, variety of techniques, and diversity of farmed species. While this productivity helps offset the pressures placed by capture fisheries on wild fish populations, it also poses important questions. The assumption that aquaculture is meeting global demand for seafood should be challenged. "I think that if there was increasing demand for fish in the world, we should be seeing higher prices," says salmon fishing captain Megan Corazza. "So I don't really buy that, to be honest, if there's a lot of demand we'd be getting more than thirty-five to forty cents a pound."[10] New York University scientist Jennifer Jacquet speaks more pointedly, avowing that "they're not trying to feed people with aquaculture, they're trying to make money."[11] In this light, perhaps carnivores like salmon and tuna should be reared mostly by hatcheries to boost wild populations, while diners concentrate on tilapia and trout and perhaps add a few new items to their menu such as farmed mussels and leafy marine algae.

Marine protected areas continue to offer the best chance to conserve wild fishes and their habitats. Ambitious targets of protecting 30 percent of the oceans by 2030 have recently gained traction, with many countries already well on their way to meeting that goal. Once marine reserves are put into place, and patrolled effectively, sea life rebounds within their borders. The resiliency of the oceans, due in part to the immense diversity of fishes and their myriad interactions, allows them to recover with amazing speed once negative impacts are halted. These reserves offer unparalleled opportunities for people to plunge into a marine world filled with fascinating wildlife, curious behaviors, and astonishing adaptations. To actually see the ocean beneath the curtain of its dappled surface, to immerse oneself in this strange and extraordinary environment, is to gain an unforgettable appreciation of our responsibility for the seas.

> Whether you eat fish or not, whether you are in the middle of America or in the middle of anywhere—the ocean touches your life.
> —Rashid Sumaila, "A Conversation with Rashid Sumaila"

The oceans have withstood radical transformations and even planetwide extinction events, and they will withstand the admittedly clumsy first centuries of our stewardship. Fortunately we have the capacity to investigate new solutions, with curiosity and creativity, to cure our errors. We are capable of dreaming loftily and adopting bold reforms when pressed by an undeniable need to change. Curiosity and ingenuity have driven human cultural evolution. Ancient ancestors figured out how to walk upright, and to scour coastlines for nutritious marine foods. Others invented canoes and paddles and circumnavigated the Pacific Ocean. We have conceived boats and steam power, winches and purse seines, refrigerated containers, and aquaculture ponds. Each was devised to resolve a dilemma, and each was successful, to a point. Today we are tasked with devising a new solution to a new impasse, managing our planet's oceans so they can provide blue foods, now and in the future, while continuing to sustain wild fish populations. Like the humble tuskfish, we will have to apply new tools to these problems. Mostly, though, what we need is to remember to enjoy the small things, the delicate intimacy between cleaner fish and groomed client, or the sonorous hooting of an amorous toadfish, and rejoice that our planet and its oceans can provide such delightful spectacles. So long as we truly care for the sea, the sea and its fishes will take care of us.

MARINE CONSERVATION AND SUSTAINABLE SEAFOOD RESOURCES

SEAFOOD PURCHASING RESOURCES

Supporting sustainable seafood, whether you make a single purchase in the supermarket or set up a supply chain for your business, is something everyone should prioritize. The following organizations offer helpful information about sustainable seafood purchasing.

Marine Stewardship Council: International nonprofit organization dedicated to assessing and certifying sustainable fisheries; look for their blue-check ecolabel on seafood. *msc.org*

Seafood Watch: Monterey Bay Aquarium has provided informed recommendations for sustainable seafood sources for more than twenty years; use their website or downloadable guides. *seafoodwatch.org*

Good Fish Guide: Marine Conservation Society is the UK's version of the Marine Stewardship Council, providing seafood sustainability ratings for consumers and businesses. *mcsuk.org*

Ocean Wise Seafood: Program that helps seafood purchasing businesses make sustainable choices; find seafood recommendations on their website and look for their Ocean Wise symbol. *ocean.org*

Local Catch Network: Information hub for community-based seafood options, including community-supported fisheries and other harvest-to-table sources. *localcatch.org*

Fish Choice: Nonprofit organization providing sustainability guidance to the seafood purchasing and sourcing industry and retail customers; explore their online buying guides. *fishchoice.com*

Sustainable Seafood Guides: World Wide Fund for Nature publishes country-specific seafood purchasing guides (for European and Asian nations). *seafoodsustainability.org*

Conservation Alliance for Seafood Solutions: International collaboration that improves the responsibility and sustainability of seafood production systems. *solutionsforseafood.org*

MARINE CONSERVATION ORGANIZATIONS

From a mere handful just two decades ago, the number of marine conservation organizations has ballooned like a swell shark to encompass hundreds of committed, hardworking, and inspiring groups. Policy and protest, science and advocacy, education and capacity-building are just some of the tools they employ to help save our seas. A few of the most influential and effective organizations are highlighted below, followed by a more comprehensive list for you to investigate. Consider joining their campaigns, educating yourself with their resources, and supporting their work with your time or by making a donation.

Antarctic and Southern Ocean Coalition: Advocates for protection of the southern polar seas, along with partners, campaigners, and supporters. *asoc.org*

Black Girls Dive: Provides young and underserved girls with science skills and underwater experiences through innovative programming. *blackgirlsdivefoundation.squarespace.com*

Conservation International Center for Oceans: Supports integrated management of marine protected areas and publishes the ocean health index for more than 200 countries. *conservation.org*

Deep Sea Conservation Coalition: Promotes policies to halt the harmful impacts of deep-water trawling, particularly on seamounts and sensitive deep-sea habitats. *deep-sea-conservation.org*

Defenders of Wildlife: Promotes conservation of US national marine sanctuaries and monuments, and protection of whales and other marine wildlife. *defenders.org*

Environmental Defense Fund: Collaborates with fishing communities to improve sustainability through new technologies and supporting rights of artisanal fishers. *edf.org*

Fauna and Flora International: Focuses on reducing the harmful impact of marine plastics by preventing plastic precursors from reaching the ocean. *fauna-flora.org*

Great Barrier Reef Research Foundation: Coordinates large-scale programs to restore and protect the Great Barrier Reef and the animals that rely on the reef. *barrierreef.org*

Institute for Ocean Conservation Science: Stony Brook University applies scientific research to a range of global marine conservation challenges. *oceanconservationscience.org*

International Seafood Sustainability Foundation: Organization dedicated specifically to improving the sustainability of tuna fisheries around the world. *iss-foundation.org*

IUCN Red List of Threatened Species: Assesses the vulnerability of terrestrial and marine species and provides data on threats, ranges, population trends, and more. *iucnredlist.org*

Manta Trust: Coordinates research and conservation programs to conserve the ocean's giant rays and their habitats through research, education, collaboration. *mantatrust.org*

Mission Blue: Sylvia Earle's organization inspires action to protect and explore the oceans, focusing on vital places critical to marine health. *missionblue.org*

Ocean Conservancy: Promotes protection of ocean ecosystems and support for communities who depend on them through science, policy, and partnerships. *oceanconservancy.org*

Oceana: Advocates for marine conservation through science-based campaigns to improve fishery management, and ocean restoration. *oceana.org*

Project Seahorse: Conserves marine life with a special emphasis on seahorses, by connecting research and management at community levels and international scales. *project seahorse.org*

Reef Environmental Education Foundation: Gathers data on reef fishes and provides educational resources, to support decision making by international conservationists and local fisheries managers. *reef.org*

Sea Shepherd Conservation Society: Supports governments in countering illegal, unreported, and unregulated fishing. *seashepherd.org*

Shark Research Institute: Sponsors and carries out research on sharks to promote their protection through education, conservation, and advocacy. *sharks.org*

UNEP Protecting Coral Reefs: United Nations Environmental Programme protects coral reefs against human impacts by promoting better marine resource management. *unep.org*

American Cetacean Society: *acsonline.org*
Aquaculture Stewardship Council: *asc-aqua.org*
Australian Marine Conservation Society: *marineconservation.org.au*
Azul: *multiplier.org/project/azul/*
Bahamas Plastic Movement: *bahamasplasticmovement.org*
Billion Oyster Project: *billionoysterproject.org*
Bite-Back Shark and Marine Conservation: *bite-back.com*
Blue Frontier Campaign: *bluefront.org*
Center for Biological Diversity: *biologicaldiversity.org*
Center for Marine Biodiversity and Conservation: *cmbc.ucsd.edu*
Coastal Research and Education Society of Long Island: *cresli.org*
Coral Reef Alliance: *coral.org*
EarthEcho International: *earthecho.org*
Earthwatch Institute: *earthwatch.org*
Equilibrio Azul: *equilibrioazul.org*
Global Coral Reef Alliance: *globalcoral.org*
Great Barrier Reef Marine Park Authority: *gbrmpa.gov.au*
Greenpeace International: *greenpeace.org*
Hawaii Ocean Watch: *hawaiioceanwatch.org*
I Am Water Foundation: *iamwaterfoundation.org*
Institute for Fisheries Resources: *ifrfish.org*
Institute for Ocean Conservation Science: *oceanconservationscience.org*
International Coral Reef Initiative: *icriforum.org*
International Fund for Animal Welfare: *ifaw.org*
IUCN Global Marine Programme: *iucn.org*
Lighthouse Foundation: *lighthouse-foundation.org*
Mare Nostrum: *joinmnf.org*
Marine Biological Association: *mba.ac.uk*
Marine Connection: *marineconnection.org*
Marine Conservation Alliance: *marineconservationalliance.org*
Marine Conservation Institute: *marine-conservation.org*
Marine Conservation Society: *mcsuk.org*
Marine Fish Conservation Network: *conservefish.org*
Marine Life Information Network: *marlin.ac.uk*
Marine Mammal Center: *marinemammalcenter.org*
Marine Megafauna Foundation: *marinemegafauna.org*
Monterey Bay Aquarium Research Institute: *mbari.org*
Mote Marine Laboratory: *mote.org*
National Marine Sanctuary Foundation: *marinesanctuary.org*
National Marine Sanctuaries Program: *sanctuaries.noaa.gov*

National Marine Mammal Foundation: *nmmf.org*
National Oceanographic and Atmospheric Association: *noaa.gov*
Natural Resources Defense Council: *nrdc.org*
Nature Conservancy: *nature.org*
Ocean Alliance: *whale.org*
Ocean Cleanup: *theoceancleanup.com*
Ocean Conservation Research: *ocr.org*
Oceanic Preservation Society: *opsociety.org*
The Ocean Project: *theoceanproject.org*
Ocean Research and Conservation Association: *teamorca.org*
OceanCare: *oceancare.org*
Oceanswell Sri Lanka: *oceanswell.org*
Point Blue Conservation Science: *pointblue.org*
Project Aware Foundation: *projectaware.org*
Reef Check: *reefcheck.org*
Reef Relief: *reefrelief.org*
Sea Turtle Conservancy: *conserveturtles.org*
Seal Conservation Society: *pinnipeds.org*
Shark Stewards: *sharkstewards.org*
Shark Trust: *sharktrust.org*
Society for Conservation Biology: *conbio.org*
Surfrider Foundation: *surfrider.org*
Take 3 For The Sea: *take3.org*
TRAFFIC Trade In Wild Species: *traffic.org*
Whale and Dolphin Conservation: *whales.org*
WildAid: *wildaid.org*
Woods Hole Oceanographic Institution: *whoi.edu*
World Resources Institute: *wri.org*
World Wildlife Fund: *worldwildlife.org*

NOTES

Introduction

1. D. E. Facey et al., eds., *The Diversity of Fishes: Biology, Evolution and Ecology, 3rd Edition* (New York: Wiley, 2022).

2. Food and Agriculture Organization (FAO), *The State of the World Fisheries and Aquaculture 2016: Contributing to Food Security and Nutrition for All* (Rome: FAO, 2016).

3. J. A. Koslow et al., "Continental Slope and Deep-Sea Fisheries: Implications for a Fragile Ecosystem," *ICES Journal of Marine Science* 57 (2000): 548–557.

4. G. Barone et al., "Estimated Dietary Intake of Trace Metals from Swordfish Consumption: A Human Health Problem," *Toxics* 6, no. 2 (2018): 22.

5. I. Urbina, "Palau vs. the Poachers," *New York Times*, February 17, 2016, https://www.nytimes.com/2016/02/21/magazine/palau-vs-the-poachers.html.

Part I

1. G. P. Harris, *Phytoplankton Ecology: Structure, Function and Fluctuation* (Berlin: Springer, 1986).

2. M. J. Dagg and G. A. Breed, "Biological Effects of Mississippi River Nitrogen on the Northern Gulf of Mexico—A Review and Synthesis," *Journal of Marine Systems* 43 (2003): 133–152.

3. E. I. Stiefel, "Molybdenum Bolsters the Bioinorganic Brigade," *Science* 272, no. 5268 (1996): 1599–1600.

4. P. Greenberg, *The Omega Principle: Seafood and the Quest for a Long Life and a Healthier Planet* (New York: Penguin Press, 2018).

5. D. Pauly and I. Tsukayama, *The Peruvian Anchoveta and Its Upwelling Ecosystem: Three Decades of Change* (Callao, Peru: Instituto del Mar del Peru, 1987).

6. J. Tarazon and W. Arntz, "The Peruvian Coastal Upwelling System," in *Coastal Marine Ecosystems of Latin America. Ecological Studies, Vol. 144*, ed. U. Seeliger and B. Kjerfve (Berlin: Springer, 2001), 229-244.

1. All Together Now

1. C. L. Bartlett, *Guano: A Treatise on the History, Economy, as a Manure, and Modes of Applying Peruvian Guano, in the Cultivation of the Various Crops of the Farm and the Garden* (Boston: C. L. Bartlett, 1860).

2. E. Schnug, F. Jacobs, and K. Stöven, "Guano: The White Gold of the Seabirds," in *Seabirds*, ed. H. Mikkola (London: Intech Open, 2018), 89.

3. Bartlett, *Guano: A Treatise*, title page.

4. Schnug, Jacobs, and Stöven, "Guano: White Gold."

5. Guano Islands Act, 48 U.S.C. Ch. 8 § 1411–1419 (1856).

6. E. M. Leal, L. R. Castro, and G. Claramunt, "Variability in Oocyte Size and Batch Fecundity in Anchoveta (*Engraulis Ringens*, Jenyns 1842) from Two Spawning Areas off the Chilean Coast," *Scientia Marina* 73, no. 1 (2009): 59–66.

7. G. S. Helfman et al., *The Diversity of Fishes: Biology, Evolution, and Ecology, 2nd Edition* (Oxford: Wiley-Blackwell, 2009).

8. D. Pauly and I. Tsukayama, *The Peruvian Anchoveta and Its Upwelling Ecosystems: Three Decades of Change* (Callao, Peru: Instituto del Mar del Peru, 1987).

9. A. Levy-Lior et al., "Guanine-Based Biogenic Photonic-Crystal Arrays in Fish and Spiders," *Advanced Functional Materials* 20 (2010): 320–329.

10. F. Crick and J. D. Watson, "Molecular Structure of Nucleic Acids: A Structure for Deoxyribose Nucleic Aid," *Nature* 171 (1953): 737–738.

11. R. M. Berquist et al., "The Coelacanth Rostral Organ Is a Unique Low-Resolution Electro-Detector that Facilitates the Feeding Strike," *Scientific Reports* 5, no. 1 (2015): 1–5.

12. R. C. Eaton, R.K.K. Lee, and M. B. Foreman, "The Mauthner Cell and Other Identified Neurons of the Brainstem Escape Network of Fish," *Progress in Neurobiology* 63, no. 4 (2001): 467–485.

13. J-G. J. Godin and M. J. Morgan, "Predator Avoidance and School Size in a Cyprinodontid Fish, the Banded Killifish (*Fundulus diaphanus* Lesueur)," *Behavioral Ecology & Sociobiology* 16 (1985): 105–110.

14. T. J. Pitcher and J. K. Parrish, "Functions of Shoaling Behaviour in Teleosts," in *The Behaviour of Teleost Fishes, 2nd Edition*, ed. T. J. Pitcher (London: Chapman & Hall, 1993), 294–337.

15. H. Kutschale, "Long-Range Sound Transmission in the Arctic Ocean," *Journal of Geophysical Research* 66, no. 7 (1961): 2189–2198.

16. O. Sand, "Lateral-Line System," in *Comparative Physiology of Sensory Systems*, ed. L. Bolis, R. D. Keynes, and S.H.P. Maddrell (Cambridge: Cambridge University Press, 1984), 3–32.

17. E. S. Hassan, "On the Discrimination of Spatial Intervals by the Blind Cave Fish (*Anoptichthys jordani*)," *Journal of Comparative Physiology* 159A (1986): 701–710.

18. T. J. Pitcher, B. L. Partridge, and C. S. Wardle, "A Blind Fish Can School," *Science* 194, no. 4268 (1976): 963–965.

19. R. R. Stephens, "A Description of the Cephalic Lateralis System of *Anchoa mitchilli* (Valenciennes) (Clupeomorpha: Engraulidae) with Identification of Synapomorphies for the Engraulidae," *Proceedings of the Biological Society of Washington* 123, no. 1 (2010): 8–16.

20. D. E. Hoss and J.H.S. Blaxter, "Development and Function of the Swimbladder-Inner Ear-Lateral Line System in the Atlantic Menhaden, *Brevoortia tyrannus* (Latrobe)," *Journal of Fish Biology* 20 (1982): 131–142.

21. P.J.P. Whitehead and J.H.S. Blaxter, "Swimbladder Form in Clupeoid Fishes," *Zoological Journal of the Linnean Society* 97 (1989): 299–372.

22. F. DiDario, "Homology between the *recessus lateralis* and Cephalic Sensory Canals, with the Proposition of Additional Synapomorphies for the Clupeiformes and the Clupeoidei," *Zoological Journal of the Linnean Society* 141 (2004): 257–270.

23. T. J. Pitcher, "Shoaling and Schooling in Fishes," in *Comparative Psychology: A Handbook*, ed. G. Greenberg and M. M. Hararway (New York: Garland, 1998), 748–760.

24. N. O. Handegard et al., "The Dynamics of Coordinated Group Hunting and Collective Information Transfer among Schooling Prey," *Current Biology* 22 (2012): 1213–1217.

25. J. E. Herbert-Read et al., "Proto-cooperation: Group Hunting Sailfish Improve Hunting Success by Alternating Attacks on Grouping Prey," *Proceedings of the Royal Society B* 283 (2016): 20161671.

26. P. Domenici et al., "How Sailfish Use their Bills to Capture Schooling Prey," *Proceedings of the Royal Society B* 281 (2014): 20140444.

27. S. P. Oliver et al., "Thresher Sharks Use Tail-Slaps as a Hunting Strategy," *PLoS ONE* 8, no. 7 (2013): e67380.

28. R. P. Wilson et al., "Conspicuous Coloration May Enhance Prey Capture in Some Piscivores," *Animal Behaviour* 35, no. 5 (1987): 1559–1560.

29. U. Kils, "The ecoSCOPE and dynIMAGE: Microscale Tools for In Situ Studies of Predator-Prey Interactions," *Ergebnisse der Limnologie* 36 (1992): 83–96.

30. S. Yamaguchi and K. Endo, "Molecular Phylogeny of Ostracoda (Crustacea) Inferred from 18S Ribosomal DNA Sequences: Implication for Its Origin and Diversification," *Marine Biology* 143 (2003): 23–38.

31. W. Root, *Food* (New York: Simon & Schuster, 1980).

32. B. Wilson, R. S. Batty, and L. M. Dill, "Pacific and Atlantic Herring Produce Burst Pulse Sounds," *Proceedings of the Royal Society B (Supplement)* 271 (2004): S95–S97.

33. P. Greenberg, *The Omega Principle: Seafood and the Quest for a Long Life and a Healthier Planet* (New York: Penguin Press, 2018).

34. A. D. Thaler, "Herring Wars: Quotas, Conflicts, and Climate Change in the North Atlantic," *Southern Fried Science*, July 25, 2013, https://www.southernfriedscience.com/herring-wars-quotas-conflicts-and-climate-change-in-the-north-atlantic.

35. Greenberg, *The Omega Principle*.

36. "Atlantic Herring, *Clupea harengus* Linnaeus 1758," FishBase: A Global Information System on Fishes, accessed August 11, 2020, https://www.fishbase.se/summary/24.

37. M. S. Love, *Certainly More than You Want to Know about the Fishes of the Pacific Coast: A Postmodern Experience* (Santa Barbara, CA: Really Big Press, 2011).

38. J. Radovich, "The Collapse of the California Sardine Fishery: What Have We Learned?" in *Resource Management and Environmental Uncertainty: Lessons from Coastal Upwelling Fisheries*, ed. M. H. Glantz and J. D. Thompson (New York: John Wiley & Sons, 1981), 56–78.

39. T. R. Baumgartner, A. Soutar, and F. Ferreira-Bartrina, "Reconstruction of the History of the Pacific Sardine and Northern Anchovy Populations over the Past Two Millennia from Sediments of the Santa Barbara Basin, California," *California Cooperative Oceanic Fisheries Investigations Reports* 33 (1992): 24–40.

40. Baumgartner, Soutar, and Ferreira-Bartrina, "Reconstruction of the History."

41. A. M. Kaltenberg and K. J. Benoit-Bird, "Diel Behavior of Sardine and Anchovy Schools in the California Current System," *Marine Ecology Progress Series* 394 (2009): 247–262.

42. S. M. Bollens and B. W. Frost, "Zooplanktivorous Fish and Variable Diel Vertical Migration in the Marine Planktonic Copepod *Calanus pacificus*," *Limnology and Oceanography* 34, no. 6 (1989): 1072–1083.

43. Kaltenberg and Benoit-Bird, "Diel Behavior," 247–262.

44. Love, *Certainly More*.

45. V. G. Russell, "Pearl Essence in San Pedro," *California Fish & Game* 13, no. 3 (1927): 216–217.

46. M. Verhaegen, P. F. Puech, and S. Munro, "Aquarboreal Ancestors?" *TRENDS in Ecology & Evolution* 17, no. 5 (2002): 212–217.

47. A. Hardy, "Was Man More Aquatic in the Past?" *New Scientist* 17 (1960).

48. J. Zilhão et al., "Last Interglacial Iberian Neandertals as Fisher-Hunter-Gatherers," *Science* 367, no. 6485 (2020).

49. N. H. Yuval, *Sapiens: A Brief History of Humankind* (New York: Harper, 2015).

50. Greenberg, *The Omega Principle*.

51. S. Kalmijn et al., "Dietary Fat Intake and the Risk of Incident Dementia in the Rotterdam Study," *Annals of Neurology* 42, no. 5 (1997): 776–782.

52. P. Barberger-Gateau et al., "Fish, Meat, and Risk of Dementia: Cohort Study," *British Medical Journal (Clinical Research Edition)* 325, no. 7370 (2002): 932–933.

53. D. Swanson, R. Block, and S. A. Mousa, "Omega-3 Fatty Acids EPA and DHA: Health Benefits throughout Life," *Advances in Nutrition* 3, no. 1 (2012): 1–7.

54. J. Frye, *The Men All Singing: The Story of Menhaden Fishing* (Virginia Beach: Donning, 1978), 19.

55. Frye, *The Men All Singing*, 19.

56. Frye, *The Men All Singing*, 20.

57. H. B. Franklin, *The Most Important Fish in the Sea: Menhaden and America* (Washington, DC: Island Press, 2007).

58. W. J. Terry, "Menhaden Movements—A New Theory," *Forest and Stream*, September 30 (1880): 176.

59. Greenberg, *The Omega Principle*.

60. D. C. Duffy, "Seabirds and the 1982–1984 El Niño-Southern Oscillation," *Elsevier Oceanography Series* 52 (1990): 395–415.

2. Hot Blooded

1. "$3.1 Million: Giant Tuna Auctioned Off at Record High Price in Japan," *Korea Times*, January 6, 2019.

2. L. E. Lilly et al., "Effect of Temperature Acclimation on Red Blood Cell Oxygen Affinity in Pacific Bluefin Tuna (*Thunnus orientalis*) and Yellowfin Tuna (*Thunnus albacares*)," *Comparative Biochemistry and Physiology, Part A* 181 (2015): 36–44.

3. M. Miya et al., "Evolutionary Origin of the Scombridae (Tunas and Mackerels): Members of a Paleogene Adaptive Radiation with 14 Other Pelagic Fish Families," *PLoS ONE* 8, no. 9 (2013): e73535.

4. V. Walters and H. L. Fierstine, "Measurements of Swimming Speeds of Yellowfin Tuna and Wahoo," *Nature* 202 (1964): 208–209.

5. R. Claro, "Características Generales de la Ictiofauna," in *Ecología de los Peces Marinos de Cuba*, ed. R. Claro (Chetumal, Mexico: Instituto de Oceanología Academia de Ciencias de Cuba and Centro de Investigaciones de Quintana Roo, 1994), 55–71.

6. M. M. Walker et al., "A Candidate Magnetic Sense Organ in the Yellowfin Tuna, *Thunnus albacares*," *Science* 224 (1984): 751–753.

7. J. Willis et al., "Spike Dives of Juvenile Southern Bluefin Tuna (*Thunnus maccoyii*): A Navigational Role?" *Behavioral Ecology and Sociobiology* 64 (2009): 57–68.

8. B. A. Block, D. Booth, and F. G. Carey, "Direct Measurement of Swimming Speeds and Depth of Blue Marlin," *Journal of Experimental Biology* 166 (1992): 264–284.

9. Peter S. Davie, *Pacific Marlins: Anatomy and Physiology* (Massey, New Zealand: Massey University Printery, 1990).

10. James L. Squire, *Cooperative Marine Game Fish Tagging Program 1985 Billfish Newsletter* (Washington, DC: National Oceanic and Atmospheric Association, National Marine Fisheries Service, 1985).

11. K. A. Fritsches et al., "Cone Visual Pigments and Retinal Mosaics in the Striped Marlin," *Journal of Fish Biology* 63 (2003): 1347–1351.

12. T. Corson, *The Story of Sushi: An Unlikely Saga of Raw Fish and Rice* (New York: Harper Collins, 2007).

13. J. Atema, K. Holland, and W. Ikehara, "Olfactory Responses of Yellowfin Tuna (*Thunnus albacares*) to Prey Odors: Chemical Search Image," *Journal of Chemical Ecology* 6 (1980): 457–465.

14. D.-E. Nilsson et al., "A Unique Advantage for Giant Eyes in Giant Squid," *Current Biology* 22 (2012): 683–688.

15. Ben Goldfarb, "A Leading Marine Biologist Works to Create a 'Wired Ocean'," *Yale Environment 360*, March 20, 2013.

16. Block, Booth, and Carey, "Direct Measurement," 264–284.

17. Janaki Lenin, "Conserving Bluefin Tuna and Sharks: An interview with Barbara Block," *Current Conservation*, June 2, 2013.

18. B. A. Block et al., "Migratory Movements, Depth Preferences, and Thermal Biology of Atlantic Bluefin Tuna," *Science* 293, no. 5533 (2001): 1310–1314.

19. J. D. Altringham and B. A. Block, "Why Do Tuna Maintain Elevated Slow Muscle Temperatures? Power Output of Muscle Isolated from Endothermic and Ectothermic Fish," *The Journal of Experimental Biology* 200 (1997): 2617–2627.

20. K. N. Holland et al., "Physiological and Behavioral Thermoregulation in Bigeye Tuna (*Thunnus obesus*)," *Nature* 358 (1992): 410–412.

21. J. B. Graham and K. A. Dickson, "Physiological Thermoregulation in the Albacore *Thunnus alalunga*," *Physiological and Biochemical Zoology* 54 (1981): 470–486.

22. K. M. Schaefer, D. W. Fuller, and B. A. Block, "Movements, Behavior, and Habitat Utilization of Yellowfin Tuna (*Thunnus albacares*) in the Northeastern Pacific Ocean, Ascertained through Archival Tag Data," *Marine Biology* 152 (2007): 503–525.

23. A. J. Williams et al., "Vertical Behavior and Diet of Albacore Tuna (*Thunnus alalunga*) Vary with Latitude in the South Pacific Ocean," *Deep-Sea Research II* 113 (2015): 154–169.

24. M. K. Musyl et al., "Vertical Movements of Bigeye Tuna (*Thunnus obesus*) Associated with Islands, Buoys, and Seamounts near the Main Hawaiian Islands from Archival Tagging Data," *Fisheries Oceanography* 12 (2003): 152–169.

25. R. T. Kraus and J. R. Rooker, "Patterns of Vertical Habitat Use by Atlantic Blue Marlin (*Makaira nigricans*) in the Gulf of Mexico," *Gulf and Caribbean Research* 19 (2007): 89–97.

26. J. B. Graham and K. A. Dickson, "Tuna Comparative Physiology," *Journal of Experimental Biology* 207 (2004): 4015–4024.

27. R. W. Hill, G. A. Wyse, and M. Anderson, *Animal Physiology, 4th Edition* (Sunderland, MA: Sinauer Associates, 2004).

28. Lilly et al., "Effect of Temperature Acclimation," 36–44.

29. H. A. Shiels, A. Di Maio, S. Thompson, and B. A. Block, "Warm Fish with Cold Hearts: Thermal Plasticity of Excitation-Contraction Coupling in Bluefin Tuna," *Proceedings of the Royal Society B* 278 (2010): 18–27.

30. B. B. Collette et al., "*Xiphias gladius*," *The IUCN Red List of Threatened Species* (2022): e.T23148A46625751.

31. F. G. Carey, "A Brain Heater in the Swordfish," *Science* 216 (1982): 1327–1329.

32. K. A. Fritsches, R. W. Brill, and E. J. Warrant, "Warm Eyes Provide Superior Vision in Swordfishes," *Current Biology* 15 (2005): 55–58.

33. W. A. Wilcox, "A Man Killed by a Swordfish," *Bulletin of the United States Fish Commission* 6 (1887): 417–418.

34. "Five Men and the Sea: Huge Marlin Sinks Filipino Fishing Boat," *Agence France-Presse*, October 10, 2018.

35. R. Ellis, *Swordfish: The Life of an Ocean Gladiator* (Chicago: University of Chicago Press, 2013).

36. J. G. Frazier et al., "Impalement of Marine Turtles by Billfishes," *Fisheries Science* 39 (2014): 85–96.

37. E.F.K. Zarudzki, "Swordfish Rams the *Alvin*," *Oceanus* 13 (1967): 14–18.

38. S. Marras et al., "Not So Fast: Swimming Behavior of Sailfish During Predator–Prey Interactions Using High-Speed Video and Accelerometry," *Integrative and Comparative Biology* 55 (2015): 719–727.

39. I. Nakamura, "Xiphiidae," in *Fishes of the North-Eastern Atlantic and the Mediterranean, Vol. 2*, ed. P.J.P. Whitehead, M. L. Bauchot, J. C. Hureau, J. Nielsen, and E. Tortonese (Paris: UNESCO, 1986), 1006–1007.

40. J. E. Randall, *Coastal Fishes of Oman* (Honolulu: University of Hawaii Press, 1995).

41. B. B. Collette et al., "High Value and Long Life—Double Jeopardy for Tunas and Billfishes," *Science* 333 (2011): 291–292.

42. S. Marikar and L. Ferran, "Jeremy Piven Defends Play Departure due to Mercury Poisoning," *ABC News*, January 15, 2008.

43. M. Harada, "Minamata Disease: Methylmercury Poisoning in Japan Caused by Environmental Pollution," *Critical Reviews in Toxicology* 25 (1995): 1–24.

44. Harada, "Minamata Disease," 1.

45. D. Borum et al., *Water Quality Criterion for the Protection of Human Health: Methylmercury* (Washington, DC: US Environmental Protection Agency, January 2001).

46. "Mercury Concentrations in Fish: FDA Monitoring Program (1990–2010)" (Washington, DC: US Food and Drug Administration, October 25, 2017).

47. J. Burgera and M. Gochfeld, "Mercury in Canned Tuna: White versus Light and Temporal Variation. *Environmental Research* 96 (2004): 239–249.

48. G. Barone et al., "Estimated Dietary Intake of Trace Metals from Swordfish Consumption: A Human Health Problem," *Toxics* 6 (2018).

49. R. Carson, *Silent Spring* (Boston: Houghton Mifflin, 1962).

50. L. R. Vieira et al., "Acute Effects of Copper and Mercury on the Estuarine Fish *Pomatoschistus microps*: Linking Biomarkers to Behaviour," *Chemosphere* 76 (2009): 416–142.

51. E. Ries, *Tales of the Golden Years of California Ocean Fishing 1900–1950* (Laguna Hills, CA: Monterey Publications, 2007), 55.

52. F. A. Parrish et al., "Foraging Interaction between Monk Seals and Large Predatory Fish in the Northwestern Hawaiian Islands," *Endangered Species Research* 4 (2008): 299–308.

53. D. M. Rowe and E. J. Denton, "The Physical Basis for Reflective Communication between Fish, with Special Reference to the Horse Mackerel, *Trachurus trachurus*," *Philosophical Transactions of the Royal Society B* 352 (1997): 531–549.

54. M. S. Love, *Certainly More than You Want to Know about the Fishes of the Pacific Coast: A Postmodern Experience* (Santa Barbara, CA: Really Big Press, 2011).

55. Corson, *Story of Sushi*.

56. M. Bricklin, *Nutrition Advisor: The Ultimate Guide to Health-Boosting and Health-Harming Factors in Your Diet* (Emmaus, PA: Rodale Press, 1993).

57. Corson, *Story of Sushi*, 143.

58. G. Lean, "Why Mackerel Has Been Taken Off the Ethical 'Fish to Eat' List," *The Telegraph*, January 22, 2013.

59. J. D. Morrow et al., "Evidence that Histamine is the Causative Toxin of Scombroid-Fish Poisoning," *New England Journal of Medicine* 324 (1991): 716–720.

60. J. M. Hungerford, "Scombroid Poisoning: A Review," *Toxicon* 56 (2010): 231–243.

61. "Anisakiasis FAQs," Washington, DC: Centers for Disease Control and Prevention, November 12, 2012, https://www.cdc.gov/parasites/anisakiasis/faqs.html.

62. K. M. MacKenzie et al., "Parasites as Biological Tags for Stock Identification of Atlantic Horse Mackerel *Trachurus trachurus* L.," *Fisheries Research* 89 (2008): 136–145.

63. B. R. Moore et al., "Stock Discrimination and Movements of Narrow-Barred Spanish Mackerel across Northern Australia as Indicated by Parasites," *Journal of Fish Biology* 63 (2003): 765–779.

64. P. Escolar, "Escolar: The World's Most Dangerous Fish," *Medellitin Blog*, July 22, 2010, http://blog.medellitin.com/2008/12/escolar-world-most-dangerous-fish.html.

65. H. McGee, *On Food and Cooking: The Science and Lore of the Kitchen* (New York: Scribner, 2004), 187.

66. M. Burros, "A Fish Puts Chefs in a Quandary," *New York Times*, March 10, 1999. https://www.nytimes.com/1999/03/10/dining/eating-well-a-fish-puts-chefs-in-a-quandary.html.

67. Lenin, "Conserving Bluefin Tuna."

68. A. Turns, "Hook to Plate: How Blockchain Tech Could Turn the Tide for Sustainable Fishing," *The Guardian*, June 9, 2021.

69. Jorge Ramírez, MSc, interview by Joe E. Meisel, June 3, 2021.

70. C. Guinet et al., "Killer Whale Predation on Bluefin Tuna: Exploring the Hypothesis of the Endurance-Exhaustion Technique," *Marine Ecology Progress Series* 347 (2007): 111–119.

3. The Oldest Fishes in the Sea

1. M. P. Heinike, G.J.P. Naylor, and S. B. Hedges, "Cartilaginous Fishes (Chondrichthyes)," in *The Timetree of Life*, ed. S. B. Hedges and S. Kumar (Oxford: Oxford University Press, 2009), 320–327.

2. M. I. Coates, "Beyond the Age of Fishes," *Nature* 458 (2009): 413–414.

3. L. Frey et al., "The Early Elasmobranch *Phoebodus*: Phylogenetic Relationships, Ecomorphology and a New Time-Scale for Shark Evolution," *Proceedings of the Royal Society B* 286 (1029): 20191336.

4. Chris Lowe, PhD, interview by Joe E. Meisel, September 16, 2020.

5. G. W. Marshall, Jr., et al., "The Dentin Substrate: Structure and Properties Related to Bonding," *Journal of Dentistry* 25, no. 6 (1997): 441–458.

6. W. Raschi and C. Tabit, "Functional Aspects of Placoid Scales: A Review and Update," *Australian Journal of Marine and Freshwater Research* 43, no. 1 (1992): 123–147.

7. F. D. Arisoy et al., "Bioinspired Photocatalytic Shark-Skin Surfaces with Antibacterial and Antifouling Activity via Nanoimprint Lithography," *ACS Applied Materials & Interfaces* 10, no. 23 (2018): 20055–20063.

8. M. Lee, "Shark Skin: Taking a Bite out of Bacteria," in *Remarkable Natural Material Surfaces and Their Engineering Potential*, ed. M. Lee (New York: Springer, 2014), 15–27.

9. H. L. Pratt, Jr., "Reproduction in the Blue Shark *Prionace flauca*," *Fisheries Bulletin* 77 (1979): 445–470.

10. K. Gammon, "We're Seeing More than Ever: White Shark Populations Rise off California Coast," *The Guardian*, November 6, 2020.

11. C. G. Lowe, "Bioenergetics of Free-Ranging Juvenile Scalloped Hammerhead Sharks (*Sphyrna lewini*) in Kane'ohe Bay, O'ahu, HI," *Journal of Experimental Marine Biology and Ecology* 278, no. 2 (2002): 141–156.

12. D. D. Chapman et al., "The Behavioural and Genetic Mating System of the Sand Tiger Shark, *Carcharias taurus*, an Intrauterine Cannibal," *Biology Letters* 9 (2013): 20130003.

13. G. K. Ostrander et al., "Shark Cartilage, Cancer and the Growing Threat of Pseudoscience," *Cancer Research* 64 (2004): 8485–8491.

14. C. B. Fox, "Squalene Emulsions for Parenteral Vaccine and Drug Delivery," *Molecules* 14 (2009): 3286–3312.

15. M. Zasloff et al., "Squalamine as a Broad-Spectrum Systemic Antiviral Agent with Therapeutic Potential," *Proceedings of the National Academy of Sciences* 108, no. 38 (2011): 15978–15983.

16. M. Kovaleva et al., "Shark Variable New Antigen Receptor Biologics—A Novel Technology Platform for Therapeutic Drug Development," *Expert Opinion on Biological Therapy* 14, no. 10 (2014): 1527–1539

17. M. Leslie, "Mini-Antibodies Discovered in Sharks and Camels Could Lead to Drugs for Cancer and Other Diseases," *Science,* May 10, 2018.

18. G. Zhao et al., "A Novel Nanobody Targeting Middle East Respiratory Syndrome Coronavirus (MERS-CoV) Receptor-Binding Domain Has Potent Cross-Neutralizing Activity and Protective Efficacy against MERS-CoV," *Journal of Virology* 92, no. 18 (2018): e00837–18.

19. S. Parker, *The Encyclopedia of Sharks, 2nd Edition* (Buffalo, NY: Firefly Books, 2008).

20. S. Wroe et al., "Three-Dimensional Computer Analysis of White Shark Jaw Mechanics: How Hard Can a Great White Bite?" *Journal of Zoology* 276, no. 4 (2008): 323–410.

21. J. A. Cooper et al., "Body Dimensions of the Extinct Giant Shark *Otodus megalodon*: A 2D Reconstruction," *Scientific Reports* 10 (2020):14596.

22. C. R. Braekevelt, "Fine Structure of the Choroidal *tapetum lucidum* in the Port Jackson Shark (*Heterodontus phillipi*)," *Anatomy and Embryology* 190 (1994): 591–596.

23. Parker, *Encyclopedia of Sharks.*

24. M. Laskaa et al., "Detecting Danger—Or Just Another Odorant? Olfactory Sensitivity for the Fox Odor Component 2,4,5-Trimethylthiazoline in Four Species of Mammals," *Physiology & Behavior* 84, no. 2 (2005): 211–215.

25. W. McKeever, *Emperors of the Deep* (New York: Harper One, 2019).

26. A. P. Nosal et al., "Olfaction Contributes to Pelagic Navigation in a Coastal Shark," *PLoS ONE* 11, no. 1 (2016): e0143758.

27. Parker, *Encyclopedia of Sharks.*

28. S. Lorenzini, *Osservazioni Intorno Alle Torpedini* (Florence: Per l'Onofri, 1678).

29. R. Murray, "Electrical Sensitivity of the Ampullæ of Lorenzini," *Nature* 187 (1960): 957.

30. B. King, Y. Hu, and J. A. Long, "Electroreception in Early Vertebrates: Survey, Evidence and New Information," *Palaeontology* 61, no. 3 (2018): 325–358.

31. J. S. Coombs, J. C. Eccles, and P. Fatt, "The Electrical Properties of the Motoneuron Membrane," *Journal of Physiology* (London) 130 (1955): 291.

32. M. Piccolino and M. Bresadola, *Shocking Frogs: Galvani, Volta, and the Electric Origins of Neuroscience* (Oxford: Oxford University Press, 2013).

33. E. E. Josberger et al., "Proton Conductivity in Ampullae of Lorenzini Jelly," *Science Advances* 13 (2016): e1600112.

34. Parker, *Encyclopedia of Sharks.*

35. B. King and J. Long, "How Sharks and Other Animals Evolved Electroreception to Find Their Prey," Phys.org, February 13, 2018, https://phys.org/news/2018-02-sharks-animals-evolved-electroreception-theirprey.html.

36. R. M. Kempster, I. D. McCarthy, and S. P. Collin, "Phylogenetic and Ecological Factors Influencing the Number and Distribution of Electroreceptors in Elasmobranchs," *Journal of Fish Biology* 80, no. 5 (2012): 2055–2088.

37. R. M. Kempster et al., "Electrosensory-Driven Feeding Behaviours of the Port Jackson Shark (*Heterodontus portusjacksoni*) and Western Shovelnose Ray (*Aptychotrema vincentiana*)," *Marine and Freshwater Research* 67 (2016): 187–194.

38. A. J. Kalmijn, "Electric and Magnetic Field Detection in Elasmobranch Fishes," *Science* 218, no. 4575 (1982): 916–918.

39. A. P. Klimley, "Highly Directional Swimming by Scalloped Hammerhead Sharks, *Sphyrna lewini* and Subsurface Irradiance, Temperature, Bathymetry, and Geomagnetic Field," *Marine Biology* 117 (1993): 1–22.

40. J. M. Morson and J. F. Morrissey, "Morphological Variation in the Electric Organ of the Little Skate (*Leucoraja erinacea*) and Its Possible Role in Communication during Courtship," in *Biology of Skates. Developments in Environmental Biology of Fishes 27*, ed. D. A. Ebert and J. A. Sulikowski (Dordrecht, Netherlands: Springer, 2007), 161–169.

41. N. W. Bellono, D. B. Leitch, and D. Julius, "Molecular Tuning of Electroreception in Sharks and Skates," *Nature* 558, no. 7708 (2018): 122–126.

42. B. O. Bratton and J. L. Ayers, "Observations on the Electric Organ Discharge of Two Skate Species (Chondrichthyes: Rajidae) and its Relationship to Behavior," *Environmental Biology of Fishes* 20 (1987): 241–254.

43. J. A. Sisneros, T. C. Tricas, and C. A. Luer, "Response Properties and Biological Function of the Skate Electrosensory System during Ontogeny," *Journal of Comparative Physiology A* 183 (1998): 87–99.

44. J. Mortenson and R. H. Whitaker, "Electric Discharge of Free Swimming Female Winter Skates (*Raja ocellata*)," *American Zoologist* 13 (1973): 1266.

45. C. W. Coates and R. T. Cox, "Observations on the Electric Discharge of *Torpedo occidentalis*," *Zoologica* 27 (1942): 25–28.

46. C. G. Lowe, R. N. Bray, and D. R. Nelson, "Feeding and Associated Electrical Behavior of the Pacific Electric Ray *Torpedo californica* in the Field," *Marine Biology* 120 (1994): 161–169.

47. L. M. Mir, "Electrochemotherapy and Electric Pulses," in *Electromagnetic Fields, Environment and Health*, ed. A. Perrin and M. Souques (Paris: Springer, 2012), 25–34.

48. Y. Tanaka et al., "An Electric Generator Using Living *Torpedo* Electric Organs Controlled by Fluid Pressure-Based Alternative Nervous Systems," *Scientific Reports* 6, no. 25899 (2016).

49. C. M. Pedroso et al., "Morphological Characterization of the Venom Secretory Epidermal Cells in the Stinger of Marine and Freshwater Stingrays," *Toxicon* 50 (2007): 688–697.

50. K. Baumann et al., "A Ray of Venom: Combined Proteomic and Transcriptomic Investigation of Fish Venom Composition Using Barb Tissue from the Blue-Spotted Stingray (*Neotrygon kuhlii*)," *Journal of Proteomics* 109 (2014): 188–198.

51. R. F. Clark et al., "Stingray Envenomation: A Retrospective Review of Clinical Presentation and Treatment in 119 Cases," *Journal of Emergency Medicine* 33, no. 1 (2007): 33–37.

52. N. K. Dulvy and J. D. Reynolds, "Evolutionary Transitions among Egg-Laying, Live-Bearing and Maternal Inputs in Sharks and Rays," *Proceedings of the Royal Society B* 264 (1997): 1309–1315.

53. H. Jung et al., "The Ancient Origins of Neural Substrates for Land Walking," *Cell* 172, no. 4 (2018): 667–682.

54. R. E. Ball, M. K. Oliver, and A. B. Gill, "Early Life Sensory Ability—Ventilatory Responses of Thornback Ray Embryos (*Raja clavata*) to Predator-Type Electric Fields," *Developmental Neurobiology* 76, no. 7 (2015): 721–729.

55. C. R. Robins and G. C. Ray, *A Field Guide to Atlantic Coast Fishes of North America* (Boston: Houghton Mifflin, 1986).

56. K. L. Omori and R. A. Fisher, "Summer and Fall Movement of Cownose Ray, *Rhinoptera Bonasus*, along the East Coast of United States Observed with Pop-Up Satellite Tags," *Environmental Biology of Fishes* 100 (2017): 1435–1449.

57. M. E. Porter et al., "Aquatic Walking and Swimming Kinematics of Neonate and Juvenile Epaulette Sharks," *Integrative and Comparative Biology* 62, no. 6 (2022): 1710–1724.

58. G. C. Grigg, "Use of the First Gill Slits for Water Intake in a Shark," *Journal of Experimental Biology* 52, no. 3 (1970): 569–574.

59. J. E. Thomerson, T. B. Thorson, and R. L. Hempel, "The Bull Shark, *Carcharhinus leucas*, from the Upper Mississippi River near Alton, Illinois," *Copeia* 1977 (1977): 166–168.

60. S. E. Smith, D. W. Au, and C. Show, "Intrinsic Rebound Potentials of 26 Species of Pacific Sharks," *Marine and Freshwater Research* 49, no. 7 (1998): 663–678.

61. N. M. Whitney and G. L. Crow, "Reproductive Biology of the Tiger Shark (*Galeocerdo cuvier*) in Hawaii," *Marine Biology* 151 (2007): 63–70.

62. Parker, *Encyclopedia of Sharks*.

63. H. Nakano and J. D. Stevens, "The Biology and Ecology of the Blue Shark, *Prionace glauca*," in *Sharks of the Open Ocean: Biology, Fisheries, and Conservation*, ed. M. D. Camhi, E. K. Pikitch, and E. A. Babcock (Oxford: Blackwell, 2008), 140–151.

64. Lowe, interview.

65. Joshua Williams, "Shark Snippets Episode 7: The Shortfin Mako (*Isurus oxyrinchus*)," July 29, 2018, in *Josh's Wave Shark Snippets Podcast*, https://radiopublic.com/shark-snippets-8jBDB7/s1!5518f.

66. Parker, *Encyclopedia of Sharks*.

67. F. G. Carey et al., "Temperature and Activities of a White Shark, *Carcharodon carcharias*," *Copeia* 1982, no. 2 (1982): 254–260.

68. N. G. Wolf, P. R. Swift, and F. G. Carey, "Swimming Muscle Helps Warm the Brain of Lamnid Sharks," *Journal of Comparative Physiology B* 157 (1988): 709–715.

69. K. C. Newton, J. Wraith, and K. A. Dickson, "Digestive Enzyme Activities Are Higher in the Shortfin Mako Shark, *Isurus oxyrinchus*, than in Ectothermic Sharks as a Result of Visceral Endothermy," *Fish Physiology and Biochemistry* 41 (2015): 887–898.

70. C. W. Speed et al., "Heat-Seeking Sharks: Support for Behavioural Thermoregulation in Reef Sharks," *Marine Ecology Progress Series* 463 (2012): 231–244.

71. S. P. Oliver et al., "Thresher Sharks Use Tail-Slaps as a Hunting Strategy," *PLoS ONE* 8, no. 7 (2013): e67380.

72. B. Worm et al., "Global Catches, Exploitation Rates, and Rebuilding Options for Sharks," *Marine Policy* 40 (2013): 194–204.

73. J. A. Forrester, T. G. Weiser, and J. D. Forrester, "An Update on Fatalities Due to Venomous and Nonvenomous Animals in the United States (2008–2015)," *Wilderness & Environmental Medicine* 29 (2018): 36–44.

74. N. K. Dulvy et al., "You Can Swim but You Can't Hide: The Global Status and Conservation of Oceanic Pelagic Sharks and Rays," *Aquatic Conservation: Marine and Freshwater Ecosystems* 18 (2008): 459–482.

75. N. K. Dulvy et al., "Extinction Risk and Conservation of the World's Sharks and Rays," *eLife* 3 (2014): e00590.

76. N. Pacoureau et al., "Half a Century of Global Decline in Oceanic Sharks and Rays," *Nature* 589 (2021): 567–571.

77. S. C. Clarke et al., "Global Estimates of Shark Catches Using Trade Records from Commercial Markets," *Ecology Letters* 9 (2006): 1115–1126.

78. "Shark Finning," *Humane Society International*, accessed November 18, 2020, https://www.hsi.org/issues/shark-finning/.

79. L. L. Hamady et al., "Vertebral Bomb Radiocarbon Suggests Extreme Longevity in White Sharks," *PLoS ONE* 9, no. 1 (2014): e84006.

80. C. A. Ward-Paige and B. Worm, "Global Evaluation of Shark Sanctuaries," *Global Environmental Change* 47 (2017): 174–189.

81. G.M.S. Vianna et al., "Socio-Economic Value and Community Benefits from Shark-Diving Tourism in Palau: A Sustainable Use of Reef Shark Populations," *Biological Conservation* 145, no. 1 (2012): 267–277.

82. G.M.S. Vianna et al., "Shark-Diving Tourism as a Financing Mechanism for Shark Conservation Strategies in Malaysia," *Marine Policy* 94 (2018): 220–226.

83. "Shark Fin Trade: Why it Should Be Banned in the United States," *Oceana Reports*, June 2016, accessed November 5, 2020, https://usa.oceana.org/publications/reports/shark-fin-trade-why-it-should-be-banned-united-states.

84. R. G. Dwyer et al., "Individual and Population Benefits of Marine Reserves for Reef Sharks," *Current Biology* 30 (2020): 480–489.

85. Gammon, "We're Seeing More than Ever."

86. Julia Mueller, "House Bans Shark Fin Trade, Curbs Illegal Fishing," *The Hill,* December 8, 2022.

4. Greats of the Great Blue

1. B. M. Norman et al., "Undersea Constellations: The Global Biology of an Endangered Marine Megavertebrate Further Informed through Citizen Science," *BioScience* 67, no. 12 (2017): 1029–1043.

2. P. J. Motta et al., "Feeding Anatomy, Filter-Feeding Rate, and Diet of Whale Sharks *Rhincodon typus* during Surface Ram Filter Feeding off the Yucatan Peninsula, Mexico," *Zoology* 113 (2010): 199–212.

3. C. R. McClain et al., "Sizing Ocean Giants: Patterns of Intraspecific Size Variation in Marine Megafauna," *PeerJ* 3 (2015): e715.

4. A. P. Martin, "Metabolic Rate and Directional Nucleotide Substitution in Animal Mitochondrial DNA," *Molecular Biology and Evolution* 12, no. 6 (1995): 1124–1131.

5. L. C. Strotz et al., "Metabolic Rates, Climate and Macroevolution: A Case Study Using Neogene Molluscs," *Proceedings of the Royal Society B* 285 (2018): 20181292.

6. S. J. Kiraly, J. A. Moore, and P. H. Jasinski, "Deepwater and Other Sharks of the U.S. Atlantic Ocean Exclusive Economic Zone," *Marine Fisheries Review* 65, no. 4 (2003): 1–64.

7. J. Nielsen et al., "Eye Lens Radiocarbon Reveals Centuries of Longevity in the Greenland Shark (*Somniosus microcephalus*)," *Science* 353, no. 6300 (2016): 702–704.

8. R. de la Parra Venegas et al., "An Unprecedented Aggregation of Whale Sharks, *Rhincodon typus*, in Mexican Coastal Waters of the Caribbean Sea," *PLoS ONE* 6, no. 4 (2011): e18994.

9. Motta, "Feeding Anatomy," 199–212.

10. J. J. Jambeck et al., "Plastic Waste Inputs from Land into the Ocean," *Science* 347, no. 6223 (2015): 768–771.

11. Law et al., "The United States' Contribution of Plastic Waste to Land and Ocean," *Science Advances* 6 (2020): eabd0288.

12. N.A.S. Abreo, D. Blatchley, and M. D. Superio, "Stranded Whale Shark (*Rhincodon typus*) Reveals Vulnerability of Filter-Feeding Elasmobranchs to Marine Litter in the Philippines," *Marine Pollution Bulletin* 141 (2019): 79–83.

13. Motta, "Feeding Anatomy," 199–212.

14. H. M. Guzman et al., "Longest Recorded Trans-Pacific Migration of a Whale Shark (*Rhincodon typus*)," *Marine Biodiversity Records* 11 (2018): 8.

15. S. A. Eckert and B. S. Stewart, "Telemetry and Satellite Tracking of Whale Sharks, *Rhincodon typus*, in the Sea of Cortez, Mexico, and the North Pacific Ocean," *Environmental Biology of Fishes* 60 (2001): 299–308.

16. R. E. Hueter, J. P. Tyminski, and R. de la Parra, "Horizontal Movements, Migration Patterns, and Population Structure of Whale Sharks in the Gulf of Mexico and Northwestern Caribbean Sea," *PLoS ONE* 8, no. 8 (2013): e71883.

17. A. R. Hearn et al., "Adult Female Whale Sharks Make Long-Distance Movements Past Darwin Island (Galapagos, Ecuador) in the Eastern Tropical Pacific," *Marine Biology* 163 (2016): 214.

18. C. T. Perry et al., "Comparing Length-Measurement Methods and Estimating Growth Parameters of Free-Swimming Whale Sharks (*Rhincodon typus*) near the South Ari Atoll, Maldives," *Marine and Freshwater Research* 69, no. 10 (2018): 1487–1495.

19. A.L.F. Castro et al., "Population Genetic Structure of Earth's Largest Fish, the Whale Shark (*Rhincodon typus*)," *Molecular Ecology* 16 (2007): 5183–5192.

20. Jennifer Schmidt, PhD, interview by Joe E. Meisel, March 23, 2021.

21. D. Rowat and K. S. Brooks, "A Review of the Biology, Fisheries and Conservation of the Whale Shark *Rhincodon typus*," *Journal of Fish Biology* 80 (2012): 1019–1056.

22. C.J.A. Bradshaw et al., "Decline in Whale Shark Size and Abundance at Ningaloo Reef over the Past Decade: The World's Largest Fish Is Getting Smaller," *Biological Conservation* 141, no. 7 (2008): 1894–1905.

23. C. A. Rohner et al., "Trends in Sightings and Environmental Influences on a Coastal Aggregation of Manta Rays and Whale Sharks," *Marine Ecology Progress Series* 482 (2013): 153–168.

24. M. B. Kaplan and S. Solomon, "A Coming Boom in Commercial Shipping? The Potential for Rapid Growth of Noise from Commercial Ships by 2030," *Marine Policy* 73 (2016): 119–121.

25. UNCTAD (UN Conference on Trade and Development), *Review of Maritime Transport 2016* (Geneva, Switzerland: UNCTAD, 2016).

26. L. Fox et al., "Quantifying the Effect of Anthropogenic Climate Change on Calcifying Plankton," *Scientific Reports* 10 (2020): 1620.

27. J. A. McKinney et al., "Feeding Habitat of the Whale Shark *Rhincodon typus* in the Northern Gulf of Mexico Determined Using Species Distribution Modelling," *Marine Ecology Progress Series* 458 (2012): 199–211.

28. B. M. Norman and D. L. Morgan, "The Return of 'Stumpy' the Whale Shark: Two Decades and Counting," *Frontiers in Ecology and the Environment* 14, no. 8 (2016): 449–450.

29. E. F. Cagua et al., "Whale Shark Economics: A Valuation of Wildlife Tourism in South Ari Atoll, Maldives," *PeerJ* 2 (2014): e515.

30. Z. Anna and D. S. Saputra, "Economic Valuation of Whale Shark Tourism in Cenderawasih Bay National Park, Papua, Indonesia," *Biodiversitas* 18, no. 3 (2017): 1026–1034.

31. F. Womersley et al., "Wound-Healing Capabilities of Whale Sharks (*Rhincodon typus*) and Implications for Conservation Management," *Conservation Physiology* 9, no. 1 (2021): coaa120.

32. J. I. Fasick et al., "The Retinal Pigments of the Whale Shark (*Rhincodon typus*) and their Role in Visual Foraging Ecology," *Visual Neuroscience* 36 (2019): E011.

33. M. G. Meekan et al., "Swimming Strategy and Body Plan of the World's Largest Fish: Implications for Foraging Efficiency and Thermoregulation," *Frontiers in Marine Science* 2 (2015): 64. doi: 10.3389/fmars.2015.00064.

34. J. P. Tyminski et al., "Vertical Movements and Patterns in Diving Behavior of Whale Sharks as Revealed by Pop-Up Satellite Tags in the Eastern Gulf of Mexico," *PLoS ONE* 10, no. 11 (2015): e0142156.

35. Meekan et al., "Swimming Strategy."

36. A. C. Gleiss, B. Norman, and R. P. Wilson, "Moved by that Sinking Feeling: Variable Diving Geometry Underlies Movement Strategies in Whale Sharks," *Functional Ecology* 25 (2011): 595–607.

37. Meekan et al., "Swimming Strategy."

38. M. Thums et al., "Evidence for Behavioural Thermoregulation by the World's Largest Fish," *Journal of the Royal Society Interface* 10 (2013): 20120477.

39. D. W. Sims and V. A. Quayle, "Selective Foraging Behaviour of Basking Sharks on Zooplankton in a Small-Scale Front," *Nature* 393 (1998): 460–464.

40. G. B. Skomal et al., "Transequatorial Migrations by Basking Sharks in the Western Atlantic Ocean," *Current Biology* 19 (2009): 1019–1022.

41. S. D'Hondt, "Consequences of the Cretaceous/Paleogene Mass Extinction for Marine Ecosystems," *Annual Review of Ecology, Evolution, and Systematics* 36 (2005): 295–317

42. M. Friedman et al., "100-Million-Year Dynasty of Giant Planktivorous Bony Fishes in the Mesozoic Seas," *Science* 327 (2010): 990–993.

43. C. Pimiento et al., "Evolutionary Pathways Toward Gigantism in Sharks and Rays," *Evolution* 73, no. 3 (2019): 588–599.

44. A. P. Summers, "Stiffening the Stingray Skeleton—An Investigation of Durophagy in Myliobatid Stingrays (Chondrichthyes, Batoidea, Myliobatidae)," *Journal of Morphology* 243 (2000): 113–126.

45. C. Pimiento et al., "Geographical Distribution Patterns of *Carcharocles megalodon* over Time Reveal Clues about Extinction Mechanisms," *Journal of Biogeography* 43, no. 8 (2016): 1645–1655.

46. O. Lambert et al., "The Giant Bite of a New Raptorial Sperm Whale from the Miocene Epoch of Peru," *Nature* 466 (2010): 105–108.

47. D. Stevens et al., *Guide to the Manta and Devil Rays of the World* (Princeton, NJ: Princeton University Press, 2018).

48. F. E. Fish et al., "Hydrodynamic Performance of Aquatic Flapping: Efficiency of Underwater Flight in the Manta," *Aerospace* 3, no. 3 (2016): 20.

49. J. D. Swenson et al., "How the Devil Ray Got Its Horns: The Evolution and Development of Cephalic Lobes in Myliobatid Stingrays (Batoidea: Myliobatidae)," *Frontiers in Ecology and Evolution* 6, no. 181 (2018). doi: 10.3389/fevo.2018.00181.

50. K. Onimaru et al., "A Shift in Anterior–Posterior Positional Information Underlies the Fin-to-Limb Evolution," *eLife* 4 (2015): e07048.

51. Motta, "Feeding Anatomy," 199–212.

52. R. V. Divi et al., "Manta Rays Feed Using Ricochet Separation, a Novel Nonclogging Filtration Mechanism," *Science Advances* 4 (2018): eaat9533.

53. M. P. O'Malley et al., "Characterization of the Trade in Manta and Devil Ray Gill Plates in China and South-east Asia through Trader Surveys," *Aquatic Conservation: Marine and Freshwater Ecosystems* 27 (2017): 394–413.

54. C. Ari, "Encephalization and Brain Organization of Mobulid Rays (Myliobatiformes, Elasmobranchii) with Ecological Perspectives," *Open Anatomy Journal* 3 (2011): 1–13.

55. A. Fathy, H. Rezk, and D. Yousri, "A Robust Global MPPT to Mitigate Partial Shading of Triple-Junction Solar Cell-Based System Using Manta Ray Foraging Optimization Algorithm," *Solar Energy* 207 (2020): 305–316.

56. H. Lassauce et al., "Diving Behavior of the Reef Manta Ray (*Mobula alfredi*) in New Caledonia: More Frequent and Deeper Night-Time Diving to 672 Meters," *PLoS ONE* 15, no. 3 (2020): e0228815.

57. K. B. Burgess et al., "*Manta birostris*, Predator of the Deep? Insight into the Diet of the Giant Manta Ray through Stable Isotope Analysis," *Royal Society Open Science* 3 (2016): 160717.

58. R. L. Alexander, "Evidence of Brain-Warming in the Mobulid Rays, *Mobula tarapacana* and *Manta birostris* (Chondrichthyes: Elasmobranchii: Batoidea: Myliobatiforrnes)," *Zoological Journal of the Linnean Society* 118 (1996): 151–164.

59. C. Ari, "Rapid Coloration Changes of Manta Rays (Mobulidae)," *Biological Journal of the Linnean Society* 113 (2014): 180–193.

60. C. Ari and D. P. D'Agostino, "Contingency Checking and Self-Directed Behaviors in Giant Manta Rays: Do Elasmobranchs Have Self-Awareness?" *Journal of Ethology* 34 (2016): 167–174.

61. R.J.Y. Perryman et al., "Social Preferences and Network Structure in a Population of Reef Manta Rays," *Behavioral Ecology and Sociobiology* 73 (2019): 114.

62. G.M.W. Stevens, "Conservation and Population Ecology of Manta Rays in the Maldives" (doctoral thesis, University of York, 2016).

63. D. A. Croll et al., "Vulnerabilities and Fisheries Impacts: The Uncertain Future of Manta and Devil Rays," *Aquatic Conservation: Marine and Freshwater Ecosystems* 26 (2016): 562–575.

64. J. V. Schmidt et al., "Paternity Analysis in a Litter of Whale Shark Embryos," *Endangered Species Research* 12 (2010): 117–124.

65. F.H.V. Hazin et al., "Reproduction of the Blue Shark *Prionace glauca* in the South-Western Equatorial Atlantic Ocean," *Fisheries Science* 60 (1994): 487–491.

66. Australian Museum, "One of the World's Largest Fish Develops from a Tiny Larval Mola Sunfish," media release, July 22, 2020.

67. E. Sawai et al., "Redescription of the Bump-Head Sunfish *Mola alexandrini* (Ranzani 1839), Senior Synonym of *Mola ramsayi* (Giglioli 1883), with Designation of a Neotype for *Mola mola* (Linnaeus 1758) (Tetraodontiformes: Molidae)," *Ichthyological Research* 65 (2018): 142–160.

68. E. W. Gudger, "From Atom to Colossus," *Natural History* 38 (1936): 26–30.

69. J. Schmidt, "New Studies of Sun-Fishes Made during the 'Dana' Expedition, 1920," *Nature* 107 (1921): 76–79.

70. E. C. Pope et al., "The Biology and Ecology of the Ocean Sunfish *Mola mola*: A Review of Current Knowledge and Future Research," *Reviews in Fish Biology and Fisheries* 20 (2010): 471–487.

71. M. Nyegaard et al., "Hiding in Broad Daylight: Molecular and Morphological Data Reveal a New Ocean Sunfish Species (Tetraodontiformes: Molidae) that Has Eluded Recognition," *Zoological Journal of the Linnean Society* 182, vol. 3 (2017): 631–658.

72. Y. Watanabe and K. Sato, "Functional Dorsoventral Symmetry in Relation to Lift-Based Swimming in the Ocean Sunfish *Mola mola*," *PLoS ONE* 3, no. 10 (2008): e3446.

73. D. P. Cartamil and C. G. Lowe, "Diel Movement Patterns of Ocean Sunfish *Mola mola* off Southern California," *Marine Ecology Progress Series* 266 (2004): 245–253.

74. N. C. Wegner et al., "Whole-Body Endothermy in a Mesopelagic Fish, the Opah, *Lampris guttatus*," *Science* 348, no. 6236 (2015): 786–789.

75. O. Gon, "Lampridae," in *Fishes of the Southern Ocean*, ed. O. Gon and P. C. Heemstra (Grahamstown, South Africa: J.L.B. Smith Institute of Ichthyology, 1990), 215–217.

76. J. Blum, "Massachusetts Coastal Town Orders Locals to Stop Calling 911 about Giant Fish," *Huffington Post*, October 14, 2020, https://www.huffpost.com/entry/sunfish-massachusetts-911_n_5f875826c5b6c5eccffcf2da.

77. L. Silvani, M. Gazo, and A. Aguilar, "Spanish Driftnet Fishing and Incidental Catches in the Western Mediterranean," *Biological Conservation* 90 (1999): 79–85.

78. S. Tudela et al., "Driftnet Fishing and Biodiversity Conservation: The Case Study of the Large-Scale Moroccan Driftnet Fleet Operating in the Alboran Sea (SW Mediterranean)," *Biological Conservation* 121 (2005): 65–78.

79. Cartamil and Lowe, "Diel Movement Patterns," 245.

80. L. Brotz et al., "Increasing Jellyfish Populations: Trends in Large Marine Ecosystems," *Hydrobiologia* 690 (2012): 3–20.

81. R. H. Condona et al., "Recurrent Jellyfish Blooms Are a Consequence of Global Oscillations," *Proceedings of the National Academy of Sciences* 110, vol. 3 (2013): 1000–1005.

82. J. F. Purcell, "Jellyfish and Ctenophore Blooms Coincide with Human Proliferations and Environmental Perturbations," *Annual Review of Marine Science* 4 (2012): 209–235.

83. B. S. Halpern et al., "A Global Map of Human Impact on Marine Ecosystems," *Science* 319 (2008): 948–952.

84. A. J. Richardson et al., "The Jellyfish Joyride: Causes, Consequences and Management Responses to a More Gelatinous Future," *Trends in Ecology and Evolution* 24, vol. 6 (2009): 312–322.

85. J. Roman and J. J. McCarthy, "The Whale Pump: Marine Mammals Enhance Primary Productivity in a Coastal Basin," *PLoS ONE* 5, no. 10 (2010): e13255.

86. J. J. Williams et al., "Mobile Marine Predators: An Understudied Source of Nutrients to Coral Reefs in an Unfished Atoll," *Proceedings of the Royal Society B* 285 (2018): 20172456.

87. L. R. Peel et al., "Stable Isotope Analyses Reveal Unique Trophic Role of Reef Manta Rays (*Mobula alfredi*) at a Remote Coral Reef," *Royal Society Open Science* 6 (2019): 190599.

Part II

1. J. Verne, *Twenty Thousand Leagues Under the Sea* (New York: Butler Brothers, 1887).

2. "Know Your Ocean: Twilight Zone," Woods Hole Oceanographic Institution, accessed November 14, 2022, https://www.whoi.edu/know-your-ocean/ocean-topics/how-the-ocean-works/ocean-zones/twilight-zone/.

3. L. Pellissier et al., "Quaternary Coral Reef Refugia Preserved Fish Diversity," *Science* 344, no. 6187 (2014): 1016–1019.

5. An Oasis of Abundance

1. J. M. de Goeij et al., "Surviving in a Marine Desert: The Sponge Loop Retains Resources within Coral Reefs," *Science* 342, no. 6154 (2013): 108–110.

2. L. Rix et al., "Coral Mucus Fuels the Sponge Loop in Warm- and Cold-Water Coral Reef Ecosystems," *Scientific Reports* 6 (2016): 18715.

3. J. R. Pawlik et al., "Chemical Warfare on Coral Reefs: Sponge Metabolites Differentially Affect Coral Symbiosis in Situ," *Limnology and Oceanography* 52, no. 2 (2007): 907–911.

4. J. L. Wulff, "Trade-offs in Resistance to Competitors and Predators, and Their Effects on the Diversity of Tropical Marine Sponges," *Journal of Animal Ecology* 74 (2005): 313–321.

5. E. G. Ball, C. F. Strittmatter, and O. Cooper, "Metabolic Studies on the Gas Gland of the Swim Bladder," *Biological Bulletin* 108, no. 1 (1955): 1–17.

6. B. Pelster, "The Generation of Hyperbaric Oxygen Tensions in Fish," *News in Physiological Science* 16 (2001): 287–291.

7. M. Berenbrink et al., "Evolution of Oxygen Secretion in Fishes and the Emergence of a Complex Physiological System," *Science* 307, no. 5716 (2005): 1752–1757.

8. M. D. Tietbohl, D. K. Ngugi, and M. L. Berumen, "A Unique Bellyful: Extraordinary Gut Microbes Help Herbivorous Fish Eat Seaweeds," *Biodiversity Frontiers for Young Minds* 8, article 58 (2020).

9. J. H. Choat and L. Axe, "Growth and Longevity in Acanthurid Fishes: An Analysis of Otolith Increments," *Marine Ecology Progress Series* 134 (1996): 15–36.

10. D. R. Robertson, N.V.C. Polunin, and K. Leighton, "The Behavioral Ecology of Three Indian Ocean Surgeonfishes (*Acanthurus lineatus, A. leucosternon* and *Zebrasoma scopas*): Their Feeding Strategies, and Social and Mating Systems," *Environmental Biology of Fishes* 4 (1979): 125–170.

11. P. C. Craig et al., "Population Biology and Harvest of the Coral Reef Surgeonfish *Acanthurus lineatus* in American Samoa," *Fishery Bulletin* 95 (1997): 680–693.

12. R. A. Horner, D. L. Garrison, and F. G. Plumley, "Harmful Algal Blooms and Red Tide Problems on the U.S. West Coast," *Limnology and Oceanography* 425 no. 5,2 (1997): 1076–1088.

13. A. Ludka, "Hitchcock's 'The Birds': Mystery Solved," *ABC News*, December 28, 2011, https://abcnews.go.com/blogs/entertainment/2011/12/hitchcocks-the-birds-mystery-solved.

14. C. R. Wylie and V. J. Paul, "Feeding Preferences of the Surgeonfish *Zebrasoma flavescens* in Relation to Chemical Defenses of Tropical Algae," *Marine Ecology Progress Series* 45 (1988): 23–32.

15. L. Fishelson, L. W. Montgomery, and A. H. Myrberg, Jr., "Biology of Surgeonfish *Acanthurus nigrofuscus* with Emphasis on Changeover in Diet and Annual Gonadal Cycles," *Marine Ecology Progress Series* 39 (1987): 37–47.

16. G. R. Allen, *Damselfishes of the World* (Melle, Germany: Mergus, 1991).

17. J. G. Eurich et al., "Stable Isotope Analysis Reveals Trophic Diversity and Partitioning in Territorial Damselfishes on a Low-Latitude Coral Reef," *Marine Biology* 166 (2019): 17.

18. M. A. Hixon and W. N. Brostoff, "Succession and Herbivory: Effects of Differential Fish Grazing on Hawaiian Coral-Reef Algae," *Ecological Monographs* 66, no. 1 (1996): 67–90.

19. H. Hata, K. Watanabe, and M. Kato, "Geographic Variation in the Damselfish–Red Alga Cultivation Mutualism in the Indo-West Pacific," *BMC Evolutionary Biology* 10 (2010): 185.

20. R. M. Brooker et al., "Domestication via the Commensal Pathway in a Fish-Invertebrate Mutualism," *Nature Communications* 11 (2020): 6253.

21. Eurich et al., "Stable Isotope Analysis," 17.

22. F. Cortesi et al., "Phenotypic Plasticity Confers Multiple Fitness Benefits to a Mimic," *Current Biology* 25, no. 7 (2015): 949–954.

23. E. Kazancıoğlu et al., "Influence of Sexual Selection and Feeding Functional Morphology on Diversification Rate of Parrotfishes (Scaridae)," *Proceedings of the Royal Society B* 276 (2009): 3439–3446.

24. G. Roberts, Jr., "X-Rays Reveal the Biting Truth about Parrotfish Teeth" (including interview with Pupa Gilbert, PhD), *Lawrence Berkeley National Laboratory News Center*, November 15, 2017.

25. C. E. Killian et al., "Self-sharpening Mechanism of the Sea Urchin Tooth," *Advanced Functional Materials* 21 (2011): 682–690.

26. A. S. Hoey and R. M. Bonaldo. *Biology of Parrotfishes* (Boca Raton, FL: CRC Press, 2018).

27. J. D. Sartori and T. G. Bright, "A Hydrophonic Study of the Feeding Activities of Western Atlantic Parrotfishes," National Sea Grant Program, Institutional Grant 2–35213, Report TAMU-SG-72–203, 1972.

28. J. T. Streelman, "Evolutionary History of the Parrotfishes: Biogeography, Eco-morphology, and Comparative Diversity," *Evolution* 56, no. 5 (2002): 961–971.

29. M. Nakazato and A. Takemura, "Acoustical Behavior of Japanese Parrot Fish *Oplenathus fasciatus*," *Nippon Suisan Gakkaishi* 53, no. 6 (1987): 967–973.

30. J. R. Cardwell and N. R. Liley, "Hormonal Control of Sex and Color Change in the Stoplight Parrotfish, *Sparisoma viride*," *General and Comparative Endocrinology* 81, no. 1 (1991): 7–20.

31. R. C. Muñoz and R. R. Warner, "Alternative Contexts of Sex Change with Social Control in the Bucktooth Parrotfish, *Sparisoma radians*," *Environmental Biology of Fishes* 68 (2003): 307–319.

32. A. S. Grutter et al., "Fish Mucous Cocoons: The 'Mosquito Nets' of the Sea," *Biology Letters* 7, no. 2 (2011): 292–294.

33. H. E. Winn and J. E. Bardach, "Formation of a Mucous Envelope at Night by Parrot Fishes," *Zoologica* 40 (1955): 145–148.

34. V. C. Sambilay, Jr., "Interrelationships between Swimming Speed, Caudal Fin Aspect Ratio and Body Length of Fishes," *Fishbyte* (December 1990): 16–20.

35. C. S. Wardle, "The Limits of Fish Swimming Speed," *Nature* 255 (1975): 725–727.

36. J. R. Grubich, A. N. Ricea, and M. W. Westneat, "Functional Morphology of Bite Mechanics in the Great Barracuda (*Sphyraena barracuda*). *Zoology* 111, no. 17 (2008): 16–29.

37. A. C. O'Toole et al., "Spatial Ecology and Residency Patterns of Adult Great Barracuda (*Sphyraena barracuda*) in Coastal Waters of The Bahamas," *Marine Biology* 158 (2011): 2227–2237.

38. N.A.J. Graham, S. J. Purkis, and A. Harris, "Diurnal, Land-Based Predation on Shore Crabs by Moray Eels in the Chagos Archipelago," *Coral Reefs* 28 (2009): 397.

39. S. C. Barley et al., "To Knot or Not? Novel Feeding Behaviours in Moray Eels," *Marine Biodiversity* 46 (2106): 703–705.

40. R. Bshary et al., "Interspecific Communicative and Coordinated Hunting between Groupers and Giant Moray Eels in the Red Sea," *PLoS Biology* 4, no. 12 (2006): e431.

41. R. G. Turingan and P. C. Wainwright, "Morphological and Functional Bases of Durophagy in the Queen Triggerfish, *Balistes vetula* (Pisces, Tetraodontiformes)," *Journal of Morphology* 215 (1993): 101–118.

42. K. E. Korsmeyer, J. F. Steffensen, and J. Herskin, "Energetics of Median and Paired Fin Swimming, Body and Caudal Fin Swimming, and Gait Transition in Parrotfish (*Scarus schlegeli*) and Triggerfish (*Rhinecanthus aculeatus*)," *Journal of Experimental Biology* 205 (2002): 1253–1263.

43. F. H. Berry and L. E. Vogle, *Filefishes (Monacanthidae) of the Western North Atlantic* (Washington, DC: US Fish and Wildlife Service, 1961).

44. Y. Cai et al., "Filefish-Inspired Surface Design for Anisotropic Underwater Oleophobicity," *Advanced Functional Materials* 24, no. 6 (2013): 809–816.

45. W. Gladstone, "Lek-Like Spawning, Parental Care and Mating Periodicity of the Triggerfish *Pseudobalistes flavimarginatus* (Balistidae)," *Environmental Biology of Fishes* 39 (1994): 249–257.

46. Karen P. Maruska, PhD, interview by Joe E. Meisel, September 15, 2021.

47. M.C.P. Amorim, "Diversity of Sound Production in Fishes," in *Sound Communication in Fishes*, ed. F. Ladich (Vienna: Springer, 2015), 71–104.

48. F. Ladich, "Fish Bioacoustics," *Current Opinion in Neurobiology* 28 (2014.):121–127.

49. J. M. Moulton, "Acoustic Behaviour of Fishes," in *Acoustic Behaviour of Animals*, ed. R. G. Busnel (Amsterdam: Elsevier, 1963), 655–693.

50. F. Ladich and A. A. Myrberg, Jr., "Agonistic Behavior and Acoustic Communication," in *Communication in Fishes*, ed. F. Ladich et al. (Enfield, NH: Science Publishers, 2006).

51. M.C.P. Amorim, Y. Stratoudakis, and A. D. Hawkins, "Sound Production during Competitive Feeding in the Grey Gurnard," *Journal of Fish Biology* 65 (2004): 182–194.

52. Amorim, "Diversity of Sound Production."

53. M. A. Connaughton, "Sound Generation in the Searobin (*Prionotus carolinus*), a Fish with Alternate Sonic Muscle Contraction," *The Journal of Experimental Biology* 207 (2004): 1643–1654.

54. J. F. Barimo and M. L. Fine, "Relationship of Swim-Bladder Shape to the Directionality Pattern of Underwater Sound in the Oyster Toadfish," *Canadian Journal of Zoology* 76 (1998): 134–143.

55. Maruska, interview.

56. R. A. Kastelein et al., "Food Intake and Body Measurements of Atlantic Bottlenose Dolphins (*Tursiops truncates*) in Captivity," *Marine Environmental Research* 53, no. 2 (2002): 199–218.

57. L. Remage-Healey, D. P. Nowacek, and A. H. Bass, "Dolphin Foraging Sounds Suppress Calling and Elevate Stress Hormone Levels in a Prey Species, the Gulf Toadfish," *The Journal of Experimental Biology* 209 (2006): 4444–4451.

58. J. Tournadre, "Anthropogenic Pressure on the Open Ocean: The Growth of Ship Traffic Revealed by Altimeter Data Analysis," *Geophysical Research Letters* 41, no. 22 (2014): 7924–7932.

59. D. Alves et al., "Boat Noise Interferes with Lusitanian Toadfish Acoustic Communication," *Journal of Experimental Biology* 224, no. 11 (2021): jcb234849.

60. J. J. Luczkovich et al., "The Lombard Effect in Fishes: How Boat Noise Impacts Oyster Toadfish Vocalization Amplitudes in Natural Experiments," *Proceedings of Meetings on Acoustics* 27 (2016): 010035.

61. A.O.H.C. Leduc et al., "Land-Based Noise Pollution Impairs Reef Fish Behavior: A Case Study with a Brazilian Carnival" *Biological Conservation* 253 (2021): 108910.

62. Maruska, interview.

6. Weird and Wonderful

1. J. Mann, *Murder, Magic, and Medicine* (Oxford: Oxford University Press, 1992), 35.

2. W. Davis, *The Serpent and the Rainbow: A Harvard Scientist's Astonishing Journey into the Secret Societies of Haitian Voodoo, Zombies, and Magic* (New York: Simon & Schuster, 1985), 119.

3. Mann, *Murder, Magic*. 35.

4. S. Lohr, "One Man's Fugu Is Another's Poison," *New York Times*, November 29, 1981.

5. J. Li et al., "Awareness on Tetrodotoxin of Illegal Activity: Forensic Issue from a Rare Homicide Case Report and Literature Review," *Journal of Forensic and Legal Medicine* 79 (2021): 102152.

6. J. B. Cole et al., "Tetrodotoxin Poisoning Outbreak from Imported Dried Puffer Fish—Minneapolis, Minnesota, 2014," *Centers for Disease Control and Prevention Morbidity and Mortality Weekly Report* 63, no. 51 (2015): 1222–1225.

7. Centers for Disease Control and Prevention, "Tetrodotoxin: Biotoxin," Emergency Response Safety and Health Database, National Institute for Occupational Safety and Health (NIOSH), accessed September 22, 2021, https://www.cdc.gov/niosh/ershdb/emergencyresponsecard_29750019.html

8. T. Yasumoto et al., "Bacterial Production of Tetrodotoxin and Anhydrotetrodotoxin," *Agricultural and Biological Chemistry* 50 (1986): 793–795.

9. M. J. Lee et al., "A Tetrodotoxin-Producing Vibrio Strain, LM-1, from the Puffer Fish *Fugu vermicularis radiatus*," *Applied and Environmental Microbiology* 66, no. 4 (2000): 1698–1701.

10. T. Saito et al., "Toxicity of the Cultured Pufferfish *Fugu rubripes rubripes* along with Their Resistibility against Tetrodotoxin," *Bulletin of the Japanese Society of Scientific Fisheries* 50 (1984): 1573–1575.

11. M. J. Caley and D. Schluter, "Predators Favor Mimicry in a Tropical Reef Fish," *Proceedings of the Royal Society B* 270 (2003): 667–672.

12. P. M. Vaelli et al., "The Skin Microbiome Facilitates Adaptive Tetrodotoxin Production in Poisonous Newts," *eLife* 2020 (2003):9:e53898.

13. E. D. Brodie III and E. D. Brodie, Jr., "Tetrodotoxin Resistance in Garter Snakes: An Evolutionary Response of Predators to Dangerous Prey," *Evolution* 44, no. 3 (1990): 651–659.

14. J. E. Randall et al., "Grammistin, the Skin Toxin of Soapfishes, and its Significance in the Classification of the Grammistidae," *Publications of the Seto Marine Biological Laboratories* 19, no. 2/3 (1971): 157–190.

15. V. R. Liguori et al., "Antibiotic and Toxic Activity of the Mucous of the Pacific Golden Striped Bass *Grammistes selineatus*," *American Zoologist* 3, no. 4 (1963): 302.

16. Randall et al., "Grammistin," 158.

17. E. Kalmanzon and E. Zlotkin, "An Ichthyotoxic Protein in the Defensive Skin Secretion of the Red Sea Trunkfish *Ostracion cubicus*," *Marine Biology* 136 (2000): 471–476.

18. K. Tachibana, M. Sakaitani, and K. Nakanishi, "Pavoninins: Shark-Repelling Ichthyotoxins from the Defense Secretion of the Pacific Sole," *Science* 226, no. 4675 (1984): 703–705.

19. V. S. Ramachandran et al., "Rapid Adaptive Camouflage in Tropical Flounders," *Nature* 379 (1996): 815–818.

20. R. A. Patzner et al., *The Biology of Gobies* (Boca Raton, FL: CRC Press, 2011).

21. J. Herler, P. L. Munday, and V. Hernaman, "Gobies on Coral Reefs," in *The Biology of Gobies*, ed. R. A. Patzner et al. (Jersey, British Isles: Science Publishers, 2011), 493–529.

22. V. Hernaman and P. L. Munday, "Life-History Characteristics of Coral Reef Gobies. II. Mortality Rate, Mating System and Timing of Maturation," *Marine Ecology Progress Series* 290 (2005): 223–237.

23. M. Depczynski and D. R. Bellwood, "The Role of Cryptobenthic Reef Fishes in Coral Reef Trophodynamics," *Marine Ecology Progress Series* 256 (2003): 183–191.

24. M. Dirnwoeber and J. Herler, "Toxic Coral Gobies Reduce the Feeding Rate of a Corallivorous Butterflyfish on *Acropora* Corals," *Coral Reefs* 32 (2013): 91–100.

25. G. E. Nilsson et al., "Hypoxia Tolerance and Air-Breathing Ability Correlate with Habitat Preference in Coral-Dwelling Fishes," *Coral Reefs* 26 (2007): 241–248.

26. I. Karplus, "The Association between Gobiid Fishes and Burrowing Alpheid Shrimps," *Oceanography and Marine Biology* 25 (1987): 507–562.

27. J. L. Preston, "Communication Systems and Social Interactions in a Goby-Shrimp Symbiosis," *Animal Behavior* 26 (1978): 791–802.

28. N. Oshima and R. Fujii, "Motile Mechanism of Blue Damselfish (*Chrysiptera cyanea*) Iridophores," *Cytoskeleton* 8, no. 1 (1987): 85–90.

29. R. Fuller and A. Berglund, "Behavioral Responses of a Sex-Role Reversed Pipefish to a Gradient of Perceived Predation Risk," *Behavioral Ecology* 7 (1996): 69–75.

30. H. N. Sköld, S. Aspengren, and M. Wallin, "Rapid Color Change in Fish and Amphibians – Function, Regulation, and Emerging Applications," *Pigment Cell Melanoma Research* 26 (2012): 29–38.

31. V. S. Ramachandran et al., "Rapid Adaptive Camouflage in Tropical Flounders," *Nature* 379 (1996): 815–818.

32. E. K. Tyrie et al., "Coral Reef Flounders, *Bothus lunatus*, Choose Substrates on Which They Can Achieve Camouflage with Their Limited Body Pattern Repertoire," *Biological Journal of the Linnean Society* 114 (2015): 629–638.

33. J. E. Randall, "A Review of Mimicry in Marine Fishes," *Zoological Studies* 44, no. 3 (2005): 299–328.

34. B. H. Nagareda and J. M. Shenker, "Evidence for Chemical Luring in the Polkadot Batfish *Ogcocephalus cubifrons* (Teleostei: Lophiiformes: Ogcocephalidae)," *Florida Scientist* 72, no. 1 (2009): 11–17.

35. B. H. Nagareda and J. M. Schenker, "Dietary Analysis of Batfishes (Lophiiformes: Ogcocephalidae) in the Gulf of Mexico," *Gulf of Mexico Science* 26, no. 1 (2008): 28–35.

36. C. L. Combs, "Structure and Probable Feeding Function of the Batfish Esca" (PhD diss., Florida State University, 1973).

37. P. Teske and L. C. Beheregaray, "Evolution of Seahorses' Upright Posture Was Linked to Oligocene Expansion of Seagrass Habitats," *Biology Letters* 5, no. 4 (2009): 521–523.

38. Q. Lin et al., "The Seahorse Genome and the Evolution of Its Specialized Morphology," *Nature* 540 (2016): 395–399.

39. M. M. Porter et al., "Why the Seahorse Tail Is Square," *Science* 349, no. 6243 (2015): aaa6683–1.

40. J. Yeung, "Robotic Tail for Humans was Inspired by Seahorses," CNN, 2019, accessed October 7, 2021, https://edition.cnn.com/style/article/japan-robotic-tail-intl-hnk-trnd/index.html.

41. B. J. Gemmell, J. Sheng, and E. J. Buskey, "Morphology of Seahorse Head Hydrodynamically Aids in Capture of Evasive Prey," *Nature Communications* 4 (2013): 2840.

42. A.C.J. Vincent and L. M. Sadler, "Faithful Pair Bonds in Wild Seahorses, *Hippocampus whitei*," *Animal Behaviour* 50 (1995): 1557–1569.

43. H. D. Masonjones and S. M. Lewis, "Courtship Behavior in the Dwarf Seahorse, *Hippocampus zosterae*," *Copeia* 3 (1996): 634–640.

44. T.P.R. Oliveira et al., "Sounds Produced by the Longsnout Seahorse: a Study of their Structure and Functions," *Journal of Zoology* 294 (2014): 114–121.

45. C. M. Whittington et al., "Seahorse Brood Pouch Transcriptome Reveals Common Genes Associated with Vertebrate Pregnancy," *Molecular Biology and Evolution* 32, no. 12 (2015): 3114–3131.

46. Q. Lin, J. Lin, and D. Zhang, "Breeding and Juvenile Culture of the Lined Seahorse, *Hippocampus erectus* Perry, 1810," *Aquaculture* 277 (2008): 287–292.

47. J. M. Lawson, S. J. Foster, and A.C. J. Vincent, "Low Bycatch Rates Add up to Big Numbers for a Genus of Small Fishes," *Fisheries* 42, no. 1 (2017): 19–33.

48. M. Grey, A.-M. Blais, and A. C. Vincent, "Magnitude and Trends of Marine Fish Curio Imports to the USA," *Oryx* 39 (2005): 413–420.

49. A.C.J. Vincent and H. J. Hall, "The Threatened Status of Marine Fishes," *Trends in Ecology and Evolution* 11 (1996): 360–361.

50. M. Dahlberg, "A Review of Survival Rates of Fish Eggs and Larvae in Relation to Impact Assessments," *Marine Fisheries Review* 41, no. 3 (1979): 1–12.

51. J. Tariel et al., "Alloparental Care in the Sea: Brood Parasitism and Adoption within and between Two Species of Coral Reef *Altrichthys* Damselfish?," *Molecular Ecology* 28, no. 20 (2019): 4680–4691.

52. W. F. Figueira and S. J. Lyman, "Context-Dependent Risk Tolerance of the Bicolour Damselfish: Courtship in the Presence of Fish and Egg Predators," *Animal Behaviour* 74, no. 2 (2007): 329–336.

53. T. Amundsen and E. Forsgren, "Male Mate Choice Selects for Female Coloration in a Fish," *Proceedings of the National Academy of Sciences* 98, no. 23 (2001.): 13155–13160.

54. N. L. Ratterman, G. G. Rosenthal, and A. G. Jones. "Sex Recognition via Chemical Cues in the Sex-Role-Reversed Gulf Pipefish (*Syngnathus scovelli*)," *Ethology* 115 (2009): 339–346.

55. G. E. MacGinitie, "The Natural History of the Blind Goby, *Typhlogobius californensis* Steindachner," *American Midland Naturalist* 21, no. 2 (1939): 489–505.

56. A.F.H. Ros et al., "The Role of Androgens in the Trade-Off between Territorial and Parental Behavior in the Azorean Rock-Pool Blenny, *Parablennius parvicornis*," *Hormones and Behavior* 46 (2004): 491–497.

57. E. Giacomello, D. Marchini, and M. B. Rasotto, "A Male Sexually Dimorphic Trait Provides Antimicrobials to Eggs in Blenny Fish," *Biology Letters* 2, no. 3 (2006): 330–333.

58. J. Atema, "Chemical Senses, Chemical Signals, and Feeding Behavior in Fishes," in *Fish Behavior and Its Use in the Capture and Culture of Fishes*, ed. J. E. Bardach et al. (ICLARM Conference Proceedings 5, International Center for Living Aquatic Resource Management, Manila, Philippines, 1980), 57–101.

59. N. Stacey, "Hormones, Pheromones and Reproductive Behavior," *Fish Physiology and Biochemistry* 28 (2003): 229–235.

60. K. P. Maruska and T. C. Tricas, "Gonadotropin-Releasing Hormone (GnRH) Modulates Auditory Processing in the Fish Brain," *Hormones and Behavior* 59 (2011): 451–464.

61. Karen Maruska, PhD, interview by Joe E. Meisel, September 15, 2021.

62. J. C. Montgomery, N. Tolimieri, and O. S. Haine, "Active Habitat Selection by Pre-Settlement Reef Fishes," *Fish and Fisheries* 2 (2001): 261–277.

63. N. Tolimieri, A. Jeffs, and J. Montgomery, "Ambient Sound as a Cue for Navigation in Reef Fish Larvae," *Marine Ecology Progress Series* 207 (2000.): 219–224.

64. J. Atema, M. J. Kingsford, and G. Gerlach, "Larval Reef Fish Could Use Odour for Detection, Retention and Orientation to Reefs," *Marine Ecology Progress Series* 241 (2002): 151–190.

65. A. G. Coppock, N. M. Gardiner, and G. P. Jones, "Sniffing out the Competition? Juvenile Coral Reef Damselfishes Use Chemical Cues to Distinguish the Presence of Conspecific and Heterospecific Aggregations," *Behavioural Processes* 125 (2016): 43–50.

66. D.L. Dixson et al., "Coral Reef Fish Smell Leaves to Find Island Homes," *Proceedings of the Royal Society B*: 275 (2008): 2831–2839.

67. G. E. Forrester, "Competition in Reef Fishes," in *Ecology of Fishes on Coral Reefs*, ed. C. Mora (Cambridge: Cambridge University Press, 2015), 34–40.

68. R. Lubbock, "Why are Clownfishes not Stung by Sea Anemones?," *Proceedings of the Royal Society B* 207, no. 1166 (1980): 35–61.

69. Z.A. Pratte et al., "Association with a Sea Anemone Alters the Skin Microbiome of Clownfish," *Coral Reefs* 37 (2018): 1119–1125.

70. R. Goldshmid et al., "Aeration of Corals by Sleep-Swimming Fish," *Limnology and Oceanography* 49, no. 5 (2004): 1832–1839.

71. M. Kremien, U. Shavit, T. Massa, and A. Genin, "Benefit of Pulsation in Soft Corals," *Proceedings of the National Academy of Sciences* 110, no. 22 (2013): 8978–8983.

72. N. Garcia-Herrera et al., "Mutualistic Damselfish Induce Higher Photosynthetic Rates in Their Host Coral," *Journal of Experimental Biology* 220 (2017): 1803–1811.

73. T. Liberman, A. Genin, and Y. Loya, "Effects on Growth and Reproduction of the Coral *Stylophora pistillata* by the Mutualistic Damselfish *Dascyllus marginatus*," *Marine Biology* 121 (1995): 741–746.

74. E. Sampaio et al., "Octopuses Punch Fishes during Collaborative Interspecific Hunting Events," *Ecology* 102, no. 3 (2021): e03266.

75. P. L. Munday et al., "Skin Toxins and External Parasitism of Coral-Dwelling Gobies," *Journal of Fish Biology* 62 (2003): 976–981.

76. Alexandra S. Grutter, PhD, interview by Joe E. Meisel, October 20, 2021.

77. K. L. Cheney, A. S. Grutter, S. P. Blomberg, and N. J. Marshall, "Blue and Yellow Signal Cleaning Behavior in Coral Reef Fishes," *Current Biology* 19 (2009): 1283–1287.

78. A. S. Grutter, "Parasite Removal Rates by the Cleaner Wrasse *Labroides dimidiatus*," *Marine Ecology Progress Series* 130 (1996): 61–70.

79. A. S. Grutter et al., "Indirect Effects of an Ectoparasite Reduce Successful Establishment of a Damselfish at Settlement," *Functional Ecology* 25 (2011): 586–594.

80. K. L. Cheney, R. Bshary, and A. S. Grutter, "Cleaner Fish Cause Predators to Reduce Aggression toward Bystanders at Cleaning Stations," *Behavioral Ecology* 19, no. 5 (2008): 1063–1067

81. A. Bshary and R. Bshary, "Interactions between Sabre-Tooth Blennies and their Reef Fish Victims: Effects of Enforced Repeated Game Structure and Local Abundance on Victim Aggression," *Ethology* 116, no. 8 (2010): 681–690.

82. L. H. Salwiczek et al., "Adult Cleaner Wrasse Outperform Capuchin Monkeys, Chimpanzees and Orang-Utans in a Complex Foraging Task Derived from Cleaner-Client Reef Fish Cooperation," *PLoS ONE* 7, no. 11 (2012): e49068.

83. D. Souter et al., eds., *Status of Coral Reefs of the World: 2020. Executive Summary* (Global Coral Reef Monitoring Network of the International Coral Reef Initiative, 2020).

7. Slow Food

1. *Cod Wars*, directed by Magnus Viðar Sigurðsson (BBC Four Documentary Films, 2001), https://www.youtube.com/watch?v=FsOytZMRXo0.

2. *Cod Wars*.

3. *Cod Wars.*

4. *Cod Wars.*

5. *Cod Wars.*

6. *Cod Wars.*

7. *Cod Wars.*

8. National Federation of Fish Friers. Accessed November 18, 2021, https://www.nfff.co.uk/pages/fish-and-chips.

9. D. A. Righton et al., "Thermal Niche of Atlantic Cod *Gadus morhua*: Limits, Tolerance and Optima. *Marine Ecology Progress Series* 420 (2010): 1–13.

10. R. M. Love, "Dark Colour in White Fish Flesh (Torry Advisory Note No. 76)," *Food and Agriculture Organization (FAO) Support Unit for International Fisheries and Aquatic Research,* 2001.

11. Bjarte Bogstad, PhD, interview by Joe E. Meisel, June 11, 2019.

12. A. Thorsen et al., "Fecundity and Growth of Atlantic Cod (*Gadus morhua* L.) along a Latitudinal Gradient," *Fisheries Research* 104, no. 1–3 (2010): 45–55.

13. D. Mountain et al., "Growth and Mortality of Atlantic Cod *Gadus morhua* and Haddock *Melanogrammus aeglefinus* Eggs and Larvae on Georges Bank, 1995 to 1999," *Marine Ecology Progress Series* 353 (2008): 225–242.

14. Mountain et al., "Growth and Mortality," 225–242.

15. B. Ellertsen et al., "Relation between Temperature and Survival of Eggs and First-Feeding Larvae of Northeast Arctic Cod (*Gadus morhua* L.)," *Rapports et procès-verbaux des réunions* 191 (1989): 209–219.

16. A. V. Dolgov, "The Role of Capelin (*Mallotus villosus*) in the Foodweb of the Barents Sea," *ICES Journal of Marine Science* 59 (2002): 1034–1045.

17. H. Gjøsæter, B. Bogstad, and S. Tjelmeland, "Ecosystem Effects of the Three Capelin Stock Collapses in the Barents Sea," *Marine Biology Research* 5, no. 1 (2009): 40–53.

18. G.A. Rose, "Capelin (*Mallotus villosus*) Distribution and Climate: A Sea 'Canary' for Marine Ecosystem Change," *ICES Journal of Marine Science* 62 (2005): 1524–1530.

19. G. Rose, "A Brief History of Canada's Iconic Northern Cod," presentation at the University of British Columbia, April 29, 2018, https://www.youtube.com/watch?v=WR-ceVRoKO0.

20. D. Ø. Hjermann, B. Bogstad, A. M. Eikeset, G. Ottersen, H. Gjøsæter, and N. C. Stenseth, "Food Web Dynamics Affect Northeast Arctic Cod Recruitment," *Proceedings of the Royal Society B* 274 (2007): 661–669.

21. J. S. Link et al., "Trophic Role of Atlantic Cod in the Ecosystem," *Fish and Fisheries* 9 (2008): 1–30.

22. N. A. Yaragina, B. Bogstad, and Y. A. Kovalev, "Variability in Cannibalism in Northeast Arctic Cod (*Gadus morhua*) during the Period 1947–2006," *Marine Biology Research* 5, no. 1 (2008): 75–85.

23. S. Mehl, "The Northeast Arctic Cod Stock's Place in the Barents Sea Ecosystem in the 1980s: An Overview," *Polar Research* 10, no. 2 (1991): 525–534.

24. D. A. Righton et al., "Thermal Niche."

25. Rose, "A Brief History."

26. Amy Willis, "National Fish and Chips Day: 24 Surprising Facts about Fish and Chips (Including How They Helped Defeat Hitler)," *Metro UK,* June 3, 2016.

27. Link et al., "Trophic Role," 1–30.

28. H. Gjøsæter et al., "Predation on Early Life Stages Is Decisive for Year-Class Strength in the Barents Sea Capelin (*Mallotus villosus*) Stock," *ICES Journal of Marine Science* 73, no. 2 (2016): 182–195.

29. Mehl, "The Northeast Arctic," 525–534.

30. D. W. Sims et al., "Hunt Warm, Rest Cool: Bioenergetic Strategy Underlying Diel Vertical Migration of a Benthic Shark," *Journal of Animal Ecology* 75 (2006): 176–190.

31. S. H. Espeland et al., "Diel Vertical Migration Patterns in Juvenile Cod from the Skagerrak Coast," *Marine Ecology Progress Series* 405 (2010): 29–37.

32. M. F. Petersen and J. F. Steffensen, "Preferred Temperature of Juvenile Atlantic Cod *Gadus morhua* with Different Haemoglobin Genotypes at Normoxia and Moderate Hypoxia," *Journal of Experimental Biology* 206 (2003): 359–364.

33. V. S. Kirpichnikov, *Genetic Bases of Fish Selection* (New York: Springer-Verlag, 1981).

34. B. Björnsson, "Thermoregulatory Behaviour in Cod: Is the Thermal Preference in Free-Ranging Adult Atlantic Cod Affected by Food Abundance?," *Canadian Journal of Fisheries and Aquatic Sciences* 76, no. 9 (2018): 1515–1527.

35. M. Kurlansky, *Cod: A Biography of the Fish That Changed the World* (New York: Penguin Books, 1997).

36. University of Exeter, "Scientists Search for Regional Accents in Cod," *ScienceDaily*, accessed December 26, 2021, www.sciencedaily.com/releases/2016/10/161006111842.htm.

37. G. A. Rose, "Cod Spawning on a Migration Highway in the North-West Atlantic," *Nature* 366 (1993): 458–461.

38. T. H. Huxley, "Inaugural Address, London Fisheries Exhibition (1885)," accessed on December 14, 2021, https://mathcs.clarku.edu/huxley/SM5/fish.html.

39. L. Z. Joncas, "Canadian Ministry of Agriculture Fisheries Report (1885)," in *Cod: A Biography of the Fish That Changed the World*, by Mark Kurlansky (New York: Penguin Books, 1997), 123.

40. Kurlansky, *Cod: A Biography*.

41. Kurlansky, *Cod: A Biography*.

42. Rose, "A Brief History."

43. R. Schijns, R. Froese, J. A. Hutchings, and D. Pauly, "Food for Thought: Five Centuries of Cod Catches in Eastern Canada," *ICES Journal of Marine Science* 78, no. 8 (2021): 2675–2683.

44. L. C. Hamilton and M. J. Butler, "Outport Adaptations: Social Indicators through Newfoundland's Cod Crisis," *Human Ecology Review* 8, no. 2 (2001): 1–11.

45. Kurlansky, *Cod: A Biography*.

46. Mark Kurlansky, *Birdseye: The Adventures of a Curious Man* (New York: Anchor Books, 2013).

47. B. Felton and M. Fowler, *Felton & Fowler's Famous Americans You Never Knew Existed* (New York: Stein and Day, 1979), 68, quoting from Clarence Birdseye Field Journals (November 9, 1910 to July 20, 1916), Amherst College Special Collection, MA.00169.

48. J. K. Baum, J. Ford, and W. Norden, "Haddock (*Melanogrammus aeglefinus*): Seafood Watch Seafood Report," *SeaChoice*, June 17, 2011.

49. J. Farkas et al., "Effects of Mine Tailing Exposure on Early Life Stages of Cod (*Gadus morhua*) and Haddock (*Melanogrammus aeglefinus*)," *Environmental Research* 200 (2021): 111447.

50. M. H. Stiasny et al., "Divergent Responses of Atlantic Cod to Ocean Acidification and Food Limitation," *Global Change Biology* 25, no. 3 (2018): 839–849.

51. E. Sørhus et al., "Unexpected Interaction with Dispersed Crude Oil Droplets Drives Severe Toxicity in Atlantic Haddock Embryos," *PLoS ONE* 10, no. 4 (2015): e0124376.

52. A. Cresci et al., "Atlantic Haddock (*Melanogrammus aeglefinus*) Larvae Have a Magnetic Compass that Guides Their Orientation," *iScience* 19 (2019): 1173–1178.

53. J. G. Pope and C. T. Macer, "An Evaluation of the Stock Structure of North Sea Cod, Haddock, and Whiting Since 1920, Together with a Consideration of the Impacts of Fisheries and Predation Effects on their Biomass and Recruitment," *ICES Journal of Marine Science* 53 (1996): 1157–1169.

54. J. Kane, "The Feeding Habits of Co-Occurring Cod and Haddock Larvae from Georges Bank," *Marine Ecology Progress Series* 16 (1984): 9–20.

55. R. E. Holt et al., "Barents Sea Cod (*Gadus morhua*) Diet Composition: Long-Term Interannual, Seasonal, and Ontogenetic Patterns," *ICES Journal of Marine Science* 76, no. 6 (2019): 1641–1652.

56. J. M. Durant et al., "Nonlinearity in Interspecific Interactions in Response to Climate Change: Cod and Haddock as an Example," *Global Change Biology* 26 (2020): 5554–5563.

57. Food and Agriculture Organization (FAO), *The State of World Fisheries and Aquaculture: Meeting the Sustainable Development Goals* (Rome: FAO, 2018).

58. S. M. Sogard and B. L. Olla, "Food Deprivation Affects Vertical Distribution and Activity of a Marine Fish in a Thermal Gradient: Potential Energy-Conserving Mechanisms," *Marine Ecology Progress Series* 133 (1996): 43–55.

59. K. M. Bailey et al., "Population Ecology and Structural Dynamics of Walleye Pollock (*Theragra chalcogramma*)," in *Dynamics of the Bering Sea: A Summary of Physical, Chemical, and Biological Characteristics, and a Synopsis of Research on the Bering Sea,* ed. T. R. Loughlin and K. Ohtani (Fairbanks: University of Alaska Sea Grant, 1999), 581–614.

60. "Fisheries of the United States, 2007," (Silver Spring, MD: Office of Science and Technology, National Marine Fisheries Service, 2007).

61. T. Taguchi et al., "Robust Sealing of Blood Vessels with Cholesteryl Group-Modified, Alaska Pollock-Derived Gelatin-Based Biodegradable Sealant Under Wet Conditions," *Journal of Biomedical Nanotechnology* 12, no. 1 (2016): 128–134.

62. "Pacific Halibut Species Profile," Alaska Department of Fish and Game, Juneau, accessed December 31, 2021, http://www.adfg.alaska.gov/index.cfm?adfg=halibut.main.

63. M. S. Love, *Certainly More Than You Want to Know about the Fishes of the Pacific Coast: A Postmodern Experience* (Santa Barbara, CA: Really Big Press, 2011).

64. N. P. Novikov, "Basic Elements of Biology of the Pacific Halibut (*Hippoglossus hippoglossus stenolepis* Schmidt) in the Bering Sea," *Soviet Fisheries Investigations in N.E. Pacific, Part II,* 1964: 175–219 (translated by US Department of Commerce and National Science Foundation).

65. L. M. Cargnelli, S. J. Griesbach, and W. W. Morse, "Essential Fish Habitat Source Document: Atlantic Halibut, *Hippoglossus hippoglossus,* Life History and Habitat Characteristics," *NOAA Technical Memorandum NMFS-NE-125* (Woods Hole, MA: National Oceanic and Atmospheric Administration Fisheries, 1999).

66. F. A. Best, "Halibut Ecology," in *Fisheries Oceanography—Eastern Bering Sea Shelf. NWAFC Processed Report 79–20* (Washington, DC: Northwest and Alaska Fisheries Center, National Marine Fisheries Service, 1979).

67. C. L. Hew and G. L. Fletcher, "The Role of Pituitary in Regulating Antifreeze Protein Synthesis in the Winter Flounder," *Federation of European Biochemical Societies (FEBS) Letters* 99, no. 2 (1979): 337–339.

68. "Pacific Halibut Species Overview," National Oceanic and Atmospheric Administration Fisheries, Washington, DC, accessed December 31, 2021, https://www.fisheries.noaa.gov/species/pacific-halibut.

69. F. H. Bell, *The Pacific Halibut, the Resource, and the Fishery* (Edmonds, WA: Alaska Northwest Publishing Company, 1981).

70. R. A. Carvalho et al., "Development of Edible Films Based on Differently Processed Atlantic Halibut (*Hippoglossus hippoglossus*) Skin Gelatin," *Food Hydrocolloids* 22 (2008): 1117–1123.

71. "New Product Will Benefit Fishing Industry—Company Introduces 'Frozen Fingers,'" *Yorkshire Post and Leeds Intelligencer,* October 4, 1955, http://www.foodsof england.co.uk/fishfingers.htm (accessed January 3, 2022).

72. "Fish Fingers Turn 60: How Britain Fell for Not-Very-Fishy Sticks of Frozen Protein," *The Guardian,* September 15, 2015, https://www.theguardian.com/lifeandstyle/wordof mouth/2015/sep/15/fish-fingers-60-britain-taste-of-childhood (accessed January 3, 2022).

73. "Fish Fingers Surprisingly Sustainable, Say Conservationists," *BBC News,* November 2, 2018, https://www.bbc.com/news/uk-46071019 (accessed January 3, 2022).

8. Into the Abyss

1. "Auguste Piccard, Explorer, Is Dead; Auguste Piccard Is Dead at 78; Stratosphere and Sea Explorer," *New York Times,* March 26, 1962, https://www.nytimes.com/1962/03/26/archives/auguste-piccard-explorer-is-dead-auguste-piccard-is-dead-at-78.html.

2. H. M. Schmeck, Jr., "Dive May Have Hit the Deepest Sea Point," *New York Times,* January 24, 1960.

3. L. P. Paine, *Ships of Discovery and Exploration* (Boston: Mariner Books, 2000).

4. "Dives of the Bathyscaph Trieste, 1958–1963: Transcriptions of Sixty-One Dictabelt Recordings in the Robert Sinclair Dietz Papers, 1905–1994," (September 2000), MC28, Robert S. Dietz Papers, Archives of the Scripps Institution of Oceanography, University of California–San Diego, La Jolla, 59.

5. "Dives of the Bathyscaph," 14.

6. "Dives of the Bathyscaph," 65.

7. "Dives of the Bathyscaph," 16.

8. "Bathyscaph's Crew Honored by the President for Their Achievement," *New York Times,* February 5, 1960.

9. "Dives of the Bathyscaph," 14.

10. "Dives of the Bathyscaph," 15.

11. A. J. Jamieson and P. H. Yancey, "On the Validity of the Trieste Flatfish: Dispelling the Myth," *Biological Bulletin* 222, no. 3 (2012): 171–175.

12. W. Beebe, *Half Mile Down* (New York: Harcourt, Brace, 1934), 169.

13. J.A.C. Nicol, "Bioluminescence," *Proceedings of the Royal Society A* 265, no. 1322 (1962): 355–359.

14. "Dives of the Bathyscaph," 17.

15. A. R. Thurber et al., "Ecosystem Function and Services Provided by the Deep Sea," *Biogeosciences* 11 (2014): 3941–3963.

16. H. J. Wagnera, E. Fröhlich, K. Negishi, and S. P. Collin, "The Eyes of Deep-Sea Fish II. Functional Morphology of the Retina," *Progress in Retinal and Eye Research* 17, no. 4 (1998): 637–685.

17. R. H. Douglas, D. M. Hunt, and J. K. Bowmaker, "Spectral Sensitivity Tuning in the Deep-Sea," in *Sensory Processing in Aquatic Environments,* ed. S. P. Collin and N. J. Marshall (New York: Springer, 2003), 323–342.

18. Z. Musilova et al., "Vision Using Multiple Distinct Rod Opsins in Deep-Sea Fishes," *Science* 364, no. 6440 (2019): 588–592.

19. E. Landgren, K. Fritsches, R. Brill, and E. Warrant, "The Visual Ecology of a Deep-Sea Fish, the Escolar *Lepidocybium flavobrunneum* (Smith, 1843)," *Philosophical Transactions of the Royal Society B* 369 (2014): 20130039.

20. B. H. Robison and K. R. Reisenbichler, "*Macropinna microstoma* and the Paradox of its Tubular Eyes," *Copeia* 4 (2008): 780–784.

21. H-J. Wagner et al., "A Novel Vertebrate Eye Using both Refractive and Reflective Optics," *Current Biology* 19 (2009): 108–114.

22. J. Case, J. Warner, A. T. Barnes, and M. Lowenstine, "Bioluminescence of Lantern Fish (Myctophidae) in Response to Changes in Light Intensity," *Nature* 265 (1977): 179–181.

23. S.H.D. Haddock, M. A. Moline, and J. F. Case, "Bioluminescence in the Sea," *Annual Review of Marine Science* 2 (2010): 443–493.

24. H. Scales, *The Brilliant Abyss: Exploring the Majestic Hidden Life of the Deep Ocean and the Looming Threat that Imperils It* (New York: Atlantic Monthly Press, 2021).

25. J. M. Claes and J. Mallefet, "Early Development of Bioluminescence Suggests Camouflage by Counter-Illumination in the Velvet Belly Lantern Shark *Etmopterus spinax* (Squaloidea: Etmopteridae)," *Journal of Fish Biology* 73 (2008): 1337–1350.

26. L. Duchatelet et al., "Etmopteridae Bioluminescence: Dorsal Pattern Specificity and Aposematic Use," *Zoological Letters* 5 (2019):9.

27. G. D. Ruxton and D. M. Bailey, "Combining Motility and Bioluminescent Signalling Aids Mate Finding in Deep-Sea Fish: A Simulation Study," *Marine Ecology Progress Series* 293 (2005): 253–262.

28. A. L. Davis et al., "Ultra-Black Camouflage in Deep-Sea Fishes," *Current Biology* 30, no. 17 (2020): 3470–3476.e3.

29. P. J. Herring and C. Cope, "Red Bioluminescence in Fishes: On the Suborbital Photophores of Malacosteus, Pachystomias and Aristostomias," *Marine Biology* 148 (2005): 383–394.

30. G. Y. Jumper and R. C. Baird, "Location by Olfaction: A Model and Application to the Mating Problem in the Deep-Sea Hatchetfish *Argyropelecus hemigymnus*," *American Naturalist* 138, no. 6 (1991.): 1431–1458.

31. P. J. Herring, "Species Abundance, Sexual Encounter and Bioluminescent Signalling in the Deep Sea," *Philosophical Transactions of the Royal Society B* 355 (2000): 1273–1276.

32. Y. L. Gagnon, T. T. Sutton, and S. Johnsen, "Visual Acuity in Pelagic Fishes and Mollusks," *Vision Research* 92 (2013): 1–9.

33. R. R. Wilson, Jr. and K. L. Smith, Jr., "Effect of Near-Bottom Currents on Detection of Bait by the Abyssal Grenadier Fishes *Coryphaenoides* spp., Recorded in Situ with a Video Camera on a Free Vehicle," *Marine Biology* 84 (1984): 83–91.

34. E. A. Ellis and T. H. Oakley, "High Rates of Species Accumulation in Animals with Bioluminescent Courtship Displays," *Current Biology* 26: 1916–1921.

35. S. J. Gould, *Hen's Teeth and Horse's Toes* (New York: W. W. Norton, 1983).

36. A. Dubin et al., "Complete Loss of the MHC II Pathway in an Anglerfish, *Lophius piscatorius*," *Biology Letters* 15 (2019): 20190594.

37. T. W. Pietsch, "Dimorphism, Parasitism, and Sex Revisited: Modes of Reproduction among Deep-Sea Ceratioid Anglerfishes (Teleostei: Lophiiformes)," *Ichthyological Research* 52 (2005): 207–236.

38. T. W. Pietsch, *Oceanic Anglerfishes: Extraordinary Diversity in the Deep Sea* (Berkeley: University of California Press, 2009).

39. M. G. Haygood and D. L. Distel, "Bioluminescent Symbionts of Flashlight Fishes and Deep-Sea Anglerfishes Form Unique Lineages Related to the Genus *Vibrio*," *Nature* 363 (1993): 110–111.

40. Y. L. Tee and P. Tran, "On Bioinspired 4D Printing: Materials, Design and Potential Applications," *Australian Journal of Mechanical Engineering* (2021): 10.1080.

41. M. Bessho-Uehara et al., "Kleptoprotein Bioluminescence: *Parapriacanthus* Fish Obtain Luciferase from Ostracod Prey," *Scientific Advances* 6, no. 2 (2020): eaax4942.

42. C. Porcu et al., "Reproductive Biology of a Bathyal Hermaphrodite Fish, *Bathypterois mediterraneus* (Osteichthyes: Ipnopidae) from the South-Eastern Sardinian Sea (Central-Western Mediterranean)," *Journal of the Marine Biological Association of the United Kingdom* 90, no. 4 (2010): 719–728.

43. J. C. Drazen and B. A. Seibel, "Depth-Related Trends in Metabolism of Benthic and Benthopelagic Deep-Sea Fishes," *Limnology and Oceanography* 52, no. 5 (2007): 2306–2316.

44. V. van Ginneken et al., "Eel Migration to the Sargasso: Remarkably High Swimming Efficiency and Low Energy Costs," *Journal of Experimental Biology* 208 (2005): 1329–1335.

45. I. G. Priede et al., "The Absence of Sharks from Abyssal Regions of the World's Oceans," *Proceedings of the Royal Society B* 273 (2006): 1435–1441.

46. N. Pinte et al., "Ecological Features and Swimming Capabilities of Deep-Sea Sharks from New Zealand," *Deep Sea Research Part I: Oceanographic Research Papers* 156 (2020): 103187.

47. A. J. Pershing et al., "The Impact of Whaling on the Ocean Carbon Cycle: Why Bigger Was Better," *PLoS ONE* 5, no. 8 (2010): e12444.

48. S. A. Mizroch et al., "Estimating the Adult Survival Rate of Central North Pacific Humpback Whales (*Megaptera novaeangliae*)," *Journal of Mammalogy* 85, no. 5 (2004): 963–972.

49. C. R. Smith and A.W.J. Demopoulos, "The Deep Pacific Ocean Floor," in *Ecosystems of the World, Volume 28: Ecosystems of the Deep Ocean*, ed. P. A. Tyler (Amsterdam: Elsevier, 2003), 179–218.

50. A. Krogh, "Conditions of Life at Great Depths in the Ocean," *Ecological Monographs* 4 (1934): 430–439.

51. C. R. Smith and A. R. Baco, "Ecology of Whale Falls at the Deep-Sea Floor," *Oceanography and Marine Biology* 41 (2003): 311–354.

52. J. Aguzzi et al., "Faunal Activity Rhythms Influencing Early Community Succession of an Implanted Whale Carcass Offshore Sagami Bay, Japan," *Scientific Reports* 8 (2018.): 11163.

53. Smith and Baco, "Ecology of Whale Falls," 311.

54. C. R. Smith et al., "Whale-Fall Ecosystems: Recent Insights into Ecology, Paleoecology, and Evolution," *Annual Review of Marine Science* 7 (2015): 571–596.

55. P. Wessel, D. T. Sandwell, and S.-S. Kim, "The Global Seamount Census," *Oceanography* 23, no. 1 (2010): 24–33.

56. A. Genin and J. F. Dower, "Seamount Plankton Dynamics," in *Seamounts: Ecology, Fisheries, and Conservation, vol. 12*, ed. T. J. Pitcher et al. (Oxford: Blackwell, 2007), 87–-100.

57. M. R. Clark et al., "The Ecology of Seamounts: Structure, Function, and Human Impacts," *Annual Review of Marine Science* 2 (2010): 253–278.

58. H. T. Pinheiro et al., "Fish Biodiversity of the Vitória-Trindade Seamount Chain, Southwestern Atlantic: An Updated Database," *PLoS ONE* 10, no. 3 (2015): e0118180.

59. D. M. Tracey and P. L. Horn, "Background and Review of Ageing of Orange Roughy (*Hoplostethus atlanticus*) from New Zealand and Elsewhere," *New Zealand Journal of Marine and Freshwater Research* 33 (1999): 67–86.

60. Clark et al., "Ecology of Seamounts."

61. Scales, *The Brilliant Abyss.*

62. M. R. Clark, "Experience with Management of Orange Roughy (*Hoplostethus atlanticus*) in New Zealand Waters, and the Effects of Commercial Fishing on Stocks over the Period 1980–1993," in *Deep-water Fisheries of the North Atlantic Oceanic Slope*, ed. A. G. Hopper (Dordrecht: Springer Netherlands, 1995), 251–266.

63. M. R. Clark et al., "The Effects of Commercial Exploitation on Orange Roughy (*Hoplostethus atlanticus*) from the Continental Slope of the Chatham Rise, New Zealand, from 1979 to 1997," *Fisheries Research* 45 (2000): 217–238.

64. P. M. Mace et al., "Growth and Productivity of Orange Roughy (*Hoplostethus atlanticus*) on the North Chatham Rise," *New Zealand Journal of Marine and Freshwater Research* 24 (1990): 105–119.

65. W. R. Allen, "Ovulation, Pregnancy, Placentation and Husbandry in the African Elephant (*Loxodonta africana*)," *Philosophical Transactions of the Royal Society B* 361(1469): 821–834.

66. K. Crane, *Sea Legs: Tales of a Woman Oceanographer* (Boulder, CO: Westview Press, 2003), 54.

67. Crane, *Sea Legs*, 112.

68. S. K. Haldar, *Introduction to Mineralogy and Petrology, 2nd Edition* (Amsterdam: Elsevier, 2020).

69. F. Gaill and S. Hunt, "Tubes of Deep Sea Hydrothermal Vent Worms *Riftia pachyptila* (Vestimentifera) and *Alvinella pompejana* (Annelida)," *Marine Ecology Progress Series* 34, no. 3 (1986): 267–274.

70. C. Arndt, F. Gaill, and H. Felbeck, "Anaerobic Sulfur Metabolism in Thiotrophic Symbioses," *Journal of Experimenal Biology* 204 (2001): 741–750.

71. Z. Minic and G. Hervé, "Biochemical and Enzymological Aspects of the Symbiosis between the Deep-Sea Tubeworm *Riftia pachyptila* and Its Bacterial Endosymbiont," *European Journal of Biochemistry* 271 (2004): 3093–3102.

72. S. Nakagawa and K. Takai, "Deep-Sea Vent Chemoautotrophs: Diversity, Biochemistry and Ecological Significance," *FEMS Microbiology Ecology* 65, no. 1 (2008): 1–14.

73. S. Okada et al., "The Making of Natural Iron Sulfide Nanoparticles in a Hot Vent Snail," *Proceedings of the National Academy of Sciences* 116, no. 41 (2019): 20376–20381.

74. J. R. Voight, "A Review of Predators and Predation at Deep-Sea Hydrothermal Vents," *Cahiers de Biologie Marine* 41 (2000.): 155–166.

75. G. Sancho et al., "Selective Predation by the Zoarcid Fish *Thermarces cerberus* at Hydrothermal Vents," *Deep-Sea Research I* 52 (2005): 837–844.

76. E. Ramirez-Llodra, T. M. Shank, and C. R. German, "Biodiversity and Biogeography of Hydrothermal Vent Species," *Oceanography* 20, no. 1 (2007): 30–41.

77. R. C. Vrijenhoek, "Genetic Diversity and Connectivity of Deep-Sea Hydrothermal Vent Metapopulations," *Molecular Ecology* 19 (2010): 4391–4411.

78. Y. Yano et al., "Adaptive Changes in Membrane Lipids of Barophilic Bacteria in Response to Changes in Growth Pressure," *Applied and Environmental Microbiology* 64, no. 2 (1998): 2572–2577.

79. P. H. Yancey et al., "Trimethylamine Oxide Counteracts Effects of Hydrostatic Pressure on Proteins of Deep-Sea Teleosts," *Journal of Experimental Zoology* 289 (2001): 172–176.

80. A. L. Samerotte et al., "Correlation of Trimethylamine Oxide and Habitat Depth within and among Species of Teleost Fish: An Analysis of Causation," *Physiological and Biochemical Zoology* 80, no. 2 (2007): 197–208.

81. D. Lu, "Scientists Find Deepest Fish Ever Recorded at 8,300 Metres Underwater Near Japan," *The Guardian*, April 3, 2023.

82. Scales, *The Brilliant Abyss*.

83. S. Shigeru and N. Kazuhiro, "Functional Morphology of Feeding Apparatus of the Cookie-cutter Shark, *Isistius hrasiliensis* (Elasmobranchii, Dalatiinae)," *Zoological Science* 9, no. 4 (1992): 811–821.

84. P. B. Best and T. Photopoulou, "Identifying the "Demon Whale-Biter": Patterns of Scarring on Large Whales Attributed to a Cookie-Cutter Shark *Isistius* sp.," *PLoS ONE* 11, no. 4 (2016): e0152643.

85. J. M. Claes and J. Mallefet, "Bioluminescence of Sharks: First Synthesis," in *Bioluminescence in Focus—A Collection of Illuminating Essays*, ed. B. Meyer-Rochow (Kerala, India: Research Signpost, 2009), 51-65.

86. S. J. Kiraly, J. A. Moore, and P. H. Jasinski, "Deepwater and Other Sharks of the U.S. Atlantic Ocean Exclusive Economic Zone," *Marine Fisheries Review* 65, no. 4 (2003): 1–64.

87. R. Proud et al., "From Siphonophores to Deep Scattering Layers: Uncertainty Ranges for the Estimation of Global Mesopelagic Fish Biomass," *ICES Journal of Marine Science* 76, no. 3 (2019): 718–733.

88. P. A. Hulley, "Myctophidae," in *Check-List of the Fishes of the Eastern Tropical Atlantic (CLOFETA), Vol. 1*, ed. J. C. Quero et al. (Paris: UNESCO, 1990), 429–483.

89. T. Valinassab, G. J. Pierce, and K. Johannesson, "Lantern Fish (*Benthosema pterotum*) Resources as a Target for Commercial Exploitation in the Oman Sea," *Journal of Applied Ichthyology* 23 (2007): 573–577.

90. J.B.S. Haldane, "The Origin of Life," *Rationalist Annual* 148 (1929): 3–10.

91. M. J. Russell, "The Importance of Being Alkaline," *Science* 302 (2003): 580–581.

92. S. F. Jordan, E. Nee, and N. Lane, "Isoprenoids Enhance the Stability of Fatty Acid Membranes at the Emergence of Life Potentially Leading to an Early Lipid Divide," *Interface Focus* 9 (2019): 20190067.

93. Scales, *The Brilliant Abyss*, 124.

Part III

1. J. A. Burns, "Vertebrate Paleontology and the Alleged Ice-Free Corridor: The Meat of the Matter," *Quaternary International* 32 (1996): 107–112.

2. T. D. Dillehay et al., "New Archaeological Evidence for an Early Human Presence at Monte Verde, Chile," *PLoS ONE* 10, no. 11 (2015): e0141923.

3. J. M. Erlandson et al., "The Kelp Highway Hypothesis: Marine Ecology, the Coastal Migration Theory, and the Peopling of the Americas," *Journal of Island & Coastal Archaeology* 2 (2007): 161–174.

4. J. M. Erlandson, "Anatomically Modern Humans, Maritime Adaptations, and the Peopling of the New World," in *The First Americans*, ed. N. Jablonski (San Francisco: California Academy of Sciences, 2002), 59–92.

5. Robert Steneck, PhD, interview by Joe E. Meisel, July 8, 2022.

6. J. M. Erlandson, T. C. Rick, and T. J. Braje, "Fishing up the Food Web?: 12,000 Years of Maritime Subsistence and Adaptive Adjustments on California's Channel Islands," *Pacific Science* 63, no. 4 (2009): 711–724.

7. J. M. Erlandson et al., "Ecology of the Kelp Highway: Did Marine Resources Facilitate Human Dispersal from Northeast Asia to the Americas?," *Journal of Island & Coastal Archaeology* 10, no. 3 (2015): 392–411.

8. E. J. Duffy, "Biodiversity and the Functioning of Seagrass Ecosystems," *Marine Ecology Progress Series* 311 (2006): 23–250.

9. Flowing River, Pounding Surf

1. Preceding text draws on several sources: M. Burke, *Lords of the Fly: Madness, Obsession, and the Hunt for the World-Record Tarpon* (New York: Pegasus, 2021); D. Simmons, "Swimming for Tarpon . . . and Staying Alive," Key West Fly Fishing, April 30, 2020, https://keywestflyfishing.com/swimming-for-tarponand-staying-alive; E. Johnston, "Fly Fishing the Homosassa River," Leisure Time Travel, March 1, 2011, https://www.leisuretimetravel.com/locklear.htm.

2. P. L. Forey et al., "Interrelationships of Elopomorph Fishes," in *Interrelationships of Fishes*, ed. M.L.J. Stiassny et al. (San Diego: Academic Press, 1996), 175–191

3. P.J.P. Whitehead and R. Vergara, "Megalopidae," in *FAO Species Identification Sheets for Fishery Purposes: Western Central Atlantic (Fishing Area 31), Vol. 3*, ed. W. Fischer (Rome: FAO, 1978).

4. A. P. Salati et al., "Effect of Different Levels of Salinity on Gill and Kidney Function in Common Carp *Cyprinus carpio* (Pisces: Cyprinidae)," *Italian Journal of Zoology* 78, no. 3 (2011): 298–303.

5. T. D. Clark et al., "Changes in Cardiac Output during Swimming and Aquatic Hypoxia in the Air-Breathing Pacific Tarpon," *Comparative Biochemistry and Physiology, Part A* 148 (2007): 562–571

6. R. S. Seymour et al., "Partitioning of Respiration between the Gills and Air-Breathing Organ in Response to Aquatic Hypoxia and Exercise in the Pacific Tarpon, *Megalops cyprinoides*," *Physiological and Biochemical Zoology* 77, no. 5 (2004): 760–767.

7. R. S. Seymour et al., "Continuous Measurement of Oxygen Tensions in the Air-Breathing Organ of Pacific Tarpon (*Megalops cyprinoides*) in Relation to Aquatic Hypoxia and Exercise," *Journal of Comparative Physiology B* 177 (2007): 579–587.

8. Seymour et al., "Partitioning of Respiration."

9. S. M. Taylor, E. R. Loew, and M. S. Grace, "A Rod-Dominated Visual System in Leptocephalus Larvae of Elopomorph Fishes (Elopomorpha: Teleostei)," *Environmental Biology of Fishes* 92 (2011): 513–523.

10. A. V. Zale and S. G. Merrifield, "Species Profiles: Life History and Environmental Requirements of Coastal Fishes and Invertebrates (South Florida): Ladyfish and Tarpon," *Oklahoma Cooperative Fish and Wildlife Research Unit*, Biological Report 82(11.104), TR El-82–4 (1989).

11. J. Luo et al., "Migrations and Movements of Atlantic Tarpon Revealed by Two Decades of Satellite Tagging," *Fish and Fisheries* 21, no. 2 (2020): 290–318.

12. J.-J. Albaret, "Mugilidae," in *Faune des Poissons d'Eaux Douce et Saumâtres de l'Afrique de l'Ouest, Tome 2. Coll. Faune et Flore Tropicales 40*, ed. C. Lévêque et al. (Paris: Museum National d'Histoire Naturalle and Institut de Recherche pour le Développement, 2003), 601–611.

13. C. R. Robins and G. C. Ray, *A Field Guide to Atlantic Coast Fishes of North America* (Boston: Houghton Mifflin, 1986).

14. K. J. Murchie et al., "Movement Patterns of Bonefish (*Albula vulpes*) in Tidal Creeks and Coastal Waters of Eleuthera, The Bahamas," *Fisheries Research* 147 (2013): 404–414.

15. R. Humston et al., "Movements and Site Fidelity of the Bonefish *Albula vulpes* in the Northern Florida Keys Determined by Acoustic Telemetry," *Marine Ecology Progress Series* 291 (2005): 237–248.

16. M. Hewage, "Mangrove Bounty," *Sunday Observer*, April 25, 2021, https://www.sundayobserver.lk/2021/04/25/feature/mangrove-bounty.

17. P. B. Tomlinson, *The Botany of Mangroves* (Cambridge: Cambridge University Press, 2016).

18. S.J.M. Blaber, *Tropical Estuarine Fishes: Ecology, Exploitation and Conservation* (Oxford: Blackwell, 2000).

19. E. F. Granek, J. E. Compton, and D. L. Phillips, "Mangrove-Exported Nutrient Incorporation by Sessile Coral Reef Invertebrates," *Ecosystems* 12 (2009): 462–472.

20. M.E.M. Walton et al., "Outwelling from Arid Mangrove Systems Is Sustained by Inwelling of Seagrass Productivity," *Marine Ecology Progress Series* 507 (2014): 125–137.

21. J. M. Fourqurean et al., "Seagrass Ecosystems as a Globally Significant Carbon Stock," *Nature Geoscience* 5 (2012): 505–509.

22. K. L. Heck et al., "Trophic Transfers from Seagrass Meadows Subsidize Diverse Marine and Terrestrial Consumers," *Ecosystems* 11 (2008): 1198–1210.

23. S. Lai et al., "First Experimental Evidence of Corals Feeding on Seagrass Matter," *Coral Reefs* 32 (2013): 1061–1064.

24. R. D. Clark et al., "Nocturnal Fish Movement and Trophic Flow across Habitat Boundaries in a Coral Reef Ecosystem (SW Puerto Rico)," *Caribbean Journal of Science* 45, no. 2–3 (2009): 282–303.

25. J. Nelson et al., "Flux by Fin: Fish-Mediated Carbon and Nutrient Flux in the Northeastern Gulf of Mexico," *Marine Biology* 159 (2012): 365–372.

26. P. J. Mumby et al., "Mangroves Enhance the Biomass of Coral Reef Fish Communities in the Caribbean," *Nature* 427 (2004): 533–536.

27. J. E. Serafy, C. H. Faunce, and J. J. Lorenz, "Mangrove Shoreline Fishes of Biscayne Bay," *Bulletin of Marine Science* 72, no. 1 (2003): 161–180.

28. S. Schuster, "Archerfish: Quick Guide," *Current Biology* 17, no. 13 (2007): R494-R495.

29. S. Temple et al., "A Spitting Image: Specializations in Archerfish Eyes for Vision at the Interface between Air and Water," *Proceedings of the Royal Society B* 277 (2010): 2607–2615.

30. S. Shuster et al., "Archer Fish Learn to Compensate for Complex Optical Distortions to Determine the Absolute Size of Their Aerial Prey," *Current Biology* 14 (2004): 1565–1568.

31. Schuster, "Archerfish: Quick Guide."

32. S. Rossel, J. Corlija, and S. Schuster, "Predicting Three-Dimensional Target Motion: How Archer Fish Determine Where to Catch Their Dislodged Prey," *Journal of Experimental Biology* 205 (2002): 3321–3326.

33. S. Shuster et al., "Animal Cognition: How Archer Fish Learn to Down Rapidly Moving Targets," *Current Biology* 16 (2006): 378–383.

34. A. Vergés and A. H. Campbell, "Kelp Forests: Quick Guide," *Current Biology* 30 (2020): R905–R931.

35. K. Minami et al., "Estimation of Kelp Forest, *Laminaria* spp., Distributions in Coastal Waters of the Shiretoko Peninsula, Hokkaido, Japan, Using Echosounder and Geostatistical Analysis," *Fisheries Science* 76 (2010): 729–736.

36. K. A. Krumhansl et al., "Global Patterns of Kelp Forest Change over the Past Half-Century," *Proceedings of the National Academy of Sciences* 113, no. 48 (2016): 13785–13790.

37. Robert Steneck, PhD, interview by Joe E. Meisel, July 8, 2022.

38. "Design & Construction Stats," Golden Gate Bridge Highway and Transportation District, accessed June 10, 2022, https://www.goldengate.org/bridge/history-research/statistics-data/design-construction-stats/.

39. D. O. Duggins, C. A. Simenstad, and J. A. Estes, "Magnification of Secondary Production by Kelp Detritus in Coastal Marine Ecosystems," *Science* 245, no. 4914 (1989): 170–173.

40. A. Ortega et al., "Important Contribution of Macroalgae to Oceanic Carbon Sequestration," *Nature Geoscience* 12 (2019): 748–754.

41. H. Christie et al., "Species Distribution and Habitat Exploitation of Fauna Associated with Kelp (*Laminaria hyperborea*) along the Norwegian Coast," *Journal of the Marine Biological Association of the United Kingdom* 83 (2003): 4181.1–4181.13.

42. K. M. Schoenrock et al., "An Ecological Baseline for *Laminaria hyperborea* Forests in Western Ireland," *Limnology and Oceanography* 66 (2021): 3439–3454.

43. Vergés and Campbell, "Kelp Forests."

44. C. Darwin, *The Voyage of the Beagle* (1839, released by Project Gutenberg, June 15, 1997), chapter 11.

45. A. L. Shanks and G. L. Eckert, "Population Persistence of California Current Fishes and Benthic Crustaceans: A Marine Drift Paradox," *Ecological Monographs* 75 (2005): 505–524.

46. R. P. Dunn, "Tool Use by a Temperate Wrasse, California sheephead *Semicossyphus pulcher*," *Journal of Fish Biology* 88, no. 2 (2015): 805–810.

47. D. Helvarg, "Send in the Sea Otters to Help Save California's North Coast," *New York Times*, August 8, 2021.

48. Steneck, interview.

49. R. S. Steneck, "Regular Sea Urchins as Drivers of Shallow Benthic Marine Community Structure," in *Sea Urchins: Biology and Ecology, 4th Edition*, ed. J. M. Lawrence (Amsterdam: Elsevier, 2020), 255–279.

50. J. S. Stephens, "California Sheephead," in *California's Marine Living Resources: A Status Report*, ed. W. S. Leet et al. (Sacramento: California Department of Fish and Game, 2001), 155–156.

51. S. L. Hamilton et al., "Size-Selective Harvesting Alters Life Histories of a Temperate Sex-Changing Fish," *Ecological Applications* 17 (2007): 2268–2280.

52. S. L. Hamilton and J. E. Caselle, "Exploitation and Recovery of a Sea Urchin Predator Has Implications for the Resilience of Southern California Kelp Forests," *Proceedings: Biological Science* 282, no. 1799 (2015): 20141817.

53. A. C. Davenport and T. D. Anderson, "Positive Indirect Effects of Reef Fishes on Kelp Performance: The Importance of Mesograzers," *Ecology* 88, no. 6 (2007): 1548–1561.

54. M. S. Love, *Certainly More than You Want to Know about the Fishes of the Pacific Coast: A Postmodern Experience* (Santa Barbara, CA: Really Big Press, 2011).

55. C. A. Stepien, "Color Pattern and Habitat Differences Between Male, Female and Juvenile Giant Kelpfish (Blennoidei: Clinidae)," *Bulletin of Marine Science* 41, no. 1 (1987): 45–58.

56. C. F. Holder, *The Channel Islands of California* (Chicago: A. C. McClurg, 1910).

57. P. H. House, B.L.F. Clark, and L. G. Allen, "The Return of the King of the Kelp Forest: Distribution, Abundance, and Biomass of Giant Sea Bass (*Stereolepis gigas*) off Santa Catalina Island, California, 2014–2015," *Bulletin of the Southern California Academy of Sciences* 115, no. 1 (2016): 1–14.

58. J. E. Fitch and R. J. Lavenberg, *Marine Food and Game Fishes of California* (Berkeley: University of California Press, 1971).

59. E. Clark, "Notes on the Inflating Power of the Swell Shark, *Cephaloscyllium uter*," *Copeia* 4 (1947): 279–280.

10. Sweet and Salty

1. Aristotle, *The History of Animals, Book VI, Part 14*, trans. D. W. Thompson (MIT, Internet Classics Archive, 2000), 15–17.

2. M. Syrski, "On the Reproductive Organs of the Eels," *Annals and Magazine of Natural History* 15 (1875): 304–305.

3. S. Freud, *The Letters of Sigmund Freud to Eduard Silberstein, 1871–1881*, ed. W. Boehlich (Cambridge, MA: Belknap Press of Harvard University Press, 1990).

4. S. Freud, "Beobachtungen über Gestaltung und feineren Bau der als Hoden beschriebenen Lappenorgane des Aals," *Sitzungsberichte der kaiserlichen Akademie der Wissenschaften (Wien)* 75 (1877): 419–431. Translated by Google Translate.

5. Freud, "Beobachtung über Gestaltung."

6. P. Svensson, *The Book of Eels: Our Enduring Fascination with the Most Mysterious Creature in the Natural World*, trans. Agnes Broomé (New York: Harper Collins, 2020).

7. J. Schmidt, "The Breeding Place of the Eel," *Philosophical Transactions of the Royal Society of London* 211 (1923): 179–208.

8. M. P. Fish, "Preliminary Note on the Egg and Larva of the American Eel (*Anguilla rostrata*)," *Science* 64, no. 1662 (1926): 455–456.

9. "Much Mirth as Sweden 'Mourns' its Oldest Eel," *BBC News*, August 9, 2014, https://www.bbc.com/news/world-europe-28721701.

10. R. L. Carson, *Under the Sea-Wind* (London: Penguin, 1941), 107.

11. P. Verhelst et al., "Downstream Migration of European Eel (*Anguilla anguilla*) in an Anthropogenically Regulated Freshwater System: Implications for Management," *Fisheries Research* 199 (2018): 252–262.

12. F.-A. Weltzien et al., "Androgen-Dependent Stimulation of Brain Dopaminergic Systems in the Female European Eel (*Anguilla anguilla*)," *Endocrinology* 147, no. 6 (2006): 2964–2973.

13. S. Chow et al., "Japanese Eel *Anguilla japonica* Do Not Assimilate Nutrition during the Oceanic Spawning Migration: Evidence from Stable Isotope Analysis," *Marine Ecology Progress Series* 402 (2010): 233–238.

14. B. Pelster, "Swim Bladder Function and the Spawning Migration of the European Eel *Anguilla anguilla*," *Frontiers in Ecology* 5, no. 486 (2015): 1–10.

15. S. Kalujnaia et al., "Salinity Adaptation and Gene Profiling Analysis in the European Eel (*Anguilla anguilla*) Using Microarray Technology," *General and Comparative Endocrinology* 152 (2007): 274–280.

16. T. Hirano and N. Mayer-Gostan, "Eel Esophagus as an Osmoregulatory Organ," *Proceedings of the National Academy of Sciences USA* 73 (1976): 1348–1350.

17. J. Isaia, "Water and Nonelectrolyte Permeation," in *Fish Physiology. Volume 10. Gills. Part B. Ion and Water Transfer*, ed. W. S. Hoar and D. J. Randall (London: Academic, 1984), 1–38.

18. V.J.T. van Ginneken and G. E. Maes, "The European Eel (*Anguilla anguilla*, Linnaeus), Its Lifecycle, Evolution and Reproduction: A Literature Review," *Reviews in Fish Biology and Fisheries* 15 (2005): 367–398.

19. R. MacNamara and T. K. McCarthy, "Size-Related Variation in Fecundity of European Eel (*Anguilla anguilla*)," *ICES Journal of Marine Science* 69, no. 8 (2012): 1333–1337.

20. K. Aarestrup et al., "Oceanic Spawning Migration of the European Eel (*Anguilla anguilla*)," *Science* 235 (2009): 1660.

21. L. C. Naisbett-Jones et al., "A Magnetic Map Leads Juvenile European Eels to the Gulf Stream," *Current Biology* 27 (2017): 1236–1240.

22. V. E. van Ginneken et al., "Eel Migration to the Sargasso: Remarkably High Swimming Efficiency and Low Energy Costs," *Journal of Experimental Biology* 208 (2005): 1329–1335.

23. P. Bryant, T. Nunes, and R. Snaith, "Eel Fat Stores Are Enough to Reach the Sargasso," *Nature* 403 (2000): 156–157.

24. J. Dannewitz et al., "Panmixia in the European Eel: A Matter of Time . . .," *Proceedings of the Royal Society B* 272 (2005): 1129–1137.

25. W. Dekker, "Worldwide Decline of Eel Resources Necessitates Immediate Action," *Fisheries* 28 (2003): 28–30.

26. "European Eel," The Fisheries Secretariat, accessed July 28, 2022, https://www.fishsec.org/eel/.

27. J. Prosek, *Eels: An Exploration, from New Zealand to the Sargasso, of the World's Most Mysterious Fish* (New York: Harper Perennial, 2011).

28. P. Verhelst et al., "Downstream Migration."

29. V. E. van Ginneken et al., "PCBs and the Energy Cost of Migration in the European Eel (*Anguilla anguilla* L.)," *Aquatic Toxicology* 92, no. 4 (2009): 213–220.

30. D. D. Bloom and N. R. Lovejoy, "The Evolutionary Origins of Diadromy Inferred from a Time-Calibrated Phylogeny for Clupeiformes (Herring and Allies)," *Proceedings of the Royal Society B* 281 (2014): 20132081.

31. M. Gross, "Evolution of Diadromy in Fishes," *American Fisheries Society Symposium* 1 (1987): 14–25.

32. Bloom and Lovejoy, "Evolutionary Origins."

33. R. M. McDowall, "Driven by Diadromy: Its Role in the Historical and Ecological Biogeography of the New Zealand Freshwater Fish Fauna," *Italian Journal of Zoology* 65 Supplement (1998): 73–85.

34. D. J. Jellyman et al., "First Record of the Australian Longfinned Eel, *Anguilla reinhardtii*, in New Zealand," *Marine and Freshwater Research* 47, no. 8 (1996): 1037–1040.

35. R. O'Gorman and T. J. Stewart, "Ascent, Dominance, and Decline of the Alewife in the Great Lakes: Food Web Interactions and Management Strategies," in *Great Lakes Fisheries Policy and Management: A Binational Perspective*, ed. W. W. Taylor and C. P. Ferrari (East Lansing: Michigan State University Press, 1999), 489–514.

36. N. J. Reigle, Jr., "Bottom Trawl Explorations in Southern Lake Michigan, 1962–65," *U.S. Fish and Wildlife Circular* 301.

37. M. R. Greenwood, *1968 State-Federal Lake Michigan Alewife Die-Off Control Investigation* (Ann Arbor, MI: Bureau of Commercial Fisheries, Fish and Wildlife Service, 1970).

38. N. S. Baldwin et al., "Commercial Fish Production in the Great Lakes 1867–1977," *Great Lakes Fishery Commission Technical Report* 3 (1979).

39. M. L. Jones et al., "Limitations to Lake Trout Rehabilitation in the Great Lakes Imposed by Biotic Interactions Occurring at Early Life Stages," *Journal of Great Lakes Research* 21, Supplement 1 (1995): 505–517.

40. R. M. Claramunt, C. P. Madenjian, and D. Clapp. "Pacific Salmonines in the Great Lakes Basin," in *Great Lakes Fisheries Policy and Management: A Binational Perspective*, ed. W. W. Taylor and C. P. Ferrari (East Lansing: Michigan State University Press, 1999), 455–488.

41. G. R. Jacobs et al., "Chinook Salmon Foraging Patterns in a Changing Lake Michigan," *Transactions of the American Fisheries Society* 142 (2013): 362–372.

42. S. C. Riley et al., "Deepwater Demersal Fish Community Collapse in Lake Huron," *Transactions of the American Fisheries Society* 137 (2008): 1879–1890.

43. J. C. Pearson, *The Life History of the Striped Bass, or Rockfish, Roccus saxatilis (WALBAUM), Bulletin of the Bureau of Fisheries* 49 (Washington, DC: US Department of Commerce, 1938), 825.

44. W. N. Eschmeyer, E. S. Herald, and H. Hammann, *A Field Guide to Pacific Coast Fishes of North America* (Boston: Houghton Mifflin, 1983).

45. C. R. Robins and G. C. Ray, *Peterson Field Guides: Atlantic Coast Fishes* (Boston: Houghton Mifflin, 1986).

46. W. B. Scott and E. J. Crossman, *Freshwater Fishes of Canada* (Ottawa: Fisheries Research Board of Canada, 1973).

47. T. Morton, *New English Canaan or New Canaan; Containing an Abstract of New England* (Amsterdam, Iacob Frederick Stam, 1637; reprinted by Prince Society, Boston, 1883), 87.

48. L. W. Miller and R. J. McKechnie, "Observation of Striped Bass Spawning in the Sacramento River," *California Fish and Game* 54 (1968): 306–307.

49. Pearson, *Life History*, 826.

50. Morton, *New English Canaan*, 87.

51. D. C. Gwinn et al., "Rethinking Length-Based Fisheries Regulations: The Value of Protecting Old and Large Fish with Harvest Slots," *Fish and Fisheries* 16 (2015): 259–281.

52. "Striped Bass in the Hudson River," New York State Department of Environmental Conservation, accessed October 27, 2022, https://www.dec.ny.gov/animals/108092.html.

53. H. W. Herbert, *Frank Forester's Fish and Fishing in the United States and British Provinces of North America* (London: Richard Bentley, 1849), 11.

54. W. G. Robbins, *The Great Northwest: The Search for Regional Identity* (Corvallis: Oregon State University Press, 2001).

55. J. E. Morrow, *The Freshwater Fishes of Alaska* (Anchorage: Alaska Northwest Publishing, 1980).

56. W. Clark, "April 16, 1806," in *Journals of the Lewis and Clark Expedition* (Lincoln: University of Nebraska Press), https://lewisandclarkjournals.unl.edu/item/lc.jrn.1806-04-16.

57. Clark, "April 16, 1806."

58. K. Barber, *Death of Celilo Falls* (Seattle: University of Washington Press, 2005).

59. S. Anderson, "Salmon Color and the Consumer," *International Institute of Fisheries Economics and Trade Proceedings, July 10–14* (2000).

60. M. S. Love, *Certainly More than You Want to Know about the Fishes of the Pacific Coast: A Postmodern Experience* (Santa Barbara, CA: Really Big Press, 2011).

61. M. Kurlansky, *Salmon: A Fish, the Earth, and the History of Their Common Fate* (Ventura, CA: Patagonia Works, 2020).

62. US Department of Commerce, *Bulletin of the United States Bureau of Fisheries, Volume 37: 1921–1922* (Washington, DC: Government Printing Office, 1923).

63. W. Pennell and P. Prouzet, "Salmonid Fish: Biology, Conservation Status, and Economic Importance of Wild and Cultured Stocks," *Fisheries and Aquaculture* 3 (2009): 42–65.

64. Kurlansky, *Salmon.*

65. Y. Hiyama et al., "A Preliminary Experiment on the Return of Tagged Chum Salmon to the Otsuchi River, Japan," *Bulletin of the Japanese Society of Scientific Fisheries* 33, no. 1 (1967): 18–19.

66. G. A. Nevitt et al., "Evidence for a Peripheral Olfactory Memory in Imprinted Salmon," *Proceedings of the National Academy of Sciences* 91, no. 10 (1994): 4288–4292.

67. Kurlansky, *Salmon.*

68. G. S. Helfman et al., *The Diversity of Fishes: Biology, Evolution, and Ecology, 2nd Edition* (Oxford: John Wiley & Sons, 2009).

69. P. J. Lisi and D. E. Schindler, "Spatial Variation in Timing of Marine Subsidies Influences Riparian Phenology through a Plant-Pollinator Mutualism," *Ecosphere* 2, no. 9 (2011): 101.

70. Kurlansky, *Salmon.*

71. E. D. Houde, "Differences between Marine and Freshwater Fish Larvae: Implications for Recruitment," *ICES Journal of Marine Science* 51 (1994): 91–97.

72. Kurlansky, *Salmon.*

73. S. Hodgson and T. P. Quinn, "The Timing of Adult Sockeye Salmon Migration into Fresh Water: Adaptations by Populations to Prevailing Thermal Regime," *Canadian Journal of Zoology* 80 (2002): 542–555.

74. P. Gilhousen, "Prespawning Mortalities of Sockeye Salmon in the Fraser River System and Possible Causal Factors," *International Pacific Salmon Fisheries Commission Bulletin* 26 (1990): 58.

75. S. Macdonald et al., "The Influence of Extreme Temperatures on Migrating Fraser River Sockeye Salmon during the 1998 Spawning Season," *Canadian Technical Report of Fisheries and Aquatic Sciences* 2326 (2000).

76. J. Cornish, *A View of the Present State of the Salmon and Channel-Fisheries: And of the Statute Laws By Which They Are Regulated* (London: Longman, Hurst, Rees, Orme, Brown, and Green, 1824), 19.

Part IV

1. Megan Corazza, interview by Joe E. Meisel, December 18, 2022.

11. Fish to the Rescue

1. Food and Agriculture Organization (FAO), *The State of World Fisheries and Aquaculture 2022: Towards Blue Transformation* (Rome: FAO, 2022).

2. FAO, *Fish and Seafood Consumption Per Capita*, extracted from Food Balances database, October 10, 2023, https://www.fao.org/faostat/en/#data/FBS.

3. D. Pauly, *Vanishing Fish: Shifting Baselines and the Future of Global Fisheries* (Vancouver, BC: Greystone Books, 2019).

4. Pauly, *Vanishing Fish*, 32.

5. Pauly, *Vanishing Fish*, 22.

6. R. A. Myers and B. Worm, "Rapid Worldwide Depletion of Predatory Fish Communities," *Nature* 423 (2003): 280–283.

7. "Commercial Fishing Safety: National Overview," Centers for Disease Control and Prevention, accessed August 20, 2022, https://www.cdc.gov/niosh/topics/fishing/national overview.html.

8. FAO, *State of World Fisheries*, xix.

9. F. T. Christy, Jr., "Fisherman Quotas: A Tentative Suggestion for Domestic Management," National Technical Information Service, Washington, DC (1973).

10. R. Arnason, "Management of the Icelandic Demersal Fisheries," in *Fishery Access Control Programs Worldwide (Alaska Sea Grant Report No. 86)*, ed. N. Mollet (Fairbanks: University of Alaska Press, 1986), 83–101.

11. C. Chu, "Thirty Years Later: The Global Growth of ITQs and Their Influence on Stock Status in Marine Fisheries," *Fish and Fisheries* 10, no. 2 (2009): 217–230.

12. "SEDAR 31 Stock Assessment Report: Gulf of Mexico Red Snapper," North Charleston, SC: SEDAR (Southeast Data, Assessment, and Review), 2013, https://sedar-web.org/sedar-31-stock-assessment-report-gulf-mexico-red-snapper.

13. G. M. Lange, "The Value of Namibia's Commercial Fisheries," *Namibian Directorate of Environmental Affairs Research Discussion Paper* 55 (2003).

14. FAO, *State of World Fisheries*.

15. B. Muse and K. Schelle, "Individual Fisherman's Quotas: A Preliminary Review of Some Recent Programs," Report CFEC 89–1, Alaska Commercial Fisheries Entry Commission, Juneau (1989).

16. L. Van der Voo, *The Fish Market: Inside the Big-Money Battle for the Ocean and Your Dinner Plate* (New York: St. Martin's Press, 2016).

17. Christy, "Fisherman Quotas."

18. Pauly, *Vanishing Fish*.

19. C. Costello, S. D. Gaines, and J. Lynham, "Can Catch Shares Prevent Fisheries Collapse?," *Science* 321 (2008): 1678–1681.

20. A. M. Birkenbach, D. J. Kaczan, and M. D. Smith, "Catch Shares Slow the Race to Fish," *Nature* 544 (2017): 223–226.

21. R. Hilborn et al., "Effective Fisheries Management Instrumental in Improving Fish Stock Status," *Proceedings of the National Academy of Sciences* 117, no. 4 (2020): 2218–2224.

22. M. Kurlansky, *Salmon: A Fish, the Earth, and the History of their Common Fate* (Ventura, CA: Patagonia Works, 2020).

23. FAO, *State of World Fisheries*, 43–45.

24. F. Lee, *The Chinese Fish Culture Classic*, trans. T.S.Y. Moo, Contribution no. 459, Chesapeake Biological Laboratory (Solomons: University of Maryland), 1.

25. R. L. Naylor et al., "A 20-Year Retrospective Review of Global Aquaculture," *Nature* 591 (2021): 551–563.

26. FAO, *State of World Fisheries*.

27. H. E. Froehlich et al., "Comparative Terrestrial Feed and Land Use of an Aquaculture-Dominant World," *Proceedings of the National Academy of Sciences* 115, no. 20 (2018): 5295–5300.

28. Naylor et al., "A 20-Year Retrospective."

29. "Farmed Salmon," World Wildlife Fund, accessed August 30, 2022, https://www.worldwildlife.org/industries/farmed-salmon.

30. S. Deady, S.J.A. Varian, and J. M. Fives, "The Use of Cleaner-Fish to Control Sea Lice on Two Irish Salmon (*Salmo salar*) Farms with Particular Reference to Wrasse Behaviour in Salmon Cages," *Aquaculture* 131 (1995): 73–90.

31. A. B. Skiftesvik et al., "Delousing of Atlantic Salmon (*Salmo salar*) by Cultured vs. Wild Ballan Wrasse (*Labrus bergylta*)," *Aquaculture* 402 (2013): 113–118.

32. Becca Franks, PhD, interview by Joe E. Meisel, October 24, 2022.

33. L. P. Gaffney et al., "Coho Salmon (*Oncorhynchus kisutch*) Prefer and Are Less Aggressive in Darker Environments," *PLoS ONE* 11, no. 3 (2016): e0151325.

34. Kurlansky, *Salmon*.

35. I.-J. Jensen et al., "An Update on the Content of Fatty Acids, Dioxins, PCBs and Heavy Metals in Farmed, Escaped and Wild Atlantic Salmon (*Salmo salar* L.) in Norway," *Foods* 9 (2020): 1901.

36. R. A. Hites et al., "Global Assessment of Organic Contaminants in Farmed Salmon," *Science* 303 (2004): 226–229.

37. FAO, *State of World Fisheries*.

38. A. Avadí, P. Fréon, and J. Tam, "Coupled Ecosystem/Supply Chain Modelling of Fish Products from Sea to Shelf: The Peruvian Anchoveta Case," *PLoS ONE* 9 (2014): e102057.

39. E. Bachis, "Fishmeal and Fish Oil: A Summary of Global Trends," *Proceedings of the 57th IFFO Annual Conference* (2017): 23–25.

40. R. Oliveros-Ramos et al., "Management of the Peruvian Anchoveta (*Engraulis ringens*) in the Context of Climate Change," in *Adaptive Management of Fisheries in Response to Climate Change*, ed. T. Bahri et al. (Rome: FAO, 2021), 237–244.

41. FAO, *State of World Fisheries*.

42. Naylor et al., "A 20-Year Retrospective."

43. S. Diener, C. Zurbrugg, and K. Tockner, "Conversion of Organic Material by Black Soldier Fly Larvae: Establishing Optimal Feeding Rates," *Waste Management & Research* 27, no. 6 (2009): 603–610.

44. I. Belghit et al., "Black Soldier Fly Larvae Meal Can Replace Fish Meal in Diets of Sea-Water Phase Atlantic Salmon (*Salmo salar*)," *Aquaculture* 503 (2019): 609–619.

45. "New Deep-Set Longline Is Smart Gear," in *The State of the World's Sea Turtles Report, Volume 1*, ed. R. B. Mast (Washington, DC: State of the World's Sea Turtles, 2005–2006), 25.

46. N.C.H. Lo and T. D. Smith, "Incidental Mortality of Dolphins in the Eastern Tropical Pacific, 1959–72," *Fisheries Bulletin*, 84, no.1 (1986): 27–34.

47. D. Phillips and M. Palmer, "Q&A: Dolphin Safe Tuna," *International Marine Mammal Project* (August 5, 2021), https://savedolphins.eii.org/news/q-a-dolphin-safe-tuna.

48. FAO, *State of World Fisheries*.

49. B. B. Collette et al., "High Value and Long Life—Double Jeopardy for Tunas and Billfishes," *Science* 333 (2011): 291–292.

50. FAO, *The State of World Fisheries and Aquaculture 2018—Meeting the Sustainable Development Goals* (Rome: FAO, 2018).

51. FAO, *State of World Fisheries*, 2022.

52. R. McKinney et al., *Netting Billions 2020: A Global Tuna Valuation* (Philadelphia: Pew Charitable Trusts, 2020).

53. Y. Zohar et al., "Reproduction, Broodstock Management, and Spawning in Captive Atlantic Bluefin Tuna," in *Advances in Tuna Aquaculture from Hatchery to Market*, ed. D. D. Benetti et al. (Oxford: Academic, 2016), 159–187.

54. A. Medina et al., "Stereological Assessment of the Reproductive Status of Female Atlantic Northern Bluefin Tuna during Migration to Mediterranean Spawning Grounds through the Strait of Gibraltar," *Journal of Fish Biology* 60 (2002): 203–217.

55. P. Greenberg, "Tuna's End," *New York Times Magazine*, June 22, 2010, https://www.nytimes.com/2010/06/27/magazine/27Tuna-t.html.

56. A. Buentello, W. H. Neill, and D. M. Gatlin III, "Tuna Aquaculture Faces Challenges in Continued Growth" *Global Seafood Alliance*, March 1, 2009, https://www.globalseafood.org/advocate/tuna-aquaculture-faces-challenges-in-continued-growth/.

57. D. Ellis and I. Kiessling "Ranching of Southern Bluefin Tuna in Australia," in *Advances in Tuna Aquaculture from Hatchery to Market*, ed. D. D. Benetti et al. (Oxford: Academic, 2016), 217–232.

58. C. J. Hayward et al., "Concurrent Epizootic Hyperinfections of Sea Lice (Predominantly *Caligus chiastos*) and Blood Flukes (*Cardicola forsteri*) in Ranched Southern Bluefin Tuna," *Veterinary Parasitology*, 173, no. 1–2 (2010): 107–115.

59. Ellis and Kiessling, "Ranching of Southern Bluefin."

60. A. Cheshire et al., "Investigating the Environmental Effects of Sea-Cage Tuna Farming II, the Effect of Sea-Cages: A Report to the FRDC and Tuna Boat Owners Association," *FRDC Project* 94/091 (1996).

61. F. de la Gándara, A. Ortega, and A. Buentello, "Tuna Aquaculture in Europe," in *Advances in Tuna Aquaculture from Hatchery to Market*, ed. D. D. Benetti et al. (Oxford: Academic, 2016), 115–157.

62. B. Worm "Averting a Global Fisheries Disaster," *Proceedings of the National Academy of Sciences* 113, no. 18 (2016): 4895–4897.

63. Pauly, *Vanishing Fish*, 22.

64. Pauly, *Vanishing Fish*, 22.

65. U. R. Sumaila et al., "WTO Must Ban Harmful Fisheries Subsidies," *Science* 374, no. 6567 (2021): 544.

66. D. Tickler et al., "Modern Slavery and the Race to Fish," *Nature Communications* 9 (2018): 4643.

67. World Trade Organization, "The WTO Agreement on Fisheries Subsidies: What It Does and What Comes Next," accessed November 3, 2023, https://www.wto.org/english/tratop_e/rulesneg_e/fish_e/fish_factsheet_e.pdf.

68. FAO, *State of World Fisheries.*

69. FAO, *State of World Fisheries*, 53.

70. K. M. Morrison et al., "The First Cut Is the Deepest: Trawl Effects on a Deep-Sea Sponge Ground Are Pronounced Four Years On," *Frontiers in Marine Science* 7 (2020): 605281.

71. E. Sala et al., "Protecting the Global Ocean for Biodiversity, Food and Climate," *Nature* 592 (2021): 397–402.

72. C. Bailey, "Lessons from Indonesia's 1980 Trawler Ban," *Marine Policy* 21, no. 3 (1997): 225–235.

73. D. Zeller et al., "Re-estimation of Small-Scale Fishery Catches for U.S. Flag-Associated Island Areas in the Western Pacific: The Last 50 Years," *Fishery Bulletin* 105, no. 2 (2007): 266–277.

74. S.J.M. Blaber, "Relationships between Tropical Coastal Habitats and (Offshore) Fisheries," in *Ecological Connectivity among Tropical Coastal Ecosystems*, ed. I. Nagelkerken (Dordrecht, Netherlands: Springer, 2009), 533–564.

75. V. S. Batista et al., "Tropical Artisanal Coastal Fisheries: Challenges and Future Directions," *Reviews in Fisheries Science & Aquaculture* 22, no. 1 (2014): 1–15.

76. "Marine Protected Areas," National Marine Protected Areas Center, accessed September 1, 2022, https://marineprotectedareas.noaa.gov/.

77. UN Environment Programme World Conservation Monitoring Centre, *World Database of Protected Areas*, June 2022, https://www.protectedplanet.net/en.

78. Pauly, *Vanishing Fish*, 31.

79. D. D. Eisenhower, quoted in W. Lambers, "Eisenhower's Warning about the Military-Industrial Complex Is Still Valid," *Chicago Sun-Times*, January 16, 2022.

80. J. Humphreys and R.W.E. Clark, "A Critical History of Marine Protected Areas," in *Marine Protected Areas: Science, Policy and Management*, ed. J. Humpreys and R.W.E. Clark (Amsterdam: Elsevier, 2020), 1–12.

81. E. V. Sheehan et al., "Drawing Lines at the Sand: Evidence for Functional vs. Visual Reef Boundaries in Temperate Marine Protected Areas," *Marine Pollution Bulletin* 76 (2013): 194–202.

82. S. E. Lester et al., "Biological Effects within No-Take Marine Reserves: A Global Synthesis," *Marine Ecology Progress Series* 384 (2009): 33–46.

83. C. M. Roberts et al., "Effects of Marine Reserves on Adjacent Fisheries," *Science* 294, no. 5548 (2001): 1920–1923.

84. S. Medoff, J. Lynham, and J. Raynor, "Spillover Benefits from the World's Largest Fully Protected MPA," *Science* 378, no. 6617 (2022): 313–316.

85. N. S. Barrett et al., "Changes in Fish Assemblages Following 10 Years of Protection in Tasmanian Marine Protected Areas," *Journal of Experimental Marine Biology and Ecology* 345 (2007): 141–157.

86. J. A. Bohnsack, "The Potential of Marine Fisheries Reserves for Reef Fish Management in the U.S. Southern Atlantic: Snapper-Grouper Development Team Report to the Southern Atlantic Fishery Management Council," *NOAA Technical Memorandum* NMFS-SEFC-261 (1990).

87. A. García-Rubies, B. Hereu, and M. Zabala, "Long-Term Recovery Patterns and Limited Spillover of Large Predatory Fish in a Mediterranean MPA," *PLoS ONE* 8, no. 9 (2013): e73922.

88. M. E. Bond et al., "Abundance and Size Structure of a Reef Shark Population within a Marine Reserve Has Remained Stable for More than a Decade," *Marine Ecology Progress Series* 576 (2017): 1–10.

89. T. R. McClanahan and N. A. Muthiga, "Change in Fish and Benthic Communities in Belizean Patch Reefs in and Outside of a Marine Reserve, across a Parrotfish Capture Ban," *Marine Ecology Progress Series* 645 (2020): 25–40.

90. C. Cox et al., "Establishment of Marine Protected Areas Alone Does Not Restore Coral Reef Communities in Belize," *Marine Ecology Progress Series* 563 (2017): 65–79.

91. A. J. Frisch and J. R. Rizzari, "Parks for Sharks: Human Exclusion Areas Outperform No-take Marine Reserves," *Frontiers in Ecology and the Environment* 17, no. 3 (2019): 145–150.

92. Jorge Ramírez, MSc, interview by Joe E. Meisel, June 3, 2021.

93. "Ecuador Detains Chinese Boat with Endangered Sharks," *BBC News*, August 16, 2017, https://www.bbc.com/news/world-latin-america-40944886.

94. T. B. Letessier et al., "Remote Reefs and Seamounts Are the Last Refuges for Marine Predators across the Indo-Pacific," *PLoS Biology* 17, no. 8 (2019): e3000366.

95. G.M.S. Vianna et al., "Socio-Economic Value and Community Benefits from Shark-Diving Tourism in Palau: A Sustainable Use of Reef Shark Populations," *Biological Conservation* 145 (2012): 267–277.

96. K. Whiting, "This Pacific Island Has Banned Fishing for Marine Conservation," World Economic Forum, December 11, 2019, https://www.weforum.org/agenda/2019/12/palau-pacific-marine-conservation-fishing-environment/.

97. *At What Price: The Economic, Social and Icon Value of the Great Barrier Reef* (Brisbane: Deloitte Access Economics, 2017).

98. "Unlocking Sustainable Tourism: A Challenge for our Times," Great Barrier Reef Foundation, February 19, 2022, https://www.barrierreef.org/news/project-news/unlocking-sustainable-tourism-resilient-reefs-initiative.

99. FAO, *State of World Fisheries*.

100. D. Russi et al., "Socio-Economic Benefits of the EU Marine Protected Areas," *Institute for European Environmental Policy* (2016).

101. J. Jacquet and D. Pauly, "Trade Secrets: Renaming and Mislabeling of Seafood," *Marine Policy* 32, no. 3 (2008): 309–318.

102. G. K. Luque and C. J. Donlab, "The Characterization of Seafood Mislabeling: A Global Meta-analysis," *Biological Conservation* 236 (2019): 556–570.

103. D. A. Willette et al., "Using DNA Barcoding to Track Seafood Mislabeling in Los Angeles Restaurants," *Conservation Biology* 31, no. 5 (2017): 1076–1085.

104. H. Bornatowski, R. R. Braga, and J.R.S. Vitule, "Shark Mislabeling Threatens Biodiversity," *Science* 340 (2013): 923.

105. T. Pazartzi et al., "High Levels of Mislabeling in Shark Meat – Investigating Patterns of Species Utilization with DNA Barcoding in Greek Retailers," *Food Control* 98 (2019): 179–186.

106. Brian Perkins, interview by Joe E. Meisel, October 23, 2019.

107. S. R. Bush et al., "The 'Devils Triangle' of MSC Certification: Balancing Credibility, Accessibility and Continuous Improvement," *Marine Policy* 37 (2013): 288–293.

108. Marine Stewardship Council (MSC), *Celebrating Sustainable Seafood: The Marine Stewardship Council Annual Report 2022–23* (London: MSC, 2023), 21.

109. F. Le Manach et al., "Small Is Beautiful, but Large Is Certified: A Comparison between Fisheries the Marine Stewardship Council (MSC) Features in Its Promotional Materials and MSC-Certified Fisheries," *PLoS ONE* 15, no. 5 (2020): e0231073.

110. P. Greenberg, *American Catch: The Fight for Our Local Seafood* (New York: Penguin Random House, 2015).

111. Jacquet and Pauly, "Trade Secrets," 309.

112. Perkins, interview.

113. S. Long et al., "Deep-Sea Benthic Habitats and the Impacts of Trawling on Them in the Offshore Greenland Halibut Fishery, Davis Strait, West Greenland," *ICES Journal of Marine Science* 78, no. 8 (2021): 2724–2744.

114. V. Restrepo et al., *ISSF 2019–08: Report of the International Workshop on Mitigating Environmental Impacts of Tropical Tuna Purse Seine Fisheries* (Washington, DC: International Seafood Sustainability Foundation, 2019).

115. J. Jacquet et al., "Seafood Stewardship in Crisis," *Nature* 467, no.2 (2010): 28–29.

116. Jennifer Jacquet, PhD, interview by Joe E. Meisel, September 12, 2022.

117. Perkins, interview.

118. Bush et al., " 'Devils Triangle,' " 289.

119. M. Wakamatsu and H. Wakamatsu, "The Certification of Small-Scale Fisheries," *Marine Policy* 77 (2017): 97–103.

120. S. Opitz et al., "Assessment of MSC-Certified Fish Stocks in the Northeast Atlantic," *Marine Policy* 71 (2016): 10–14.

121. R. Froese and A. Proelss, "Evaluation and Legal Assessment of Certified Seafood," *Marine Policy* 36, no. 6 (2012): 1284–1289.

122. Froese and Proelss, "Evaluation and Legal Assessment."

123. "Global Catches of Alaska Pollock (*Gadus chalcogrammus*) by EEZ," Sea Around Us: Fisheries, Ecosystems & Biodiversity, accessed November 6, 2022, https://www.seaaroundus.org/data/#/taxon/600318?chart=catch-chart&dimension=eez&measure=tonnage&limit=10.

124. "Alaska Pollock Species Overview," National Oceanic and Atmospheric Administration Fisheries, accessed November 6, 2022, https://www.fisheries.noaa.gov/species/alaska-pollock.

125. U. Eberly, "The Surprising Success Story of Fish Sticks," *Smithsonian*, April 26, 2021, https://www.smithsonianmag.com/innovation/surprising-success-story-fish-sticks-180977578/.

126. US Environmental Protection Agency, *An Inventory of Sources and Environmental Releases of Dioxin-like Compounds in the U.S. for the Years 1987, 1995, and 2000* (Washington, DC: US Environmental Protection Agency, 2006).

127. N.A.J. Graham et al., "Seabirds Enhance Coral Reef Productivity and Functioning in the Absence of Invasive Rats," *Nature* 559 (2018): 250–253.

128. D. J. McCauley et al., "From Wing to Wing: The Persistence of Long Ecological Interaction Chains in Less-Disturbed Ecosystems," *Scientific Reports* 2 (2012):409.

129. Graham et al., "Seabirds Enhance," 250.

130. D.A. Kroodsma et al., "Tracking the Global Footprint of Fisheries," *Science* 359 (2018): 904–908.

131. FAO, *State of World Fisheries*.

132. Pauly, *Vanishing Fish*, 31.

133. R. Sumaila, "A Conversation with Rashid Sumaila," Tyler Prize for Environmental Achievement, July 2023, https://tylerprize.org/laureates-laureate-conversations/a-conversation-with-rashid-sumaila-part-2/.

Epilogue

1. A. M. Jones, C. Brown, and S. Gardner, "Tool Use in the Tuskfish *Choerodon schoenleinii?*," *Coral Reefs* 30, no. 3 (2011): 865.

2. J. Coyer, "Use of a Rock as an Anvil for Breaking Scallops by the Yellowhead Wrasse, *Halichoeres garnoti* (Labridae)," *Bulletin of Marine Science* 57 (1995): 548–549.

3. G. Bernardi, "The Use of Tools by Wrasses (Labridae)," *Coral Reefs* 31 (2011): 39.

4. H. Fricke, "Behaviour as Part of Ecological Adaptation," *Helgoländer Wissenschaftliche Meeresuntersuchungen* 24 (1973): 120–144.

5. K. J. Pryor and A. M. Milton, "Tool Use by the Graphic Tuskfish *Choerodon graphicus*," *Journal of Fish Biology* 95 (2019): 663–667.

6. R. Dunn, "Tool Use by a Temperate Wrasse, California Sheephead *Semicossyphus pulcher*," *Journal of Fish Biology* 88 (2015): 805–810.

7. Shark Fin Sales Elimination Act of 2021, H.R. 2811, 117th Cong. (2021–2022), https://www.congress.gov/bill/117th-congress/house-bill/2811.

8. E. T. Sherwood et al., "Tampa Bay (Florida, USA): Documenting Seagrass Recovery since the 1980's and Reviewing the Benefits," *Southeastern Geographer* 57, no. 3 (2017): 294–319.

9. C. Costello and D. Ovando, "Status, Institutions, and Prospects for Global Capture Fisheries," *Annual Review of Environment and Resources* 44 (2019): 177–200.

10. Megan Corazza, interview by Joe E. Meisel, December 18, 2022.

11. Jennifer Jacquet, interview by Joe E. Meisel, September 12, 2022.

INDEX

Page numbers in **bold** refer to figures.